Congenital Diseases and the Environment

Environmental Science and Technology Library

VOLUME 23

Congenital Diseases and the Environment

Edited by

P. Nicolopoulou-Stamati
National and Kapodistrian University of Athens, Medical School, Department of Pathology, Athens, Greece

L. Hens
Vrije Universiteit Brussel, Human Ecology Department, Brussels, Belgium

and

C.V. Howard
Bioimaging Research Group, Centre for Molecular Biosciences, University of Ulster, Coleraine, United Kingdom

A C.I.P. Catalogue record for this book is available from the Library of Congress.

ISBN-10 1-4020-4830-0 (HB)
ISBN-13 978-1-4020-4830-2 (HB)
ISBN-10 1-4020-4831-9 (e-book)
ISBN-13 978-1-4020-4831-9 (e-book)

Published by Springer,
P.O. Box 17, 3300 AA Dordrecht, The Netherlands.

www.springer.com

Printed on acid-free paper

Desktop publishing by Dao Kim Nguyen Thuy Binh and Vu Van Hieu.

Editorial Statement

It is the policy of AREHNA, EU–SANCO project, to encourage the full spectrum of opinions to be represented at its meetings. Therefore it should not be assumed that the publication of a paper in this volume implies that the Editorial Board is fully in agreement with the contents, though we ensure that contributions are factually correct. Where, in our opinion, there is scope for ambiguity we have added notes to the text, where appropriate.

TABLE OF CONTENTS

SECTION 1: METHODS

CLINICAL TERATOLOGY

M. CLEMENTI AND E. DI GIANANTONIO

SECTION 2: TERATOGENS

DIOXINS AND CONGENITAL MALFORMATIONS

J.G. KOPPE, M. LEIJS, G. TEN TUSSCHER AND P.D. BOER

LINKS BETWEEN *IN UTERO* EXPOSURE TO PESTICIDES AND EFFECTS ON THE HUMAN PROGENY. DOES EUROPEAN PESTICIDE POLICY PROTECT HEALTH?

C. WATTIEZ

ASSOCIATION OF *INTRA-UTERINE* EXPOSURE TO DRUGS WITH CONGENITAL DEFECTS: THE THALIDOMIDE EFFECT

M. CLEMENTI, K. LUGWIG AND A. ANDRISANI

SECTION 3: CONGENITAL DISEASES

ENDOCRINE DISRUPTER EXPOSURE AND MALE CONGENITAL MALFORMATIONS

M.F. FERNÁNDEZ AND N. OLEA

TESTICULAR DYSGENESIS SYNDROME AS A CONGENITAL DISEASE

H.E. VIRTANEN AND J. TOPPARI

ENVIRONMENTAL RISK AND SEX RATIO IN NEWBORNS

M. PETERKA, Z. LIKOVSKY AND R. PETERKOVA

SECTION 4: COUNTRY REPORTS

PREFACE

For many years, interest in the prevention of diseases in children was concentrated on improvement of the postnatal environment. However, since the major problems of infectious diseases and nutrition were solved with the help of vaccinations and better feeding regimes, it became clear that new approaches were needed to prevent and treat the disorders and problems we are facing now - problems mainly arising in prenatal life.

Today's epidemics in children are prematurity; intra-uterine growth retardation; learning disabilities; Attention Deficit Hyperactivity Disorder (ADHD); asthma and allergies; auto-immune diseases such as type 1 diabetes and Crohn's disease (both increasing in recent decades, due to so far unknown environmental factors); cancer; obesity and hearing problems. All of these problems can have their roots in prenatal life. The idea that the mother is protecting the child while (s)he is developing *in utero* has been proved to be wrong. Environmental toxicants, especially the fat soluble persistent bioaccumulating chemicals, pass across the placenta and can damage the developing baby. Inhibiting influences on placental enzymes, such as by pesticides, might have long-term effects, for instance on blood pressure.

This book examines various aspects of congenital diseases and the environment.

Congenital anomalies are among the most important causes of handicap and mortality. About 4-5 % of newborns have a congenital defect, of which 1% is severe.

Already in 3000 BC, congenital abnormalities like achondroplasia were described and the Romans thought that an abnormal child was born as a predictor of the future, using the word "monstrum". Moreover, in the middle-ages many mothers, together with the midwife and the abnormal child, died at the stake on accusation of being bewitched.

After Mendel, in the first half of the twentieth century, ideas about causes pointed to genetics.

In 1941, for the first time, a publication from Australia made it clear that a rubella-infection caused a congenital cataract. In addition, the more recent thalidomide scandal underlined the effects of drugs in the early phase of pregnancy.

Congenital malformations are mostly disorders of organs that develop in the first trimester of pregnancy. However, even more important quantitatively are the functional teratological disturbances arising in the second and third trimester.

Besides drugs, other environmental stressors also became known to cause disturbances in development, resulting in anomalies when taking place in the first three months of pregnancy, and in functional problems when occurring in the second and third trimester of pregnancy or in the early postnatal period.

It is not so surprising that the prenatal period is the most vulnerable one. Between conception and birth, the fertilized ovum goes through some forty-two cycles of cell division to develop into a full-term infant; after birth, only another five are needed to attain adult size. Credit is due to David Barker, who brought attention to the early origin of adult disease. Well-known are the studies done in the Netherlands on the influence of hunger in the Dutch Hunger winter in the period of November 1944 - May 1945. Besides a high mortality and morbidity at birth, the growth restriction of the fetus during this period of hunger (mostly a protein deficiency) resulted fifty years later in an abnormal glucose metabolism with a defective glucose tolerance, abnormal lipid profile and higher blood pressure. The cohort conceived during the peak of the hunger, in the months February until May 1945, and born in November with normal or even higher birth-weights, showed later in life an increase in obesity and schizophrenia.

It is fair to say that at this moment, the causes of most congenital malformations are not known, but these are probably multi-factorial, with infinite numbers of causative factors. A single-compound approach to testing, as is done now in classical toxicology, is no longer sufficient in a world of multiple exposures. The combination of different chemicals, or the combination with an infection or stress, can result in additive, synergistic, antagonistic or neutral effects. In general, almost nothing is yet known about fetal toxicology. Disturbances in the thyroid hormone homeostasis causes impaired brain development, but in the first half of pregnancy, it is the task of the mother to provide her baby with sufficient thyroid hormone and in that period, any problem in the mother is decisive. Abnormalities in the estrogen/androgen status in the fetal period can result in abnormal sexual development of the brain, such as trans-sexualism.

Genetic causes are becoming better known, but clear-cut genetic reasons for congenital diseases are rare and clinical aspects can be confusing. The recognition of epigenetics also makes it clear that the environment plays a definitive role. Genes are activated or inactivated by environmental influences such as hunger periods or Diethylstilboestrol (DES) in pregnancy, or later, and these effects can be transferred to the next generation when the ovum or spermatogonium is susceptible. The female oocyte is susceptible for this imprinting around the 26th week of pregnancy of the female baby, while the male spermatocyte is vulnerable in the pre-pubertal period in the male.

Because the fetal period, especially the first three months, is so essential for the individual's later life, preconception counselling is very important. Happily, obstetricians have become more and more convinced of this. Because the cycle of the production of sperm is three months, a time point of three months before conception would be ideal to counsel the future parents. The father might be given advice to avoid all sorts of solvents, alcohol and smoking, while the mother can, for example, lower her body burden of mercury during these three months, by avoiding fish rich in mercury, such as the bigger tuna fish, and she can also eat a lot of foods rich in anti-oxidants, such as grapes, berries, broccoli, carrots and beetroot. Supplementation of folic acid and a control of the vitamin status, such as vitamin A, are important for the mother-to-be. Control of diseases such as diabetes, hypertension, and epilepsy is also important before conception.

This book addresses the crucial question "how environmental factors/stressors influence the intra-uterine life of the fetus" and highlights current relevant scientific knowledge.

J.G. KOPPE
Emeritus Professor of Neonatology
Ecobaby Foundation
The NETHERLANDS

ACKNOWLEDGEMENTS

All papers of this book have been peer reviewed. The Editors are most indebted to the colleagues who reviewed various chapters of this book:

Ahmed Mahmoud, Laboratory of Andrologie, University of Ghent, Gent, Belgium

Alain Dupont, Clinical Pharmacology, Faculty of Medicine and Pharmacy, Vrije Universiteit Brussel, Belgium

Antoni Duleba, Department of Obstetrics and Gynecology, Yale University School of Medicine, USA

Arnold Schecter, University of Texas, School of Public Health, Dallas, USA

Asher Ornoy, The Israeli Teratogen Information Service, Laboratory of Teratology, Department of Anatomy and Cell Biology, The Hebrew University – Hadassah Medical School and Israeli Ministry of Health, Israel

Athina Tsmamadou, Teratogen Information Center, Poison Information Center, Children's Hospital A and P Kyriakou, Athens, Greece

Barbara D. Abbott, Developmental Toxicology Division, U.S Environmental Protection Agency, USA

David Miller, Department of Obstetrics and Gynaecology, University of Leeds, UK

David Stone, Padiatric Epidemiology and Community Health, Yorkhill Hospital Glasgow, Scotland

Didima M.G. de Groot, TNO Quality of Life (Location Zeist), Zeist, The Nederlands

Emmanuel Brilakis, Paediatric Surgery Department, "TZANEIO" General Hospital of Piraeus, Piraeus, Greece

Erminio Giavini, Department of Biology, State University of Milan, Italy

Ettore Caroppo, IRCCS "S. de Bellis", UO Fisiopatologia della Riproduzione Umana, Castellana Grotte (Ba), Italy

Evangelos Fousteris, Department of Internal Medicine, General Hospital of Livadeia, Piraeus, Greece

Faith G. Davis, Epidemiology-Biostatistics, School of Public Health, University of Illinois at Chicago, USA

George P. Daston, The Procter and Gamble Company, Cincinnati, OH, USA

Gies Andreas, Department of Health Policy and Management, Mailman School of Public Health, Columbia University, USA

Ilias Maglogiannis, Department of Information and Communication Systems Engineering, University of Aegean, Greece

James Mills, Pediatric Epidemiology Section, Division of Epidemiology, Statistics and Prevention Research, UK

Janna G. Koppe, Ecobaby Foundation, The Netherlands

Jeanne Mager Stellman, Department of Health Policy and Management, Mailman School of Public Health, University of Columbia, USA

Jeremy R. Montague, School of Natural and Health Sciences, University of Barry, USA

John A. Harris, California Birth Defects Monitoring Program, USA

Jorma Toppari, Department of Physician of Pediatric Endocrinology, University of Turku, Finland

Judith Rankin, School of Population and Health Sciences, University of Newcastle, UK

Julien I.E.Hoffman, Cardiovascular Research Institute, University of California, San Francisco, USA

Le Thi Nham Tuyet, Research Centre for Gender, Family and Environment in Development (CGFED), Ha Noi, Vietnam

Linda Birnbaum, Environmental Toxicology Division U.S, Environment Protection Agency (EPA), USA

Luc Pussemier, CODA-CERVA-VAR, Brussels, Belgium

Marc Nyssen, Department of Medical Informatics, Vrije Universiteit Brussel, Brussels, Belgium

Marie-Christine Dewolf, EEN, EPHA Environmental Network, Hygiène Publique en Hainaut asbl/Provincial Institute of Hygiene and Bacteriology of the Hainaut, Belgium

Martine Vrijheid, International Agency for Research on Cancer (IARC), Lyon, France

Maurizio Clementi, CEPIG, Genetica Clinica ed Epidemiologica, Dipartimento di Pediatria, Università di Padova, Italy

Monique Ryan, Paediatric Neurologist, The Children's Hospital at Westmead, Australia

Olle Soder, Karolinska Institute and University Hospital Stockholm, Sweden

Patricia B. Hoyer, University of Arizona, Tucson, USA

Petroff D.V.M Brian, University of Kansas Medical Center, Breast Cancer Prevention Center, USA

Ramsden David, University of Birmingham, School of Bioscience, UK

Richard Nelson, Northern General Hospital, Sheffield, UK

Richard Sharpe, MRC Human Reproductive Sciences Unit, Centre for Reproductive Biology, Queen's Medical Research Institute, UK

Sándor János, University Pécs, Faculty of Health Sciences, Institute of Applied Health Sciences, Department of Public Health. Hungary

Shigetaka Katow, CDC (Centers for Disease Control and Prevention), Department of Viral Disease and Vaccine Control, National Institute of Infectious Diseases, Musashi-Murayama, Japan

Stephen Safe, Department of Veterinary Physiology and Pharmacology, College Station, University of Texas A and M, USA

Stoyan Stoyanov. University of Chemical Technology and Metallurgy, Ecology Center, Bulgaria

Sylvaine Cordier, INSERM U625, University of Rennes, France

Trent D. Stephens, Department of Biological Sciences, Idaho State University, USA

Tsung O. Cheng, Department of Medicine, George Washington University Medical Center, Washington, D.C, USA

Warren G. Foster, Reproductive Biology, Department of Obstetrics and Gynecology University of McMaster, Canada

Werner Kloas, Department of Endocrinology, Institute of Biology, Humboldt University Berlin, Germany

The editors of this book wish also to thank the EU-SANCO and the Municipality Dikaiou Kos for their support to the A.R.E.H.N.A. project (www.arehna.di.uoa.gr) which provided the main scientific bases of this book.

Language editing of the book was done by Ms. V A Mountford Chester.

The camera-ready copy of this book was produced by Vu Van Hieu and Dao Kim Nguyen Thuy Binh. Their work is most sincerely appreciated.

LIST OF CONTRIBUTORS

A. ANDRISANI
Dipartimento di Scienze ginecologiche e della riproduzione umana, Università di Padova
Via Giustiniani 3
35128 Padova
ITALY

P.D. BOER
Emeritus Professor of Neonatology
Ecobaby Foundation
Hollandstraat 6
3634 AT Loenersloot
The NETHERLANDS

E. BRILAKIS
Paediatric Surgery Department
"TZANEIO" General Hospital of Piraeus
Aristofanous street 18
185 33 Piraeus
GREECE

M. CLEMENTI
CEPIG, Genetica Clinica ed Epidemiologica
Dipartimento di Pediatria
Università di Padova
Via Giustiniani 3
35128 Padova
ITALY

E. DI GIANANTONIO
CEPIG, Genetica Clinica ed Epidemiologica
Dipartimento di Pediatria
Università di Padova
Via Giustiniani 3
35128 Padova
ITALY

H. DOLK
Faculty of Life and Health Sciences
University of Ulster
Shore Rd
BT370QB Newtownabbey
UNITED KINGDOM

M.F. FERNÁNDEZ
Radiology Department
School of Medicine
University of Granada
Av Madrid s/n
18071 Granada
SPAIN

E. FOUSTERIS
Laboratory of Experimental Pharmacology
School of Medicine
National and Kapodistrian University of Athens
Mikras Asias street 75
115 27 Athens
GREECE

L. HENS
Vrije Universiteit Brussel
Human Ecology Department
Laarbeeklaan 103
B-1090 Brussel
BELGIUM

E. HUYGHE
Human Fertility Research Group – Reproductive Health in Developing Countries
EA n°36 94
Hôpital Paule de Viguier
Avenue de Grande Bretagne 330
TSA 70034, 31059 Toulouse Cedex
FRANCE

T. KARJALAINEN
European Commission
Research Directorate General –Unit E2 (Food quality)
Squre de Mecus 8
B-1049 Brussels
BELGIUM

J. G. KOPPE
Emeritus Professor of Neonatology
Ecobaby Foundation
Hollandstraat 6
3634 AT Loenersloot
The NETHERLANDS

M. LEIJS
Department of Paediatrics and Neonatology
Emma Children's Hospital Academic Medical Centre,
University of Amsterdam
P.O. Box 22660
1100 DD Amsterdam
The NETHERLANDS

Z. LIKOVSKY
Department of Teratology
Institute of Experimental Medicine
Academy of Sciences CR
Videnská 1083
142 20 Prague 4
CZECH REPUBLIC

K. LUDWIG
CEPIG, Genetica Clinica ed Epidemiologica
Dipartimento di Pediatria
Università di Padova
Via Giustiniani 3
35128 Padova
ITALY

G. LYONS
Toxics Policy Advisor
WWF UK
17 The Avenues
NR2 3PH
Norwich
ENGLAND

A. MANTOVANI
Department Food Safety and Veterinary Public Health
Istituto Superiore di Sanità
viale Regina Elena 299
00161 Rome
ITALY

F. MARANGHI
Department Food Safety and Veterinary Public Health
Istituto Superiore di Sanità
viale Regina Elena 299
00161 Rome
ITALY

R. MIEUSSET
Human Fertility Research Group – Reproductive Health in Developing Countries
EA n°36 94
Hôpital Paule de Viguier
Avenue de Grande Bretagne 330
TSA 70034, 31059 Toulouse Cedex
FRANCE

P. NICOLOPOULOU-STAMATI
Department of Pathology
Medical School
University of Athens
M.Asias street 75
11527 Athens 11527
GREECE

N. OLEA
Radiology Department
School of Medicine
University of Granada
Av Madrid s/n
18071 Granada
SPAIN

J. PAPADOPULOS
Department of Internal Medicine
General Hospital of Livadeia
Agiou Eleutheriou street 143
185 41 Piraeus
GREECE

M. PETERKA
Department of Teratology
Institute of Experimental Medicine
Academy of Sciences CR
Videnská 1083
142 20 Prague 4
CZECH REPUBLIC

R. PETERKOVA
Department of Teratology
Institute of Experimental Medicine
Academy of Sciences CR
Videnská 1083
142 20 Prague 4
CZECH REPUBLIC

J. RANKIN
School of Population and Health Sciences
University of Newcastle upon Tyne
Framlington Place
NE2 4HH
Newcastle upon Tyne
ENGLAND

E. ROBERT-GNANSIA
Institut Européen des Génomutations
rue Edmond Locard 86
69005 Lyon
FRANCE

A. ROSANO
Italian Institute of Social Medicine
Via P.S. Mancini 29
10196 Rome
ITALY

O. SÖDER
Department of Woman and Child Health
Pediatric Endocrinology Unit
Karolinska Institute & University Hospital
Q2:08
SE17176 Stockholm
SWEDEN

S. STOYANOV
University of Chemical Technology and Metallurgy
Ecology Centre
blvd. "Kl. Ohridski" 8
1756 Sofia
BULGARIA

M. STRAND
Department of Woman and Child Health
Pediatric Endocrinology Unit
Karolinska Institute and University Hospital
Q2:08
SE17176 Stockholm
SWEDEN

I. SUPORNSILCHAI
Department of Woman and Child Health
Pediatric Endocrinology Unit
Karolinska Institute and University Hospital
Q2:08
SE17176 Stockholm
SWEDEN

K. SVECHNIKOV
Department of Woman and Child Health
Pediatric Endocrinology Unit
Karolinska Institute and University Hospital
Q2:08
SE17176 Stockholm
SWEDEN

E. TERLEMESIAN
University of Chemical Technology and Metallurgy
Ecology Centre
blvd. "Kl. Ohridski" 8
1756 Sofia
BULGARIA

P. F. THONNEAU
Human Fertility Research Group – Reproductive Health in Developing Countries
EA n°36 94
Hôpital Paule de Viguier,
Avenue de Grande Bretagne 330
TSA 70034, 31059 Toulouse Cedex
FRANCE

J. TOPPARI
Departments of Physiology and Paediatrics,
University of Turku
Kiinamyllynkatu 10
20520 Turku
FINLAND

G.T. TUSSCHER
Department of Paediatrics and Neonatology
Westfries Gasthuis
Maelsonstraat 3
1624 NP Hoorn
The NETHERLANDS

V.VECHNIKOVA
Department of Woman and Child Health
Pediatric Endocrinology Unit
Karolinska Institute & University Hospital
Q2:08
SE17176 Stockholm
SWEDEN

H.E. VIRTANEN
Departments of Physiology and Paediatrics,
University of Turku
Kiinamyllynkatu 10
20520 Turku
FINLAND

A. WAHLGREN
Department of Woman and Child Health
Pediatric Endocrinology Unit
Karolinska Institute & University Hospital
Q2:08
SE17176 Stockholm
SWEDEN

C. WATTIEZ
Consultant for Pesticides Action Network Europe
Leonard Street 56-64
EC2A 4JX
London
UNITED KINGDOM

C. ZETTERSTRÖM
Department of Woman and Child Health
Pediatric Endocrinology Unit
Karolinska Institute & University Hospital
Q2:08
SE17176 Stockholm
SWEDEN

LIST OF FIGURES

LIST OF TABLES

LIST OF BOXES

INTRODUCTION: CONCEPTS IN THE RELATIONSHIP OF CONGENITAL DISEASES WITH THE ENVIRONMENT

P. NICOLOPOULOU-STAMATI
Department of Pathology
Medical School, University of Athens
Athens
GREECE
Email: aspis@ath.forthnet.gr

Summary

In recent years, increasing attention and resources have been brought to bear on the relationship between environment and congenital diseases. These diseases were previously thought to be mostly due to genetic causes. Even though the importance of genes as factors in causation is accepted, environmental factors seem to be implicated just as strongly. This book explores some further concepts that have arisen from more recent perceptions of environmental effects and their possible interactions with living systems.

Bearing in mind the difficulty of assessing the cause and extent of congenital diseases, methods of studying environmental impacts are presented and a new approach for toxicology is advocated: assessing low dose and chronic exposure. Emphasis is placed on developmental endpoints as markers of endocrine disruption with possible teratogenic effects of compounds and mixtures of substances. The acquisition of valuable markers and sources of information, obtained by examining congenital anomalies in wildlife and tracing back exposures and contaminants is discussed.

Furthermore, this book includes details of the most recent studies on the effects of compounds such as dioxins, thalidomide, PCBs, and phthalates, in the causation of congenital anomalies. As a result, their mechanisms are now far better understood, though not yet fully elucidated. Intra-uterine exposures to pesticides causing congenital anomalies are discussed from the viewpoint of European policy.

P. Nicolopoulou-Stamati et al. (eds.), Congenital Diseases and the Environment, 1–17.

I contrast, the example of Vietnam is presented, which is known to have undergone very high exposure to dioxins three decades ago.

Some of the most recent definitions and aspects of specific congenital anomalies are also addressed in this book: examples include testicular dysgenesis syndrome, cryptorchidism, effects on the human sex ratio and congenital malformations in boys. Certain pathways are presented, such as endocrine disruption affecting normal development, but also novel ones, such as inflammation and steroidogenesis.

Reports from different countries are outlined, with recent trends of particular conditions and their incidence rates: the interest also lies in their widespread geographic distribution. Countries considered are Greece, Bulgaria, and UK.

Following these factual reports, the manner in which policy elaboration and implementation are affected by this data are discussed, as well as the importance of raising awareness, increasing education, sharing information and instigating research. The exchange of information and knowledge within the EU constitutes a priority that should be promoted using recent advances in information technology. The concerns of NGOs should be also considered in this context, as they represent intermediaries between citizens, scientists and policy makers.

1. Introduction

The environment can have a considerable impact on several stages of reproductive health, ranging from the maturation of organs and endocrine systems to indeed, the health of the developing organism as a whole. A systematic investigation and assessment of the current state of knowledge concerning congenital abnormalities is therefore important. Congenital abnormalities have traditionally been associated with inherited conditions. Recent advances in scientific knowledge allied to discoveries of environmental agents such as EDs have altered this perception. It is the aim of this book to present these new concepts and frameworks in the field of environmental effects on congenital anomalies, in order to inform both scientists and non-expert professionals of recent developments. Additionally, a strengthening of communication between all levels is advocated, from investigation to implementation of policy.

2. The changing concepts of environmental influences in the causation of congenital anomalies

Our traditional understanding about how congenital anomalies occur, what their causes are, and how to identify them have been overturned by technological and scientific advances. Examples addressing diet (Jirtle *et al.,* 2004) and exposure to environmental agents, such as toxins, and/or EDs (Newbold, 2005) have been published. Contrary to the traditional tenets of toxicology, toxins have been found to be able to exert subtle effects at concentrations several orders of magnitude lower than those found in classical high dose studies. The fact that these can be altered by individual susceptibility has resulted in the new discipline of pharmacogenomics. An example of this low dose toxicology is provided by Rajapakse *et al.* (2001). The timing of exposure during development is crucial, as evidenced by the vulnerability of the organism during critical windows of embryonic development.

The main changes that have emerged to challenge previously widely held beliefs concerning the origin and nature of congenital diseases, the current state of science concerning this issue and the directions that future enquiry are likely to take, are addressed in this book. The validity of current studies might be contested by many, due to the inability to compare the present results with ones from previous periods. No comparable records are available for environmental levels of certain substances, even from some decades ago, since they were not deemed to represent a hazard or because measurement technique changed. A view which, even though changed today, does not grant an accurate idea of past exposure. Therefore, the current studies and research methods will have to rely on screening and monitoring services as well as laboratory research. Multidisciplinary approaches and increased professional interactions are considered essential for understanding the effects of the environment on congenital disorders, and identifying areas of concern. In addition, another challenge is the necessity for the scientist to persist and develop findings, in the face of peer pressure or adverse media impact, in order to raise public awareness and advise politicians of possible risks. Establishing networks of communication between experts, and tying closer links with the media, are perceived as being essential, in order to protect future generations.

3. Methods to study congenital anomalies and their links to the environment

3.1. EUROCAT: Surveillance of environmental impact

In order to study congenital anomalies, it is essential to be able to detect them consistently. In recent years, it has been difficult to assert whether there has been a genuine rise in some congenital anomalies, or whether our screening and detection capacities have improved to the extent of detecting them earlier and more accurately than before (Dolk *et al.,* 1998). To this end, an organization that would monitor levels of occurrence of congenital anomalies would help determine whether there are any trends or fluctuations that might necessitate action or the taking of precautions. This would provide advanced warning as well as a tool for monitoring and statistically analyzing these occurrences, thus identifying risk factors (Dolk and Vrijheid, 2003; Dolk, 2004). This is reflected by the surveillance policy of EUROCAT, the main points of which can be summarized as: routine monitoring, detection and response to spatial and temporal clusters of cases, and finally evaluation of specific environmental exposure hypotheses with the data obtained. However, the current position, comprising a pan-European network of surveillance consisting of different data sets from different sources, makes a strong argument for a unified European environmental health surveillance strategy with the required stable and robust monitoring. This would have the main aim of assessing exposure from environmental factors, rather than determining definitive causes for phenomena.

3.2. Endpoints for prenatal exposures in toxicological studies

Regulatory toxicology consists primarily, though not exclusively, of product testing using toxicological studies of high dose and brief duration of exposure before being released for commercial use. Yet, it has been established that the various systems in an organism work in a homeostatic fashion, and that there are specific periods, such as critical windows of development, where these systems are likely to be disrupted. These disruptions can occur at low doses, and in long term exposures: conditions that are not tested by routine toxicology. Hence, in order to account for the variability of the organism's dynamic systems and the specific pathways that these substances take, it is essential to increase the use of developmental toxicology, with an emphasis on its importance for risk assessment. The testing of substances during *in utero* exposure in animals has been the standard for many years, and their results have been deemed useful (Newman *et al.,* 1993). The development of an organism is regulated by multiple events, which if delayed or disrupted, will affect the development of the organism for life, resulting not only in congenital anomalies, but also in functional deficits: for example, thyroid hormones (Stoker *et al.,* 2000). The use of specific tests, that calculate the variable susceptibility of specific organs and how they develop, by custom-tailoring them to specific maturation markers, will ensure that the action or influence of any potential toxicant can be detected at an early stage (Mantovani and Calamandrei, 2001). It is difficult to assess some of these using animal models or in vitro assays; certain sensitive aspects, such as reproduction (Gray *et al.,* 2000) or immunity (Nohara *et al.,* 2004), can prove challenging.

The responsibility remains to refine and develop ever more sensitive tests, with better detection in mind, and for specific end points of prenatal exposure. The future development of these tests is presented in this book (Mantovani *et al.,* 2006), together with the current issues involving their use, such as quality of detection, and validation of tests.

3.3. ***Evidence from wildlife***

The development of ever more sensitive tests and in vitro models of exposure for studying congenital abnormalities and their links to the environment are essential, even though they present various limitations to their applicability and the extent of assessment they offer. Even then, the information they produce might not be representative of the holistic manner in which exposure occurs in the environment. In this respect, an interesting complement to these methods is the studying of environmental exposure of wildlife. Congenital abnormalities can be studied together with levels of exposure of substances and mixtures that are actually present in the environment, and do not need to be artificially replicated. These observations can provide useful indications of possible pathways and modes of action of substances, as well as combinations thereof, which result in congenital abnormalities (Dunbar *et al.,* 1996). They can then be used as a direct warning for determining exposure, and also provide a foundation for more refined investigations in laboratories. An example is DDT exposure, which has been shown to present congenital abnormalities in wildlife studies: a study in humans found a similar link between DDT levels and spontaneous abortions in Chinese workers (Korrick *et al.,* 2001).

In this book, some different examples studied in wildlife are presented and the information that can be gained from environmental disasters or recorded levels of high exposures is discussed. An example is the contamination of lakes in Florida and the resulting rise in congenital abnormalities seen in alligators (Guillette *et al.*, 1994). Admittedly, it is more difficult to correlate observations to specific factors. These indications provide a valuable source of information, along with excellent models of exposure to environmental substances and factors, potentially leading to the development of more realistic hypotheses, and a wider conceptual framework of incidence of congenital abnormalities.

3.4. Epidemiology

The epidemiology of congenital abnormalities represents an essential aspect of study, in order to determine prevalence and risk factors of these conditions, with correlations to environmental factors (Dolk, 1998). By associating, in both a temporal and spatial manner, specific patterns of incidence of congenital abnormalities, possible risk factors can be isolated and extracted, with the potential to elaborate hypotheses that can be tested, in the laboratory (Game *et al.*, 2005). This process allows an improved understanding and elucidation of the causes of congenital abnormalities, permitting the determination of any fluctuations in their occurrence.

In this book, the different aspects of epidemiological studies on congenital abnormalities, with the current state of research, are presented (Rosano *et al.*, 2006). These studies face several complications, such as the accurate determination of specific causes, the possibility to isolate and correlate specific factors with specific conditions, data acquisition and validation, as well as adjustments for human factors. Presenting a population report on a congenital abnormality requires accurate records and reliable sources: if these are not available, the integrity and validity of the study, as well as the quality resulting from its conclusions, might be flawed.

3.5. Clinical teratology

Clinical observations play an important role in determining the prevalence and incidence of congenital abnormalities. Indeed, it is an essential task to be able to correlate the incidence of specific rare diseases within a population, and enquire whether there are any underlying patterns of exposure. Repeated presentation within a specific temporal interval of rare conditions that present similar symptoms or features is a powerful diagnostic tool. Congenital abnormalities can also arise due to iatrogenic interventions, by prescription, application of, or exposure to, certain drugs and substances. In this book, the importance of detecting specific indications of exposure from clinical examinations and determining possible origins and causes for teratogenesis are discussed. This, in turn, helps to assess and estimate the risk factors, whether they are limited to specific cases or might represent a threat to a wider segment of the population or area. As such, it constitutes an important complement to surveillance and epidemiology studies, representing a powerful method for detecting and determining likely causes for concern.

4. Chemicals and exposure conditions associated with congenital anomalies

4.1. Congenital diseases related to environmental exposure to dioxins

Dioxins are one group of chemicals that has been implicated in congenital diseases. These substances have a wide distribution, and represent a characteristic group of EDs with many effects. They are lipophilic and easily absorbed and stored in adipose tissue. Dioxins are known to interfere with Vitamin A metabolism, necessary for the monoglycosylation of proteins. Dioxin poisoning possesses short-term effects, such as chloracne poisoning, as seen in the recent case of the two Austrian secretaries (Geusau *et al.,* 2001; Geusau *et al.,* 2002)). It also exhibits long-term effects, characteristic of EDs, as these substances can be transferred from the mother's adipose stores to the embryo. Due to the lack of an effective BBB, together with the fact that dioxins can be taken up through the placenta by the fetus, they can affect neurological development, and cause congenital abnormalities, or disturbances of sex-specific behaviour (Vreugdenhill *et al.,* 2002). In this book, it is demonstrated (Koppe, 2006) how various forms of dioxin exposure that occurred during specific documented events correlated with a rise in several, otherwise rare, congenital diseases, such as cleft lips, and especially relating to NTDs such as hydrocephalus, or spina bifida. These can be linked to the inhibition of vitamin A, and therefore be traced back to an action of dioxin. More research is necessary to be able to determine the exact mechanisms of action of these complex substances; yet detection of exposure and counselling of pregnant women could greatly help to alleviate and control risk factors in the future.

4.2. Association of intra-uterine exposure with drugs: the thalidomide effect

EDs possess multiple forms of action that can be more long-term than anticipated by normal toxicological studies, rendering these studies inadequate for assessing the effects of some of these substances. This is particularly true in the context of congenital abnormalities, as some of these substances and other toxins can act as teratogens, resulting in damage to the embryo or fetus (Finnell *et al.,* 2002).

However, another particular danger is that drugs can potentially pass toxicology tests, and only present problems at a much later stage, such as pregnancy. Such was the case with DES, or fetal abnormalities with valproic acid; but the more notorious example in relation to congenital abnormalities is thalidomide. Thalidomide caused substantial damage to the embryo, and the main issue, which it exemplifies along with other teratogens, is that its effects occur much later. As another point of interest, animal studies might not be suited to determine intra-uterine effects of certain drugs or toxins, proving to be misleading: the case of thalidomide exposure in rats is complex (Vorhees *et al.,* 2001). A possible reason for that might be due to the different ways of some animals, of metabolizing and excreting these agents, compared to humans.

In this book, the finer clinical aspects of effects of drugs during intrauterine exposure and congenital abnormalities are discussed, while also pointing to the importance of including genetic studies within the conceptual framework necessary to assess the origin of congenital abnormalities (Clementi *et al.,* 2006). Consideration of the prevalence within a population, specific genetic individual susceptibility, and the methods to detect and test these factors, such as micro arrays and pharmacogenetic studies, represent some important aspects in clinical assessments.

4.3. Endocrine disrupter exposure and male congenital malformations

One of the major concerns about epidemiological studies is whether they are actually representative of the individual: according to certain authors, the principle of 'the ecological fallacy' relies on the repeated failure to correlate biological effects observed on a population level with the individual level (Vidaeff and Sever, 2005). The real problem actually lies in the fact that, in the environment, chemicals are rarely isolated, but they are present in multitudes, and in different low doses. Their effects may be weakly estrogenic or androgenic, but in mixtures they can show potentiated effects, in a synergistic fashion (Kortenkamp and Altenburger, 1998). Previous epidemiological studies also presented inconsistencies in their methodologies, in that they relied on questionnaires and models. In a current chapter of this book, Fernandez and Olea (2006) present an epidemiological study involving a study cohort of children, and link it to exposure with a variety of factors. The main factors considered were geographical regions and blood samples to measure the Total Effective Xenoestrogen Burden of the alpha fraction (TEXB-alpha), indicating different exposure patterns. An increase in prevalence of congenital abnormalities in the Southeast of Spain seemed linked to an increase of pesticide use. This study would indicate some links between increased use of pesticide, limited to the rural parts of the country, and prevalence of congenital abnormalities, such as genital tract malformations. One of the main benefits of these studies, noted by the author, is their hypothesis-generating ability, while highlighting the importance of epidemiological studies in order to assess harmful effects in humans. In addition, emphasis is laid on the necessity to establish a certain process for data acquisition: information from scientists should be transformed into knowledge, which then should raise concern in non-expert professionals that can in turn raise the awareness of the public. This will have an impact on decision making. Waiting for the necessary evidence might not be an option: the Precautionary Principle should be applied to prevent deleterious situations from arising before they occur.

4.4. Links between in utero exposure to pesticides and their effects on human progeny. Does European pesticide policy protect health?

Various population studies have revealed multiple incidences of congenital abnormalities which can be linked to exposure to pesticides (Nurminen, 1995; Kristensen *et al.,* 1997; Shaw *et al.,* 1999; Garry *et al.,* 2002). Toxicological safety testing in laboratories is performed on animal models, which can prove to be inadequate for extrapolation. In this book, the issue of whether pesticide exposure was covered by European directives, and whether there was a certain protocol to follow to ensure safe practice, is explored. According to the author (Wattiez, 2006), these directives are, from a European point of view, neither properly assessed, nor considered. There do not seem to be any specific tests required for ED properties nor for developmental neurotoxicity. Furthermore, there is no consideration of the possibility of combined effects. In addition, the components that are necessary, in order to cover the basic requirements for maintaining efficient criteria in monitoring pesticide exposure are presented.

4.5. Phthalates

Phthalate esters (more usually simply called phthalates) constitute an essential component of many modern industrial products, from plastics to cosmetics, such as perfumes. They are more or less ubiquitous in our environment, and research has also identified them in human tissue (Paganetto *et al.,* 2000). Connections have been made between the effects of members of this chemical family which are EDs, acting on the development of genitalia, being also correlated to increased incidence of hypospadias in a dose-dependent manner (Steinhardt, 2004). Other focused studies on the genetic effects of phthalates in animals have identified a link with Male

Dysgenesis Syndrome (MDS) (Fisher *et al.,* 2004), and another recent paper identified new gene targets of *in utero* exposure of phthalates (Liu *et al.,* 2005). One of the pathways which is thought to be affected is for the normal development of Sertoli cells, leading to MDS. This is evidenced by decrease in the anogenital distance, and other testicular anomalies, observed in rodents: a strong link with these symptoms was forged, when the same effects amongst male human infants were correlated with prenatal phthalate exposure (Swan *et al.,* 2005).

5. Environmental congenital anomalies

5.1. Testicular dysgenesis syndrome

Neonatal exposure to environmental agents can affect development. This is well illustrated by MDS. Testicular development depends on a complex feedback pattern but usually there are other symptoms expressed along with MDS, such as reduced sperm quality, or testicular cancer. This pattern is linked to the interaction between certain basic systems that compensates for fluctuations: if many of them show reduced function, or one of them ceases altogether, as seen with Leydig cells, then the effects can lead to MDS. Our knowledge of this syndrome and its underlying possible environmental causes is presented by Virtanen and Toppari (2006) in the relevant chapter of this book. A review of the current scientific perspective is coupled with an epidemiological approach to population migrations and prevalence of MDS in Finland. The incidence of various congenital abnormalities, such as cryptorchidism (Boisen *et al.,* 2004) and hypospadias (Boisen *et al.,* 2005), is considered. Mainly, the questions posed by the authors concern whether the changes observed in MDS can be quantified, and how to differentiate between environmental and genetic reasons in this context. This raises the issue of the importance of closer interaction amongst researchers, towards a common goal. In addition, compiling together of unrelated data should be avoided. In this respect, it is not possible to make a common unified outcome out of different, yet possibly linked, environmental factors.

5.2. Endocrine disrupters, inflammation and steroidogenesis

Testicular physiology is a sensitive system, depending on various factors, such as the prenatal sexual differentiation. The correct levels of hormones, especially androgens, will lead to various important changes in the development of the male fetus, such as brain differentiation. The development of male external genital organs, for instance, is represented by testicular descent. The physiological conditions prevalent in the developing organism are important, with various crucial factors occurring, such as steroidogenesis evoked by androgens seen in Leydig cells. A classic perception of EDs is that they affect the levels of androgens in the fetus by either weak or strong estrogenic effects, overwhelming the levels of natural androgen (Rajapakse *et al.,* 2001). A novel pathway, presented by Svechnikov *et al.* (2006) in this book, is that EDs can cause congenital abnormalities by affecting the expression of steroidogenesis, which will in turn affect androgen and hormonal levels, due to the interconnections of the system. Another element that is essential to consider, presented in this chapter, is the possible interactions of EDs and inflammation, by activation of cytokines in multiple cell systems. This can lead to tissue injury, with possible consequences of damage to cells involved in the synthesis of the essential factors in male differentiation: cells in the testis also produce cytokines (Strand *et al.,* 2005). Recent studies tend to validate the principles outlined above (Assmus *et al.,* 2005), and outline an interesting, and exciting, new avenue of research to understand the possible molecular and physiological pathways leading to congenital abnormalities.

5.3. Environmental impact on congenital diseases: the case of cryptorchidism

The prevalence of certain congenital diseases such as cryptorchidism is increasing in different industrialised countries. There does not seem to be any specific reason or cause to explain this: but the recent trends in male reproductive issues, such as decreasing fertility (Skakkebaek, 2003), indicate that there might be a correlation between these and congenital abnormalities such as TSD, hypospadias, cryptorchidism, and many others. The question posited by Thonneau *et al.* (2006) in this book is about the link between potent estrogens, the possible decrease of male androgens which they might cause, and the combined result of their effects in the differentiation of genital organs. There are multiple chemicals that have been identified to function as anti-androgens (Fisher *et al.,* 2004); yet the main difficulties that are encountered when attempting to apply this data to human studies are the limitations on information on environmental exposures of humans. Different countries use different chemicals, and positing a clear relationship can be particularly challenging. It is therefore important to define the exact window of exposure, and where multiple effects can affect systems that will remain unchanged even in adult life. Once again, the difficulties of epidemiological studies are rendered clear, and an emphasis is made on the necessity of collaboration between different disciplines in order to capitalise on the information acquired from these studies.

6. Policy aspects

6.1. Raising awareness of information society on impact of EU policies and deployment aspects

More and more, the exchange of information acquires a predominant role. This is also felt strongly due to the decline of the industrial and manufacturing branch of European economy, with developing countries acquiring an edge, due to cheaper labour and manufacturing costs. Europe still has the leading role in education of individuals and elaboration of high quality ideas, within the framework of a knowledge-based economy. The marketing of ideas and of high tech methods remains one of Europe's best selling exports. The problem in the future would consist in maintaining that edge, and allowing this to be developed even further. Questions remain on the prospects of the information society, and how information technology, along with electronic equipment, should be put to better use in order to facilitate dissemination of knowledge. Indeed, Europe has, so far, a record of good research, yet bad implementation and capitalisation on the level of products and marketable results. This in turn results in the 'brain drain' of our most competitive and brilliant minds, depleting the intellectual reserves of Europe when they are needed most. The emphasis in this book, due to all these factors, is on how to foster effective communication, how to implement policies for disseminating information, rendering it more accessible to the citizen, and how to promote cooperation.

6.2. The concerns of NGO's related to congenital diseases

Congenital diseases represent an issue that requires regulation on the political level, and further research by scientists to elucidate possible causes. However, it is useful to remember that they also affect the individuals concerned, their families, and also the non-expert professionals that are charged with the treatment and diagnosis of these conditions. There seems to be a distinct lack of communication between these various levels of individuals. NGOs represent organizations that usually intervene as the middle ground between these levels, allowing them to have a unique perspective on the debate at hand, while also being in closer contact to the individuals directly affected.

6.3. Environmental impacts on congenital anomalies – information for the non-expert professional

Non-expert professionals are deeply involved with congenital anomalies, as they usually represent the first level of contact, once they occur. In that sense, they are a group that is in a unique position for gathering and disseminating information about these conditions, particularly to the affected individuals, who can be bewildered by all the information displayed elsewhere: the internet is such an example of conflicting information. Hens (2006) discusses in this chapter the importance of the non-expert professional to define and communicate the risks involved, by evaluating various factors, as diverse as from employment to geographical location. This is particularly important when considering the responsibility towards public health, and the need to educate and prevent where possible. Another goal is to try and minimize the prescription behaviour, and try to avoid the use and prescription of drugs that are known to be potentially harmful during pregnancy. Finally, the search for alternatives should not be underestimated, and should be positively encouraged, along with better communication with experts. Valuable contributions can be made from both sides, as long as communication remains open.

7. Conclusions

The conclusions reached in the course of this book identify certain main issues relating to the current state of research on congenital anomalies, reflecting the difficulties they represent. Moreover, it assembles various complex features of the problems under consideration, and suggests ways in which communication can be improved and refined.

Current recent concepts, such as critical windows of development, when applied to congenital abnormalities, help in understanding the problem more accurately. The main consideration is that the range of answers that can be obtained from research, as well as the quality of knowledge derived, depends entirely on the relevance of the question asked. In that respect, the better the problem formulation, the more valuable the data obtained.

The main problem in the field of congenital diseases currently is that methods are rarely standardized, and that not everyone is using unified approaches. This can be applied to epidemiology but also to general toxicology, developmental toxicology, and clinical evidence-based medicine. As a result, it is more difficult to appreciate and compare the data, which may cause confusion to experts, non-experts, policy-makers and the public. The evidence obtained from wildlife can prove to be a useful tool for assessing risk factors, and to further improve methodology.

Furthermore, in this book the most important issue, involving communication, is addressed. This aspect has been ignored, with potential serious consequences. None of the research and scientific evidence will make any difference unless it is actively disseminated to the correct channels, a fact which should have been self-evident in the knowledge-based society that is Europe. The dissemination of information has acquired a primary role throughout all other aspects of life, which unfortunately does not seem to extend to scientific knowledge and research. This communication should involve the public, the non-expert professionals, NGOs, and policy-makers. There needs to be an efficient interplay of communication between all levels, with the message getting through efficiently, targeted at raising awareness on the environmental impact of congenital diseases.

References

Assmus, M., Svechnikov, K., von Euler, M., Setchell, B., Sultana, T., Zetterstrom, C., Holst, M., Kiess, W. and Soder, O. (2005) Single subcutaneous administration of chorionic gonadotropin to rats induces a rapid and transient increase in testicular expression of pro-inflammatory cytokines, *Pediatr. Res.* **57**, 896-901.

Boisen, K. A., Chellakooty, M., Schmidt, I. M., Kai, C. M., Damgaard, I. N., Suomi, A. M., Toppari, J., Skakkebaek, N. E. and Main, K. M. (2005) Hypospadias in a cohort of 1072 Danish newborn boys: prevalence and relationship to placental weight, anthropometrical measurements at birth, and reproductive hormone levels at three months of age, *J. Clin. Endocrinol. Metab.* **90**, 4041-4046.

Boisen, K. A., Kaleva, M., Main, K. M., Virtanen, H. E., Haavisto, A. M., Schmidt, I. M., Chellakooty, M., Damgaard, I. N., Mau, C., Reunanen, M., Skakkebaek, N. E. and Toppari, J. (2004) Difference in prevalence of congenital cryptorchidism in infants between two Nordic countries, *Lancet* **363**, 1264-1269.

Dolk, H. (1998) Rise in prevalence of hypospadias, *Lancet* **351**, 770.

Dolk, H. (2004) Epidemiologic approaches to identifying environmental causes of birth defects", *Am. J. Med. Genet. C. Semin. Med. Genet.* **125**, 4-11.

Dolk, H. and Vrijheid, M. (2003) The impact of environmental pollution on congenital anomalies, *Br. Med. Bull.* **68**, 25-45.

Dolk, H., Busby, A., Armstrong, B. G. and Walls, P. H. (1998) Geographical variation in anophthalmia and microphthalmia in England, 1988-94, *Br. Med. J.* **317**, 905-909.

Dunbar, M.R., Cunningham M.W., Wooding J.B., Roth R.P. (1996) Cryptorchidism and delayed testicular descent in Florida black bears. *J. Wildl. Dis.* **32**, 661-664.

Dwernychuk, L. W. (2005) Dioxin hot spots in Vietnam, *Chemosphere* **60**, 998-999.

Dwernychuk, L. W., Cau, H. D., Hatfield, C. T., Boivin, T. G., Hung, T. M., Dung, P. T. and Thai, N. D. (2002) Dioxin reservoirs in southern Viet Nam--a legacy of Agent Orange, *Chemosphere* **47**, 117-137.

Finnell, R. H., Waes, J. G., Eudy, J. D. and Rosenquist, T. H. (2002) Molecular basis of environmentally induced birth defects, *Annu. Rev. Pharmacol. Toxicol.* **42**, 181-208.

Fisher, J. S. (2004) Environmental anti-androgens and male reproductive health: focus on phthalates and testicular dysgenesis syndrome, *Reproduction* **127**, 305-315.

Garne, E., Loane, M., Dolk, H., De Vigan, C., Scarano, G., Tucker, D., Stoll, C., Gener, B., Pierini, A., Nelen, V., Rosch, C., Gillerot, Y., Feijoo, M., Tincheva, R., Queisser-Luft, A., Addor, M. C., Mosquera, C., Gatt, M. and Barisic, I. (2005) Prenatal diagnosis of severe structural congenital malformations in Europe, *Ultrasound Obstet. Gynecol.* **25**, 6-11.

Garry, V. F., Harkins, M. E., Erickson, L. L., Long-Simpson, L. K., Holland, S. E. and Burroughs, B. L. (2002) Birth defects, season of conception, and sex of children born to pesticide applicators living in the Red River Valley of Minnesota, USA, *Environ. Health Perspect.* **110** (Suppl 3), 441-449.

Geusau, A., Abraham, K., Geissler, K., Sator, M.O., Stingl, G. and Tschachler, E. (2001) Severe 2,3,7,8-tetrachlorodibenzo-p-dioxin (TCDD) intoxication: clinical and laboratory effects, *Environ. Health Perspect.* **109**, 865-869.

Geusau, A., Schmaldienst, S., Derfler, K., Pápke, O. and Abraham, K. (2002) Severe 2,3,7,8-tetrachlorodibenzo-pdioxin (TCDD) intoxication:kinetics and trials to enhance elimination in two patients, *Arch. Toxicol.* **76**, 316-325.

Gray, L. E., Jr., Ostby, J., Furr, J., Price, M., Veeramachaneni, D. N. and Parks, L. (2000) Perinatal exposure to the phthalates DEHP, BBP, and DINP, but not DEP, DMP, or DOTP, alters sexual differentiation of the male rat, *Toxicol. Sci.* **58**, 350-365.

Guillette, L.J., Gross, T.S. and Masson, G.R. (1994) Developmental abnormalities of the gonad and abnormal sex hormone concentrations in juvenile alligators from contaminated and control lakes in Florida, *Environ. Health Perspect.* **102**, 680-688.

Hwang, H., Fisher, S. W., Kim, K. and Landrum, P. F. (2004) Comparison of the toxicity using body residues of DDE and select PCB congeners to the midge, Chironomus riparius, in partial-life cycle tests, *Arch. Environ. Contam. Toxicol.* **46**, 32-42.

Jirtle, R. L. (2004) IGF2 loss of imprinting: a potential heritable risk factor for colorectal cancer, *Gastroenterology* **126**, 1190-1193.

Korrick, S. A., Chen, C., Damokosh, A. I., Ni, J., Liu, X., Cho, S. i., Altshul, L., Ryan, L. and Xu, X. (2001) Association of DDT with spontaneous abortion: a case-control study, *Ann. Epidemiol.* **11**, 491-496.

Kortenkamp, A. and Altenburger, R. (1998) Synergisms with mixtures of xenoestrogens: a reevaluation using the method of isoboles, *Sci. Total Environ.* **221**, 59-73.

Kristensen, P., Irgens, L. M. and Bjerkedal, T. (1997) Environmental factors, reproductive history, and selective fertility in farmers' sibships, *Am. J. Epidemiol.* **145**, 817-825.

Kunisue, T., Watanabe, M., Iwata, H., Subramanian, A., Monirith, I., Minh, T. B., Baburajendran, R., Tana, T. S., Viet, P. H., Prudente, M. and Tanabe, S. (2004) Dioxins and related compounds in human breast milk collected around open dumping sites in Asian developing countries: bovine milk as a potential source, *Arch. Environ. Contam Toxicol.* **47**, 414-426.

Le, T. N. and Johansson, A. (2001) Impact of chemical warfare with agent orange on women's reproductive lives in Vietnam: a pilot study, *Reprod. Health Matters* **9**, 156-164.

Liu, K., Lehmann, K. P., Sar, M., Young, S. S. and Gaido, K. W. (2005) Gene expression profiling following *in utero* exposure to phthalate esters reveals new gene targets in the etiology of testicular dysgenesis, *Biol. Reprod.* **73**, 180-192.

Mantovani, A. and Calamandrei, G. (2001) Delayed developmental effects following prenatal exposure to drugs, *Curr. Pharm. Res*, **7**, 859-880.

Newbold, R. R., Padilla-Banks, E., Snyder, R. J. and Jefferson, W. N. (2005) Developmental exposure to estrogenic compounds and obesity, *Birth Defects Res.A Clin. Mol. Teratol.* **73**, 478-480.

Nohara, H., Okayama, N., Inoue, N., Koike, Y., Fujimura, K., Suehiro, Y., Hamanaka, Y., Higaki, S., Yanai, H., Yoshida, T., Hibi, T., Okita, K. and Hinoda, Y. (2004) Association of the -173 G/C polymorphism of the macrophage migration inhibitory factor gene with ulcerative colitis, *J. Gastroenterol* **39**, 242-246.

Nurminen, M. (1995) Linkage failures in ecological studies, *World Health Stat. Q.* **48**, 78-84.

Nurminen, T. (1995) Maternal pesticide exposure and pregnancy outcome, *J. Occup. Environ. Med.***37**, 935-940.

Paganetto, G., Campi, F., Varani, K., Piffanelli, A., Giovannini, G. and Borea, P. A. (2000) "Endocrine-disrupting agents on healthy human tissues, *Pharmacol. Toxicol.* **86**, 24-29.

Parkes, J., Dolk, H. and Hill, N. (1998) Does the Child Health Computing System adequately identify children with cerebral palsy? *J. Public Health Med.* **20**, 102-104.

Rajapakse, N., Ong, D. and Kortenkamp, A. (2001) Defining the impact of weakly estrogenic chemicals on the action of steroidal estrogens, *Toxicol. Sci.* **60**, 296-304.

Relyea, R. A. & Hoverman, J. T. (2003) The impact of larval predators and competitors on the morphology and fitness of juvenile treefrogs, *Oecologia.* **134**, 596-604.

Shaw, G. M., Wasserman, C. R., O'Malley, C. D., Nelson, V. and Jackson, R. J. (1999) Maternal pesticide exposure from multiple sources and selected congenital anomalies, *Epidemiology* **10**, 60-66.

Skakkebaek, N. E. (2003) Testicular dysgenesis syndrome, *Horm. Res.* **60**, 49.

Steinhardt, G. F. (2004) Endocrine disruption and hypospadias, *Adv. Exp. Med. Biol.* **545**, 203-215.

Stoker, T. E., Parks, L. G., Gray, L. E. and Cooper, R. L. (2000) Endocrine-disrupting chemicals: prepubertal exposures and effects on sexual maturation and thyroid function in the male rat. A focus on the EDSTAC recommendations. Endocrine Disrupter Screening and Testing Advisory Committee, *Crit Rev. Toxicol.* **30**, 197-252.

Strand, M. L., Wahlgren, A., Svechnikov, K., Zetterstrom, C., Setchell, B. P. and Soder, O. (2005) Interleukin-18 is expressed in rat testis and may promote germ cell growth, *Mol. Cell. Endocrinol.* **240**, 64-73.

Swan, S.H., Main, K.M., Liu F., Stewart, S.L., Kruse, R.L., Calafat, A.M., Mao, C.S., Redmon, J.B., Ternand, C.L., Sullivan, S. and Teague, J.L. Study for Future Families Research Team (2005) Decrease in anogenital distance among male infants with prenatal phthalate exposure, *Environ Health Perspect.* **113**, 1056-1061.

Topp, M., Huusom, L. D., Langhoff-Roos, J., Delhumeau, C., Hutton, J. L. and Dolk, H. (2004) Multiple birth and cerebral palsy in Europe: a multicenter study", *Acta Obstet. Gynecol. Scand.* **83**, 548-553.

Toppari, J. (2003) Physiology and disorders of testicular descent, *Endocr. Dev.* **5**, 104-109.

Toppari, J., Kaleva, M. and Virtanen, H. E. (2001) Trends in the incidence of cryptorchidism and hypospadias, and methodological limitations of registry-based data, *Hum. Reprod. Update* **7**, 282-286.

Vidaeff, A. C. and Sever, L. E. (2005) *In utero* exposure to environmental estrogens and male reproductive health: a systematic review of biological and epidemiologic evidence, *Reprod. Toxicol.* **20**, 5-20.

Virtanen, H. E., Rajpert-De Meyts, E., Main, K. M., Skakkebaek, N. E. and Toppari, J. (2005) Testicular dysgenesis syndrome and the development and occurrence of male reproductive disorders, *Toxicol. Appl. Pharmacol.* **207**, 501-505.

Vorhees, C. V., Weisenburger, W. P. and Minck, D. R. (2001) Neurobehavioral teratogenic effects of thalidomide in rats, *Neurotoxicol. Teratol.* **23**, 255-264.

Vreugdenhill H.J.I., Slijper F.M.E., Mulder P.G.H. and Weisglas-Kuprus N. (2002) Effects of perinatal exposure to pcbs and dioxins on play behavior in dutch children at school age, *Environ. Health Perspect.* **110**, A593-A598.

Young, A. L., Giesy, J. P., Jones, P. D. and Newton, M. (2004) Environmental fate and bioavailability of Agent Orange and its associated dioxin during the Vietnam War, *Environ. Sci. Pollut. Res. Int.* **11**, 359-370.

Young, A. L., Giesy, J. P., Jones, P., Newton, M., Guilmartin, J. F., Jr. and Cecil, P. F., Sr. (2004) Assessment of potential exposure to Agent Orange and its associated TCDD, *Environ. Sci. Pollut. Res. Int.* **11**, 347-348.

SECTION 1:

METHODS

ENDPOINTS FOR PRENATAL EXPOSURES IN TOXICOLOGICAL STUDIES

A. MANTOVANI* AND F. MARANGHI
Department Food Safety and Veterinary Public Health
Istituto Superiore di Sanità
Rome
ITALY
(author for correspondence, Email: alberto@iss.it)*

Summary

The standard approach to developmental toxicology includes, 1) studies targeting the organogenesis/pregnancy period, to assess birth defects, minor anomalies (that may be a signal of more severe effects at higher dose levels), fetal growth and viability (OECD guideline 414); 2) one- and two-generation studies (OECD guidelines 415 and 416) that provide an overall assessment of general parameters related to postnatal development and survival. Studies investigating effects on organogenesis are important for risk assessment, since short-term exposures in susceptible developmental phases (e.g., from pharmaceuticals or workplace chemicals) may elicit birth defects and/or embryonic loss. Several *in vitro* assays also exist, using cultured rodent embryonic stem cells or whole embryos that deserve further exploitation as a possible screening battery as well as to understand embryotoxicity mechanisms.

Nevertheless, the science of risk assessment is increasingly concerned with the subtle, but potentially important, effects of low exposure levels relevant to the general population, such as dietary intakes of contaminants. In this respect, the 2-generation study receives considerable attention, as it is the only "standard" protocol whereby an organism is exposed during the whole of development, from gamete stage through sexual development.

The wide group of endocrine disrupters (ED) are a good example of a challenge to developmental toxicologists from several standpoints, including a) the potential to

P. Nicolopoulou-Stamati et al. (eds.), Congenital Diseases and the Environment, 21–36.

induce long-term effects upon exposure in susceptible developmental phases, including postnatal life up to puberty; b) the potential impact of endocrine disruption on immune, neurobehavioral and reproductive development as well as on susceptibility to cancer later in life. Therefore, tiered approaches have to be developed. Screening in vitro/in vivo assays should identify the most relevan.mechanisms/endpoints to be investigated by targeted two-generation assays; these would allow the follow-up of late outcomes and the identification of relevant NOELs for the most sensitive effects.

Other long-term effects of prenatal exposures, such as disruption of immune development and early predisposition to cancer, may be considered as likely "hot topics" for the next wave of research projects on chemical safety.

1. Introduction

The exposure of the developing organism to manufactured xenobiotics as well as natural substances can lead to developmental alterations which can be evident immediately after birth and/or during postnatal life; such effects can be partially or completely recovered but they can also result in persistent sequelae (Peters, 1998). During pregnancy, unique events occur that influence all the subsequent physiological, anatomical, and biochemical processes. The organism passes through highly dynamic sequences of developmental events that involve such processes as cellular proliferation, apoptosis, differentiation, intercellular signalling, etc., (Zakeri and Ahuja, 1997).

Organogenesis is a short phase, as it represents approximately 10 days (GD 6-15) out of a 22-day pregnancy in rat, 13 days (GD 6-19) out of a 29-day pregnancy in rabbit and 7 weeks (GD 9-50) days out of a 9-month pregnancy in humans (Kilborn *et al.,* 2002). Only this short developmental phase is generally considered to be sensitive to the induction of malformations (Mantovani *et al.,* 1993). Nevertheless, the fetal period is also susceptible to general insults that may affect growth and viability, as well as to specific effects on the histogenesis of target organs and tissues (Calamandrei and Mantovani, 2001). Furthermore, after birth, several organs and tissues show prolonged developmental periods, that can be sensitive to effects on growth and functional maturation; thus, postnatal life, up to and including adolescence, has to be regarded as a further window of potentially enhanced susceptibility and vulnerability to any insult coming from the environment (Selevan *et al.,* 2000). Development has to be considered as a continuum, showing different peaks of vulnerability to specific insults. Thus, major tasks of developmental toxicology are:

- to detect substances that can influence reproduction and/or development,
- identify target organs and tissues,
- identify processes involved, target periods and windows of specific susceptibility; and,
- possibly reveal mechanisms of action (Peters, 1998).

The standard approach to developmental toxicology, as applied in regulatory testing strategies, includes:

1) studies targeting the organogenesis/pregnancy period, to assess birth defects, minor anomalies that may be a signal of more severe effects at higher dose levels, fetal growth and viability (OECD guideline 414);
2) one- and two-generation studies (OECD guidelines 415 and 416) that provide an overall assessment of general parameters related to postnatal development and survival (Barlow *et al.,* 2002).

The importance of studies investigating effects on organogenesis should not be overlooked, since short-term, even single, exposures in susceptible developmental phases may increase the risk of birth defects and/or embryonic loss; such instances can occur with, e.g., pharmaceuticals and workplace exposures (Koren *et al.,* 1998; McMartin and Koren, 1999). Several interesting *in vitro* assays exist, using cultured rodent embryonic stem cells or whole embryos and these could be developed as a screening battery as well as being used to understand embryotoxicity mechanisms (Bremer *et al.,* 2005). Nevertheless, the science of risk assessment is increasingly concerned with the subtle, but potentially important, developmental effects of the low exposure levels that are relevant to the general population, e.g., those resulting from dietary intakes of contaminants and residues (Mantovani and Macrì, 2002). Because of this concern, the two-generation study is receiving considerable attention as a key element of toxicological risk assessment; in fact, it is the only "standard" protocol used in regulatory testing whereby an organism is exposed throughout development, from gamete stage through to sexual development (Lamb and Brown, 2000). In particular, the two-generation study is regarded as the major approach for hazard identification and dose-response assessment of the wide group of endocrine disrupters (ED), a heterogeneous group of contaminants widespread in the environment that may interfere with the endocrine system in animals and in humans

(Mantovani, 2002). As an example, two-generation studies in the rat have characterized the effects of nonylphenol, a compound with estrogen-like activity that can contaminate the environment and food chains as a by-product of certain non-ionic surfactants or consumer products; nonylphenol showed only minor effects on neurobehavioral development (Flynn *et al.,* 2002), whereas it affected the functional development of reproductive, immune and renal systems at dose levels not inducing overt signs of toxicity (Karrow *et al.,* 2004; Nagao *et al.,* 2001).

Therefore, the selection of appropriate test systems to evaluate possible damages to those developmental phases which are more vulnerable to certain insults stands out as a critical issue in developmental toxicology.

2. An overview of regulatory tests in developmental toxicology

According to the up-to-date conceptual framework of chemical testing for developmental toxicity, developmental effects have to be viewed in the whole context of reproductive function; thus, developmental toxicity does not concern only alterations induced in the intrauterine life, but also effects that can be observed and/or induced postnatally. Also, developmental toxicity of a chemical has to be assessed by considering the whole spectrum of toxicological actions and targets (Kimmel and Makris, 2001).

2.1. *Prenatal development toxicity study (OECD guideline 414)*

In this protocol, the female animals are treated with the test substance during pregnancy, mainly and at the least in the organogenesis period, and sacrificed immediately before the delivery (Barlow *et al.,* 2002). The uterus is excised and weighed. Caesarean parturition is routinely used to avoid the cannibalism of malformed or non-viable pups that occurs in rodents and rabbits (Schardein *et al.,* 1978). Cannibalizations would obviously prevent the observation of developmental damage. Measures of prenatal lethality (early and late resorptions, non-viable fetuses) are recorded; viable fetuses are removed from the uterus, weighted, sexed as well as further processed and analysed for different measures. Possible external alterations are recorded and usually half of the pups from each litter are reserved for skeletal studies and half of the pups for studies of soft tissue anomalies. In this way, the main purpose of the developmental toxicity study is to identify lethal, teratogenic or growth retardation effects both on embryo and fetus, based on the following endpoints: the ratio of embryonic and fetal resorptions or deaths to viable fetuses; the measurement of fetal weight and sex ratio; and the examination of external, visceral and skeletal morphology.

Two animal laboratory species are used for prenatal developmental toxicity testing: a rodent, more frequently the rat; and a non-rodent such as the rabbit, a lagomorph. The mouse may be used instead of the rat; the two species share many features, but the mouse is mostly used for detailed investigation of mechanisms of action, susceptible developmental phases, etc. (e.g. Tiboni and Giampietro, 2005), rather than for hazard identification testing. The use of more than one species is widely considered necessary, since it can offer a better chance of detecting a potential human developmental toxicant. The rabbit is normally chosen as the second, non-rodent species, for historical and physiological reasons. Rabbits, but not rats, are susceptible to limb-reduction deformities induced by thalidomide, the drug that caused the most serious outbreak of chemically-induced malformations yet recorded (Gordon *et al.,* 1981). Attention should be given to interspecies extrapolation of experimental results, e.g., placentation is different in the rat, rabbit and human. The early post-implantation rat embryo is completely surrounded by the yolk sac that mediates the mother-to-conceptus transfer of nutrients and other substances; on the contrary, the yolk sac only partly occurs in the rabbit conceptus. In humans, the chorioallantoic placenta plays a major role throughout organogenesis, although supplemental absorption of nutrients by the yolk sac could also occur. Such differences in placental morphology and physiology may affect toxicant uptake and metabolism (Carney *et al.,* 2004). The physiology of pregnancy also differs across species, as the rabbit has a slightly longer fetal period (approximately 11-12 days). The short, one-week fetal phase is a potential shortcoming in using the rat as the most prominent species in developmental testing, since the extrapolation to humans, who have a prolonged intrauterine phase of fetal maturation, may be difficult. As a consequence, the rat newborn is less mature than the human; for instance, the brain-blood barrier is markedly less efficient in rat newborns, allowing a higher passage of toxicants, as observed for the pesticide ivermectin (Lankas *et al.,* 1989). The risk assessment of chemicals where experimental studies provide conflicting results, may require a comparative evaluation of pharmacokinetics among different species, including humans (Nau, 2001). A major problem of the rabbit is the comparatively limited knowledge base for toxicokinetics and systemic toxicity data in this species. Indeed, developmental toxicity should be viewed in the context of all available toxicological data, as it represents more than simply effects on the embryo, but consists of effects on the mother-placenta-conceptus complex (Carney *et al.,* 2004). Accordingly, the EU criteria for classification and labelling of developmental toxicants do not include substances which damage the conceptus as a mere consequence of maternal toxicity (Sullivan, 1995). Other alternative species have been considered as non-rodent models, but the amount of available data is comparatively limited. Non-human primates can share many physiological and metabolic pathways with humans and have a similar placenta (King, 1993); however, many ethical constraints exist regarding their use as experimental animals in Europe. The miniature pig strains deserve interest because pharmacodynamic and pharmacocokinetic data are available on pigs, a species in which various compounds for veterinary use are tested. Currently, miniature pigs are used for some specific studies (Rothcotter *et al.,* 2002).

Of course, many targeted protocols may be envisaged as spin-off from the standard prenatal toxicity test, in order to fulfil specific purposes. Examples include, but are not limited to:

- treatment on single days of pregnancy to identify the critical organogenesis for inducing a certain effect (e.g., Tiboni and Giampietro, 2005);
- examination of embryos *in vivo* to investigate the pathogenesis of teratogenicity or embryolethality (e.g. Vinson and Hales, 2003);
- follow-up of pups exposed *in utero* until weaning or adulthood, in order to assess late-appearing effects on the development of target organs, especially those with a prolonged phase of functional maturation such as the kidney or nervous system (Calamandrei and Mantovani, 2001).

2.2. Two-generation reproduction toxicity study (OECD 416)

Since human exposure to chemicals found in food and in the environment may span a lifetime, multigeneration studies are usually chosen because the test substance is administered continuously, without interruption, to parental (P) and subsequent offspring generations (F1, F2). This study is normally conducted with a single laboratory species, the rat or occasionally the mouse, and it is extended over at least two generations with one or two litters per generation (Barlow *et al.,* 2002). One-generation reproductive studies (OECD 415) are generally performed by pharmaceutical industries for testing drugs with prolonged usage.

A two-generation reproductive study will provide information about effects on male and female reproductive performances, potency and fertility, on pregnancy outcomes, on maternal lactation and offspring care, on the prenatal and postnatal survival, growth and development of the offspring as well as on their reproductive capacity. Regulators have realized that the two-generation study is the only standard test protocol where a group of animals (the F1 generation) is exposed to a substance throughout all of development, i.e., from gamete stage until sexual maturity and production of offspring; in particular, it is the only protocol providing a reliable assessment of exposure of newborn and peripubertal animals.

Thus, the two-generation study is now considered as a key study in the overall assessment of toxicological risk (Lamb and Brown, 2000). Multigeneration studies are relatively flexible protocols that can include, besides the core study, satellite groups to investigate specific effects. For instance, cross-fostering (i.e., a group of litters with intrauterine exposure nursed by untreated dams and vice versa) may be used to distinguish the role of maternal and pup exposures (Luebker *et al.,* 2005a). In other instances, different sacrifice timepoints are used to identify the changing trends of certain parameters at different stages, e.g., neonatal, juvenile and adult (Laurenzana *et al.,* 2002). The core measures of multigeneration studies provide an indication of potential adverse developmental effects induced by prenatal exposure, by the observation of parameters such as reduction of birth weight, litter size or postnatal survival; reduced offspring survival at 4 days of age is a sensitive, albeit unspecific, indicator of prenatal toxicity in the rat (Mantovani, 1992).

Such protocols may also allow the identification of any major target organs for toxicity (particularly reproductive organs) directly from the histological examination of tissues derived from parental and offspring animals. The multigeneration study is generally considered to be a powerful and comprehensive tool to reveal reproductive toxicity of chemicals (Barlow *et al.,* 2002). A new and interesting aspect is the ability to investigate long-term effects on the prenatal and postnatal development of the immune, nervous and reproductive systems (Mantovani and Maranghi, 2005; Ogata *et al.,* 2001; Omura *et al.,* 2001; Tyla *et al.,* 2000).

A drawback of multigeneration tests is their complexity and cost. The investigation of all parameters relevant to reproductive function, as well as development, results in a huge quantity of data. As a result, such studies can be difficult to manage and require the use of more animals than any other type of standard toxicity test, which raises important animal welfare issues. On a practical ground, they are time consuming, space consuming and labour intensive. It is suggested that a battery of preliminary assays should target the most relevant effects (Gelbke *et al.,* 2004). Such a stepwise approach may thus augment the unique ability of two-generation tests to trace the effects of xenobiotics from early embryogenesis until the subsequent maturation and reproductive performance of the new living organism. Another, complementary, approach is to potentiate the statistical exploitation of data. A potentially interesting development is the "benchmark dose" (BMD) approach, that makes full use of the data at all dose levels for the different parameters. The estimated BMD is the lower 95% confidence limit of an area of the dose-response curve where the slope starts to change noticeably, normally in the range of an increase of 5 - 10 % compared to the background (control) value (Slob *et al.,* 2005). The BMD provides an accurate comparative assessment of the dose-response relationships for the different effects; thus it could be very useful for complex studies where multiple parameters are evaluated together. However, it has seen only a limited application to two-generation assays (e.g., Luebker *et al.,* 2005,b; Reiss and Gaylor, 2005). A possible reason is that an appropriate exploitation of BMD would require a change of current testing protocols, with more dose level groups (Slob *et al.,* 2004).

2.3. In vitro alternative tests for developmental toxicity

The development of *in vitro* assays is viewed as a European priority for both ethical and practical reasons, in order to complement *in vivo* testing strategies and to result in both improved scientific reliability and the use of reduced numbers of laboratory animals (Hareng *et al.,* 2005). To achieve this objective, validated alternative test methods are required for the different components of toxicological testing. While some *in vivo* tests for topical toxicity have been successfully replaced by alternative methods, systemic toxicities require new test strategies in order to achieve an adequate safety level for human health while reducing or replacing whole animal testing. In this respect, reproductive toxicity offers several opportunities to develop alternative tests. The number of animals currently required in the *in vivo* assays is high; however, the reproductive system can be studied by breaking down it into well-defined sub-elements covering the reproductive cycle. A number of pioneering alternatives have already been developed and used, and since basically the same protocols for animal experiments are carried out for drugs, chemicals and cosmetics, the new testing strategies could be widely applied to the different fields of chemical testing (Hareng *et al.,* 2005).

In particular, as regards developmental toxicity, several interesting assay systems do exist. Effects on organogenesis can be investigated in part through, respectively, the Rodent Embryonic Stem Cells (Pellizzer *et al.,* 2005) and the Whole Embryo Culture (Piersma, 2004). Both assays have been validated under the auspices of the European Centre for Validation of Alternative Methods (ECVAM, http://ecvam.jrc.it/index.htm). However, they cannot be considered as replacements for *in vivo* testing (Bremer *et al.,* 2005); the Whole Embryo Culture is an excellent tool for studying embryotoxicity mechanisms, but it covers only early organogenesis when the rodent embryo is supported by yolk sac nutrition; stem cells can model differentiation processes relevant to late organogenesis, but certainly the information provided is too limited for evaluating complex effects at organism level, *in vitro* assays may serve toxicological risk assessment either as components of a screening battery and/or for better understanding mechanisms of actions (Eisenbrand *et al.,* 2002). Toxicity mechanisms are no longer regarded as a sort of academic embellishment of toxicological risk assessment. Understanding mechanisms can be highly relevant to such issues as the effects of combined exposures (Andrews *et al.,* 2004) and gene expression pathways, that could be related to differential susceptibility to chemical effects (e.g. Bonner *et al.,* 2003). The new developments in bioinformatics are also relevant. The increasing amount of information on molecular mechanisms also support new, predictive models based on advanced computational methodologies ("*in silico*") that may be highly cost-effective (Fielden *et al.,* 2002); for instance, approaches based on quantitative structure-activity relationships may model chemical-receptor interactions of test compounds (Akahori *et al.,* 2005). There are also some interesting efforts towards *in silico* models for effects on functional maturation, e.g., of the nervous system (Gohlke *et al.,* 2004). On the other hand, the overall issue of delayed developmental effects is still in dire need of non-animal assays that can be proposed for validation (Bremer *et al.,* 2005).

Toxicokinetics presents another major challenge to *in vitro* testing strategies. The many mechanisms relevant to developmental toxicity may involve multiple metabolic pathways. It is necessary to understand the ability of different test systems to perform metabolic transformation of xenobiotics, as well to implement the use of metabolic activation systems in alternative models (Combes, 2004). Rapid progress in molecular biology has led to an increasing interest in the incorporation of such tools as transcriptomics and proteomics in toxicological testing (Klemm and Schrattenholz, 2004; Pellizzer *et al.,* 2004). These innovative approaches may play a useful role in the determination of subtle effects, but there are also problems concerning reproducibility, repeatability and scientific relevance that deserve careful attention.

It is important to stress that no *in vitro* system can mimic the complexity of *in vivo* development. Nevertheless, they may be efficiently combined to compose testing strategies which might offer a realistic opportunity to achieve a substantial reduction in animal experiments in assessments of reproductive and development toxicity. Therefore, the validation of test batteries emerges as a critical issue, together with and beyond the validation of individual assays. In this respect, it is noteworthy that the Integrated Project ReProTect, funded within the EU 6th Framework Programme, targets primarily the development of existing *in vitro* assays on male/female fertility and teratogenicity as an integrated testing battery (Hareng *et al.,* 2005).

2.4. ***Developmental toxicity testing of environmental contaminants. The example of endocrine disrupters***

EDs are heterogeneous compounds sharing the potential of interacting with the endocrine system by means of different mechanisms of action, e.g., binding with nuclear receptors; alterations of the steroid and/or thyroid metabolism, etc. (Neubert, 1997). Since the homeostasis of steroid and thyroid hormones are the main targets of ED effects, it is apparent that susceptibility is associated to life-cycle phases and that it is enhanced throughout the prenatal and postnatal development of offspring.

EDs include several agrochemicals, veterinary antiparasite drugs and biocides (chlorinated insecticides, organotins, imidazoles, triazoles); compounds present in consumer products (bisphenol A, phthalates, polybrominated flame retardants); environmental contaminants (dioxins, polychlorinated biphenyls), as well as so-called "natural" products (phytoestrogens) (Neubert, 1997; Mantovani, 2002). EDs are often classified according to their mechanisms, as estrogenic, antiandrogenic, aromatase inhibitors, thyreostatic, etc.. However, such distinctions should not be considered too rigid, as the effects of interference with the nuclear receptor pathways may vary with the endocrine status (e.g., sex and/or age) of the target organism, that in turn influence the regulation patterns of receptor-mediated transcription (Beischlag and Perdew, 2005). New *in vivo* models are now available to investigate such patterns and they are, in fact, increasingly exploited in toxicology, including transgenic mice with human receptors and reporter genes indicating receptor modulation (Villa *et al.,* 2004).

Although the number of studies is steadily increasing, EDs still present a challenge to developmental toxicologists from several standpoints. EDs have the potential to induce long-term effects upon exposure during susceptible developmental phases, including postnatal life up to puberty (e.g., Maranghi *et al.,* 2003; Traina *et al.,* 2003). Thus, the two-generation test is recognized as the key study for hazard identification of EDs (Mantovani, 2002).

Due to the regulatory role of endocrine homeostasis, mechanisms of endocrine disruption may impact on immune system (e.g. AhR agonists), neurobehavioural (e.g., thyroid inhibitors), or reproductive development (e.g. antiandrogens), as well as on susceptibility to cancer (e.g., mammary or testicular germ-cell tumours) (Donahue *et al.,* 2005; Fenton *et al.,* 2002; Gray *et al.,* 1999; Inouye *et al.,* 2005;

Mahood *et al.,* 2005). In turn, this means that a number of measures are potentially relevant to certain critical effects. As an example, indicators of impaired male reproductive development include: anogenital distance; hypospadias; age and weight of offspring on the day of balanopreputial separation; weight of testis and accessory glands; sperm parameters (count, motility, morphology, genetic integrity).

In rodents, fertility is quite an insensitive parameter, since severe damage to reproductive functional development is required to detect an appreciable effect on measures of reproductive performance such as gestational index. On the contrary, the qualitative and quantitative histological examination of reproductive tissues provides effect markers detectable at lower dose levels than reduced fertility (Mantovani and Maranghi, 2005). Therefore, the development of tiered approaches is particularly relevant (Gelbke *et al.,* 2004). Screening *in vitro*/*in vivo* assays should identify the relevant mechanisms/endpoints to be investigated; in the final tier, targeted two-generation assays will allow the follow-up of late outcomes and the identification of relevant NOELs for the critical effects.

Several screening assays are being validated. *In vitro* tests include, for example: the activation of oestrogen and androgen receptors for binding/transcriptional activity; and assays for steroidogenesis. However, few or no reliable assays are yet available for such important actions as interference with thyroid function or with hypothalamic-pituitary axis (Bremer *et al.,* 2005). *In vivo* short-term assays have received much interest, since they appear to mimic the complexity of the endocrine system in a more reliable way. The uterotrophic assay and Hershberger assay are developed to identify interference with estrogen- or androgen-related pathways respectively. Both are based on weight changes of reproductive tissues, as well as relying on existing knowledge of endocrinology (Gelbke *et al.,* 2004). The peripubertal assays are more complex tests and include thyroid assessment. In addition to screening for relevant endocrine actions, they may serve as specific tests for toxicity in juvenile stages (Kim *et al.,* 2002).

3. Recommendations for further research in developmental toxicology

Based on the above considerations, the available toxicological assays can be considered as reliable tools to identify several kinds of developmental hazards. Nevertheless, some major issues for research in developmental toxicology are also apparent, such as:

- the implementation of integrated *in vitro* batteries, making full use of metabolic activation systems as well as of the new tools provided by molecular biology;
- *in vitro* models as well as *in vivo* models, to evaluate effects on the fetal phase of the development, with special attention to systems showing a complex and prolonged maturation period, such as the nervous system;
- the assessment of long term effects of exposures during early life, such as disruption of immune development and early predisposition to cancer.

These may be considered as likely "hot topics" for a next wave of research projects on chemical safety.

Besides a major role in reducing the chance for deleterious chemicals to enter our lives, toxicological assays may also support the understanding of complex, multifactorial events leading to birth defects or endocrine disruption. The use of up-to-date approaches, such as transgenic animal animals or transcriptomics may, indeed, be useful in dissecting mechanisms of gene-nutrient-chemical interactions (e.g. Spiegelstein *et al.,* 2005). Such information may also characterize possible markers of enhanced susceptibility as well as pinpoint potentially protective factors towards environmental chemicals. As the European Environment and Health Strategy points out (e.g. Kotzias, 2005), improved knowledge on toxicological effects and mechanisms is key to supporting the scientific basis of public health strategies.

Acknowledgements

The paper has been prepared within the framework of the Italian National Public Health grant (4A/F3) "Risk Analysis of Compounds present in Foods of Animal Origin – SARA".

References

Akahori, Y., Nakai, M., Yakabe, Y., Takatsuki, M., Mizutani, M., Matsuo, M. and Shimohigashi, Y. (2005) Two-step models to predict binding affinity of chemicals to the human estrogen receptor alpha by three-dimensional quantitative structure-activity relationships (3D-QSARs) using receptor-ligand docking simulation, *SAR QSAR Environ. Res.* **16**, 323-337.

Andrews, J.E., Nichols, H.P., Schmid, J.E., Mole, L.M., Hunter, E.S. 3rd. and Klinefelter, G.R. (2004) Developmental toxicity of mixtures: the water disinfection by-products dichloro-, dibromo- and bromochloro acetic acid in rat embryo culture, *Reprod. Toxicol.* **19**, 111-116.

Barlow, S.M., Greig, J.B., Bridges, J.W., Carere, A., Carpy, A.J.M., Galli, C.L., Kleiner, J., Knudsen, I., Koeter, H.B.W.M., Levy, L.S., Madsen, C., Mayerj, S., Narbonne, J-F., Pfannkuch, F., Prodanchuk M.G., Smith M.R. and Steinberg P. (2002) Hazard identification by methods of animal-based toxicology, *Food Chem. Toxicol.* **40**, 145–149.

Beischlag, T.V. and Perdew G.H. (2005) ER alpha-AHR-ARNT protein-protein interactions mediate estradiol-dependent transrepression of dioxin-inducible gene transcription, *J. Biol. Chem.* **280**, 2607-2611.

Bonner, A.E., Lemon, W.J. and You, M. (2003) Gene expression signatures identify novel regulatory pathways during murine lung development: implications for lung tumorigenesis, *J. Med. Genet.* **40**, 408-417.

Bremer, S., Cortvrindt, R., Daston, G., Eletti, B., Mantovani, A., Maranghi, F., Pelkonen, O., Ruhdel, I. and Spielmann, H. (2005) Reproductive and developmental toxicity, *ATLA* **33** (Suppl 1), 183–220.

Calamandrei, G. and Mantovani, A. (2001) Delayed developmental effects following prenatal exposure to drugs, *Curr. Phar. Design* **7**, 859-882.

Carney, E.W., Scialli, A.R., Watson, R.E. and DeSesso, J.M. (2004) Mechanisms regulating toxicant disposition to the embryo during early pregnancy: an interspecies comparison, *Birth Defects Res. C. Embryo. Today* **72**, 345-360.

Combes, R.D. (2004) The case for taking account of metabolism when testing for potential endocrine disrupters in vitro, *Altern. Lab. Anim.* **32**, 121-135.

Donahue, D.A, Dougherty, E.J. and Meserve, L.A. (2004) Influence of a combination of two tetrachlorobiphenyl congeners (PCB 47; PCB 77) on thyroid status, choline acetyltransferase (ChAT) activity, and short- and long-term memory in 30-day-old Sprague-Dawley rats, *Toxicology* **203**, 99-107.

Eisenbrand, G., Pool-Zobel, B., Baker, V., Balls, M., Blaauboer, B.J., Boobis, A., Carere, A., Kevekordes, S., Lhuguenot, J.C., Pieters, R. and Kleiner, J. (2002) Methods of in vitro toxicology, *Food Chem. Toxicol.* **40**, 193-236.

Fenton, S.E., Hamm, J.T., Birnbaum, L.S. and Youngblood, G.L. (2002) Persistent abnormalities in the rat mammary gland following gestational and lactational exposure to 2,3,7,8-tetrachlorodibenzo-p-dioxin (TCDD), *Toxicol. Sci.* **67**, 63-74.

Fielden, M.R., Matthews, J.B., Fertuck, K.C., Halgren, R.G., Zacharewski, T.R. (2002) In silico approaches to mechanistic and predictive toxicology: an introduction to bioinformatics for toxicologists, *Crit. Rev. Toxicol.* **32**, 67-112.

Flynn, K.M., Newbold, R.R. and Ferguson, S.A. (2002) Multigenerational exposure to dietary nonylphenol has no severe effects on spatial learning in female rats, *Neurotoxicology* **23**, 87-94.

Gelbke, H.P., Kayser, M. and Poole, A. (2004) OECD test strategies and methods for endocrine disrupters, *Toxicology* **205**, 17-25.

Gohlke, J.M., Griffith, W.C. and Faustman, E.M. (2004) The role of cell death during neocortical neurogenesis and synaptogenesis: implications from a computational model for the rat and mouse, *Brain Res. Dev. Brain Res.* **151**, 43-54.

Gordon, G.B., Spielberg, S.P., Blake, D.A. and Balasubramanian, V. (1981) Thalidomide teratogenesis: evidence for a toxic arene oxide metabolite, *PNAS* **78**, 2545-2548.

Gray, L.E. Jr, Wolf, C., Lambright, C., Mann, P., Price, M., Cooper, R.L. and Ostby, J. (1999) Administration of potentially antiandrogenic pesticides (procymidone, linuron, iprodione, chlozolinate, p,p'-DDE, and ketoconazole) and toxic substances (dibutyl- and diethylhexyl phthalate, PCB 169, and ethane dimethane sulphonate) during sexual differentiation produces diverse profiles of reproductive malformations in the male rat, *Toxicol. Ind. Health* **15**, 94-118.

Hareng, L., Pellizzer, C., Bremer, S., Schwarz, M. and Hartung, T. (2005) The integrated project ReProTect: a novel approach in reproductive toxicity hazard assessment, *Reprod. Toxicol.* **20,** 441-452.

Inouye, K., Pan, X., Imai, N., Ito, T., Takei, T., Tohyama, C. and Nohara, K. (2005) T cell-derived IL-5 production is a sensitive target of 2,3,7,8-tetrachlorodibenzo-p-dioxin (TCDD), *Chemosphere* **60**, 907-913.

Karrow, N.A., Guo, T.L., Delclos, K.B., Newbold, R.R., Weis, C., Germolec, D.R., White, K.L. Jr. and McCay, J.A. (2004) Nonylphenol alters the activity of splenic NK cells and the numbers of leukocyte subpopulations in Sprague-Dawley rats: a two-generation feeding study, *Toxicology* **196**, 237-245.

Kilborn, S.H., Trudel, G. and Uhthoff, H. (2002) Review of growth plate closure compared with age at sexual maturity and lifespan in laboratory animals, *Contemp. Top Lab. Anim. Sci.* **41**, 21-26.

Kim, H.S., Shin, J.H., Moon, H.J., Kim, T.S., Kang, I.H., Seok, J.H., Kim, I.Y., Park, K.L. and Han, S.Y. (2002) Evaluation of the 20-day pubertal female assay in Sprague-Dawley rats treated with DES, tamoxifen, testosterone, and flutamide, *Toxicol. Sci.* **67**, 52-62.

Kimmel, C.A. and Makris, S.L. (2001) Recent developments in regulatory requirements for developmental toxicology, *Toxicol. Lett.* **120**, 73-82.

King, B.F. (1993) Development and structure of the placenta and fetal membranes of nonhuman primates, *J. Exp. Zool.* **266**, 528-540.

Klemm, M. and Schrattenholz, A. (2004) Neurotoxicity of active compounds-establishment of hESC-lines and proteomics technologies for human embryo- and neurotoxicity screening and biomarker identification, *ALTEX* **21** (Suppl 3), 41-48.

Koren, G., Pastuszak, A. and Ito, S. (1998) Drugs in pregnancy, *N. Engl. J. Med.* **338**, 1128-1137.

Kotzias, D. (2005) Indoor air and human exposure assessment--needs and approaches, *Exp. Toxicol. Pathol.* **57** (Suppl 1), 5-7.

Lamb, J.C. and Brown, S.M. (2000) Chemical testing strategies for predicting health hazards to children, *Reprod. Toxicol.* **14**, 83-94.

Lankas, G.R., Minsker, D.H. and Robertson, R.T. (1989) Effects of ivermectin on reproduction and neonatal toxicity in rats, *Food Chem. Toxicol.* **27**, 523-529.

Laurenzana, E.M., Balasubramanian, G., Weis, C., Blaydes, B., Newbold, R.R. and Delclos, K.B. (2002) Effect of nonylphenol on serum testosterone levels and testicular steroidogenic enzyme activity in neonatal, pubertal, and adult rats, *Chem. Biol. Interact.* **139**, 23-41.

Luebker, D.J., Case, M.T., York, R.G., Moore, J.A., Hansen, K.J. and Butenhoff, J.L. (2005) Two-generation reproduction and cross-foster studies of perfluorooctanesulfonate (PFOS) in rats, *Toxicology* **215**, 126-148.

Luebker, D.J., York, R.G., Hansen, K.J., Moore, J.A. and Butenhoff, J.L. (2005) Neonatal mortality from *in utero* exposure to perfluorooctanesulfonate (PFOS) in Sprague-Dawley rats: dose-response, and biochemical and pharamacokinetic parameters, *Toxicology* **215**, 149-169.

Mahood, I.K., Hallmark, N., McKinnell, C., Walker, M., Fisher, J.S. and Sharpe, R.M. (2005) Abnormal Leydig Cell aggregation in the fetal testis of rats exposed to di(n-butyl) phthalate and its possible role in testicular dysgenesis, *Endocrinology* **146**, 613-623.

Mantovani, A. (1992) The role of multigeneration studies in safety assessment of residues of veterinary drugs and additives, *Ann. Ist. Super. Sanita.* **28**, 429-435.

Mantovani, A. (2002) Hazard identification and risk assessment of endocrine disrupting chemicals with regard to developmental effects, *Toxicology* **27**, 367-370.

Mantovani, A. and Macrì, A. (2002) Endocrine effects in the hazard assessment of drugs used in animal production, *J. Exp. Clin. Cancer Res.* **21**, 445-456.

Mantovani, A. and Maranghi F. (2005) Risk assessment of chemicals potentially affecting male fertility, *Contraception* **72**, 308-313.

Mantovani, A., Ricciardi, C., Macri, C. and Stazi, A.V. (1993) Prenatal risks deriving from environmental chemicals, *Ann. Ist. Super. Sanita.* **29**, 47-55.

Maranghi, F., Macrì C., Ricciardi, C., Stazi, A.V., Rescia, M. and Mantovani, A. (2003) Histological and histomorphometric alterations in thyroid and adrenals of CD rat pups exposed *in utero* to methyl thiofanate, *Reprod. Toxicol.* **17**, 617-623.

McMartin, K.I. and Koren, G. (1999) Proactive approach for the evaluation of fetal safety in chemical industries, *Teratology* **60**, 130-136.

Nagao, T., Wada, K., Marumo, H., Yoshimura, S. and Ono, H. (2001) Reproductive effects of nonylphenol in rats after gavage administration, a two-generation study, *Reprod. Toxicol.* **15**, 293-315.

Nau, H. (2001) Teratogenicity of isotretinoin revisited: species variation and the role of all-trans-retinoic acid, *J. Am. Acad. Dermatol.* **45**, 183-187.

Neubert, D. (1997) Vulnerability of the endocrine system to xenobiotic influence, *Regul. Toxicol. Pharmacol.* **26**, 9-29.

Ogata, R., Omura, M., Shimasaki, Y., Kubo, K., Oshima, Y., Aou, S. and Inoue, N. (2001) Two-generation reproductive toxicity study of tributyltin chloride in female rats, *J. Toxicol. Environ. Health A* **63**, 127-144.

Omura, M., Ogata, R., Kubo, K., Shimasaki, Y., Aou, S., Oshima, Y., Tanaka, A., Hirata, M., Makita, Y. and Inoue, N. (2001) Two-generation reproductive toxicity study of tributyltin chloride in male rats, *Toxicol. Sci.* **64**, 224-232.

Pellizzer, C., Adler, S., Corvi, R., Hartung, T. and Bremer, S. (2004) Monitoring of teratogenic effects *in vitro* by analysing a selected gene expression pattern, *Toxicol. In vitro* **18**, 325-335.

Pellizzer, C., Bremer, S. and Hartung, T. (2005) Developmental toxicity testing from animal towards embryonic stem cells, *ALTEX* **22**, 47-57.

Peters, P.W. (1998) Developmental toxicology, adequacy of current methods, *Food Addit. Contam.* **15**, 55-62.

Piersma, A.H. (2004) Validation of alternative methods for developmental toxicity testing, *Toxicol. Lett.* **149**, 147-153.

Reiss, R. and Gaylor, D. (2005) Use of benchmark dose and meta-analysis to determine the most sensitive endpoint for risk assessment for dimethoate, *Regul. Toxicol. Pharmacol.* **43**, 55-65.

Rothkotter, H.J., Sowa, E. and Pabst, R. (2002) The pig as a model of developmental immunology, *Hum. Exp. Toxicol.* **21**, 533-536.

Schardein, J.L., Petrere, J.A., Hentz, D.L., Camp, R.D. and Kurtz, S.M. (1978) Cannibalistic traits observed in rats treated with a teratogen, *Lab. Anim.* **12**, 81-83.

Selevan, S.G., Kimmel, C.A. and Mendola, P. (2000) Identifying critical windows of exposure for children's health, *Environ. Health Perspect.* **108** Suppl 3, 451-455.

Slob, W., Moerbeek, M., Rauniomaa, E. and Piersma, A.H. (2005) A statistical evaluation of toxicity study designs for the estimation of the benchmark dose in continuous endpoints, *Toxicol. Sci.* **84**, 167-185.

Spiegelstein, O., Gould, A., Wlodarczyk, B., Tsie, M., Lu, X., Le, C., Troen, A., Selhub, J., Piedrahita, J.A., Salbaum, J.M., Kappen, C., Melnyk, S., James, J. and Finnell, R.H. (2005) Developmental consequences of *in utero* sodium arsenate exposure in mice with folate transport deficiencies, *Toxicol. Appl. Pharmacol.* **203**, 18-26.

Sullivan, F.M. (1995) The European Community Directive on the classification and labeling of chemicals for reproductive toxicity, *J. Occup. Environ. Med.* **37**, 966-969.

Tiboni, G.M. and Giampietro, F. (2005) Murine teratology of fluconazole, evaluation of developmental phase specificity and dose dependence, *Pediatr. Res.* **58**, 94-99.

Traina, M.E., Rescia, M., Urbani, E., Mantovani, A., Macrì, C., Ricciardi, C., Stazi, A.V., Fazzi, P., Cordelli, E., Eleuteri, P., Leter, G. and Spano, M. (2003) Long-lasting effects of lindane on mouse spermatogenesis induced by *in utero* exposure, *Reprod. Toxicol.* **17**, 25-35.

Tyla, RW., Friedman, M.A., Losco, P.E., Fisher, L.C., Johnson, K.A., Strother, D.E. and Wolf, C.H. (2000) Rat two-generation reproduction and dominant lethal study of acrylamide in drinking water, *Reprod. Toxicol.* **14**, 385-401.

Villa, R., Bonetti, E., Penza, M.L., Iacobello, C., Bugari, G., Bailo, M., Parolini, O., Apostoli, P., Caimi, L., Ciana, P., Maggi, A. and Di Lorenzo, D. (2004) Target-specific action of organochlorine compounds in reproductive and nonreproductive tissues of estrogen-reporter male mice, *Toxicol. Appl. Pharmacol.* **201**, 137-148.

Vinson, R.K. and Hales, B.F. (2003) Genotoxic stress response gene expression in the mid-organogenesis rat conceptus, *Toxicol. Sci.* **74**, 157-164.

Zakeri, Z.F. and Ahuja, H.S. (1997) Cell death/apoptosis: normal, chemically induced, and teratogenic effect, *Mutat. R es.* **396**, 149-161.

CONGENITAL DEFECTS OR ADVERSE DEVELOPMENTAL EFFECTS IN VERTEBRATE WILDLIFE: THE WILDLIFE-HUMAN CONNECTION

GWYNNE LYONS
WWF-UK
Norwich
ENGLAND
Email: Lyonswwf@aol.com

Summary

This paper provides a review of defects reported in vertebrate wildlife, particularly including those present or already programmed at birth, which are known or suspected to be associated with pollutants. Species from each of the main classes of animals in the vertebrate sub-phylum (including bony fish, amphibians, reptiles, birds, and mammals) are now known to have been affected by endocrine disrupting chemicals (EDCs) in the environment.

The similarities in the endocrine system of all vertebrate animals are such that given the presence of EDCs in the environment, effects are likely to be occurring in more species than those currently reported. Section 1 outlines the animal – human connection, and highlights that *Homo sapiens* is also vulnerable.

Section 2 summarises the effects reported in wildlife. These particularly include: genital deformities; altered thyroid function; and suppressed immune system. Section 3 outlines how these pollutant-related effects have also been noted in humans. Birth defects, such as intersex features and male reproductive tract anomalies, and congenital effects on brain function and immune system function should be of major concern.

Furthermore, concern for the long-term health of wildlife populations and humans is enhanced because several laboratory studies have suggested that disorders such as

P. Nicolopoulou-Stamati et al. (eds.), Congenital Diseases and the Environment, 37–87.

deficits in sperm production, can be passed on to unexposed subsequent generations. This highlights the need for a far more precautionary regulatory approach to controlling chemical exposures, and the need for more research to understand the long-term implications for life on earth.

1. Introduction: The animal – human connection and epigenetic reprogramming

This paper reviews some congenital defects in vertebrate wildlife, which are suggested to result from exposure to chemicals in the environment. Many of the effects seen in wildlife will be relevant for humans, because anatomically and physiologically homologous endocrine tissues exist in all vertebrates. Moreover, all vertebrates have similar adrenal and sex steroid hormone receptors. Therefore, observations in vertebrate wildlife, including fish, may serve to highlight pollution issues of concern for humans. Mammalian wildlife, in particular, can serve as sentinels for humans, because although the timing of certain developmental events may vary, the basic mechanisms underlying development are the same in all mammals.

The animal / human connection has long been accepted in chemicals regulation, where effects seen in laboratory rodents form the foundation of human risk assessment. Effects seen in wildlife should also therefore inform clinicians and those concerned with human health policy about the possible role of environmental chemicals – and vice versa. Associations between adverse effects and exposure to some of the older organochlorine contaminants have been clearly noted; although it may well be that other pollutants also play a part. Nevertheless, it needs to be recognised that not all abnormalities will be the result of exposure to man-made chemicals. Therefore, an investigative approach is needed.

Over recent decades, humans will have been increasingly exposed to a diversity of chemicals through the use of ‘consumer articles’. Exposure occurs via the dermal route, via inhalation (particularly of indoor air), and via the diet, with the latter including not only man-made chemicals accumulated in the food-chain, but also chemicals introduced during factory processing of food, chemicals that leach from food packaging, and chemicals in drinking water, including those introduced by the disinfection process. Persistent chemicals may eventually end up in water-bodies and oceans, where elevated levels of bioaccumulating compounds are found in wildlife at the top of the food chain.

Most of the data on the adverse effects of chemicals on wildlife comes from polluted areas that have been the subject of in-depth studies. These particularly include the Great Lakes, the Baltic, and the Arctic. By the early 1990s, at least 14 species of fish and fish-eating wildlife in the Great Lakes basin had been reported with reproductive problems, population declines or other adverse health effects attributed to chemical contaminants (Gilman *et al.,* 1991). These species included snapping turtle (*Chelydra serpentine*), cormorant (*Phalacrocorax auritus*) black crowned night heron (*Nycticorax nycticorax*), bald eagle (*Haliaeetus leucocephalus*), osprey (*Pandion haliaetus*), herring gull (*Larus argentatus),* ring backed gull (*Larus delawarensis),* Caspian tern (*Hydroprogne caspia*), common tern (*Sterna hirundo),* Forster's tern (*Sterna forsteri*), wild mink (*Mustela vison*), otter (*Lutra canadensis*), and beluga whales (*Delphinapterus leucas*) from the Saint Lawrence river which is fed by water from the Great Lakes, and farmed mink fed Great Lakes fish. Levels of many persistent organochlorine contaminants have declined since the early 1990s, but concern is still high, and in 2001, Environment Canada (EC) initiated phase 1 (2001-2005) of the Fish and Wildlife Health Effects and Exposure Study, which was to explore the wildlife / human connection.

Apart from studies in highly polluted areas, most of the other data on wildlife come from fish, presumably because this vertebrate wildlife species is relatively easy to catch and is eaten by most people worldwide.

Epidemiological and field studies which seek to identify the chemicals that are instrumental in causing developmental effects are hampered by the fact that there are many confounding factors, and pollutants are ubiquitous. Therefore, controlled laboratory experiments have played a major part in determination of the potential causal agent(s). Unfortunately, however, laboratory studies are severely hampered because such a 'reductionist' approach cannot embrace the complex multi-causality and stepwise triggers that may impact on disease in the outside world. For example, the presence of several pollutants, natural ecosystem stressors, and other factors, such as ultra violet light exposure (Lyons *et al.,* 2002), may affect the toxic response.

Furthermore, both humans and wildlife will be exposed to many compounds concurrently, some of which may act with a similar mechanism of action or via mechanisms of action that converge. For example, certain phthalates inhibit testosterone synthesis during fetal life, and dioxins alter androgen dependent tissues, and also the pesticides vinclozolin, procymidone, linuron and DDT are all androgen receptor antagonists, which de-rail the process of masculinization (Gray *et al.,* 2001). Rodents exposed experimentally to such anti-androgenic chemicals show a cascade of features including shortened ano-genital distance, nipple retention, and testes-related effects (Gray *et al.,* 1999a).

Laboratory experiments have now clearly shown that additive effects can occur due to exposure to more than one chemical. For example, two phthalates with a similar mechanism of action can give rise to additive effects (Foster *et al.,* 2000). Furthermore, toxicants that induce malformations in androgen receptor dependent tissues can produce additive effects, even when two chemicals are acting via different mechanisms of action (Gray *et al.,* 2002). Interactive effects, including additivity or even synergism, also occur with other mechanisms of action. For example, additivity of effects has been shown in fish exposed to a mixture of estrogenic chemicals (Brian *et al.,* 2005), and similarly Crofton and colleagues (2005) have shown that when several thyroid disrupting chemicals are given to rodents, effects can occur even when each chemical is given at a dose level below their no-effect concentration. For thyroid disrupting chemicals, both additive and synergistic effects were reported. Several workers have shown that exposure to pesticide mixtures can cause greater effects than single exposures. For example, Moore and Lower (2001) have shown an additive response in male fish due to exposure to a weak mixture of atrazine and simazine pesticides (0.5ppb each), which caused reduced milt expression in fish because of altered olfactory function and reduced reaction to female pheromones. Similarly, Hayes and co-workers have shown effects in frogs exposed to low levels of several pesticides (Hayes *et al.,* 2006). In addition, epidemiological data suggest that some PCB and phthalate metabolites are associated with more than additive effects on sperm motility (Hauser *et al.,* 2005).

Organisms are most vulnerable to the effects of endocrine disrupting chemicals during puberty and during early life development in the uterus or in the egg, particularly when cell fate is determined and differentiation is occurring. Moreover, there may be a substantial delay before the effects from *in-utero* or *in-ovo* exposures are seen, although as such defects are programmed before birth, they are congenital. Effects from early life exposures prior to birth or hatching can therefore range from gross structural defects to subtle, but important, behavioural effects. This review considers congenital defects in a very wide sense, and embraces those present but not yet evident at birth. Therefore, this paper includes both chemically-induced overt structural abnormalities, and some programmed altered functional effects which manifest during development, both of which are important in species population viability.

Effects due to early life exposures may also result in effects on subsequent generations, as several laboratory studies have suggested transgenerational effects in both invertebrates and vertebrates. A transgenerational effect is an effect carried across generations, as a consequence of events that happen during the lifetime of the previous generation. Such effects have been seen in fish (Gray *et al.,* 1999b), and also in mammals. For example, in rodents, cancer has been noted in the granddaughters and grandsons of mice exposed to DES when pregnant (Newbold *et al.,* 1998; 2000). Similarly, an increased risk of hypospadias has been reported in the sons of women exposed *in utero* to DES (Brouwers *et al.,* 2005). Furthermore, in rats, deficits in sperm production have been shown in several subsequent generations, after an original exposure to methoxychlor and vinclozolin. It seems that this is not due to a genetic mutation, but instead due to altered DNA methylation, which then causes changes in the expression of a gene or genes, in a process termed epigenetic reprogramming (Anway *et al.,* 2005).

2. Summary of pollutant-related defects reported in wildlife

Many wildlife species are now reported to be affected by pollutants, and similarities can be seen in the effects recorded. The target sites affected are common to all vertebrates and particularly include male developmental pathways. It is clear that structural intersex features, including effects on the male reproductive tract, result from exposure before birth, and as such are congenital. On the other hand, abnormal secretion of the egg yolk protein, vitellogenin (VTG), in male fish, birds, and reptiles, can result from later adult-life exposure to pollutants. Nevertheless, studies showing such later-life effects are also included in this review, as they confirm the presence of sex hormone disrupting contaminants in the environment, and highlight a mechanism by which some of the reported congenital effects may arise. Reduced reproduction has been included because, although it is not a congenital defect per se, pollution-related decreased reproduction arises from abnormalities of the reproductive apparatus, or lack of viability of the offspring, both of which may be due to congenital defects.

Data also suggest that altered immune systems and altered thyroid function can result from exposure prior to birth, although in wildlife, effects on these systems have mostly been looked for in the adult. Many of these studies are also summarised in the following review, and table 1 outlines the similarities in effects that have been noted in wildlife and humans.

In addition, it is clear from studies of rats (Bowers *et al.,* 2004), monkeys (Rice, 2000) and humans (see 3.2), that congenital effects on brain function can result from *in utero* exposure to some chemicals found as contaminants. However, research into such effects in wildlife is limited, and therefore this is not a major focus of this review. Similarly, the life-long characteristics of the stress response in an individual can be altered, due to prenatal or antenatal exposure to raised levels of corticosteroid chemicals, resulting in both endocrine and behavioural consequences (Catalani *et al.,* 2000: O'Regan *et al.,* 2001; Welberg *et al.,* 2001). Altered stress response has indeed been noted in human infants (Stewart *et al.,* 2000; Health Canada, 1997) and several other species exposed to pollutants (for review see Pottinger, 2003), but this is also not dealt with in any detail in this review.

In fish, the following effects have been particularly noted: abnormal secretion in males of vitellogenin (VTG), the precursor to the egg yolk protein, which is normally only detected in females; altered spermatogenesis; eggs developing in testes (intersex); intersex genital apparatus; poor reproductive success; thyroid disruption; and immunosuppression.

In amphibians, abnormal production of VTG by males and intersex features has also been noted. Furthermore, limb, eye and CNS deformities, which are now occurring in America and Japan, may also be linked, at least in part, to pollution.

In the reptile class, turtles and alligators have been the subject of numerous studies. In turtles, the following effects have been noted: abnormal production of VTG by males; deformities of the reproductive tract (including ovo-testes and shorter estimated penis length); decreased hatching/reproduction; thyroid disruption; and altered immune system. In alligators, the effects include: abnormal production of VTG by males; sex hormone disruption; smaller phallus and testicular abnormalities in males; reduced clutch size; fertilisation failure; embryo mortality; and thyroid hormone disruption.

In birds, the effects include: abnormal VTG production in male birds; deformities of the reproductive tract; embryonic mortality and reduced reproductive success; deformities of the bill and bones; thyroid disruption; immunosuppression; and egg-shell thinning.

In mammalian species, the following effects have been noted. In otters and/or mink: reduced baculum length; smaller testes; impaired reproduction; eye defects (such as dysplastic changes in retina); and jaw deformities. In seals and/or sea lions: impaired reproduction (including implantation failure, sterility, abortion, premature pupping) and immunosuppression. In cetaceans: impaired reproduction; hermaphrodite organs; cancer; and immunosuppression. In polar bears: reduced cub survival; intersex features and deformed genitals; reduced testosterone levels in adult males; thyroid hormone disruption; and immunosuppression. In black and/or brown bears: undescended testes and intersex features. In the Florida panther: abnormal sperm and low sperm density; undescended testes; altered hormone levels; and immune system deficits. In deer: undescended testes; testicular abnormalities including cells predictive of testicular cancer, tumours, and calcifications; and antler dysgenesis. In eland (an antelope): abnormal testes, including impaired spermatogenesis. These findings are tabulated in table 1 and discussed in more detail in the rest of the section.

It is clear from table 1 that certain congenital defects, particularly linked to male reproductive development, appear to be commonssz to most vertebrate wildlife species. Moreover, as would be expected, these abnormalities have also been noted in humans. It can be concluded that contaminants appear to be affecting male reproductive health, thyroid function, and the immune system of many species.

2.1. Fish

Fish may be particularly at risk from pollutants, because their exposure is not only via the diet, but also via the gills and skin. The physical and chemical characteristics of many endocrine disrupting chemicals, especially their lack of ionization and lipophilicity also favour their movement from the water column to biological tissues.

In addition to the biological processes common to other vertebrates, the processes of smoltification (in salmonids) may also be susceptible to disruption. The following examples represent some of the studies that have shown a link between exposure to endocrine disrupting chemicals and effects in fish from the Osteichthyes class (the bony fish). It should be noted that vitellogenin (VTG), the precursor of the egg yolk protein, is normally not detectable in male fish, or is only present at very low levels. Therefore, detection of elevated levels in male fish is abnormal, and is an excellent biomarker of exposure to estrogenic endocrine disrupting chemicals (EDCs). Furthermore, VTG induction is generally accompanied by various degrees of reproductive interference at similar or lower ambient estrogen concentrations. This means that it can be a marker for a number of adverse effects (for review see Matthiessen, 2003).

Abnormal production of vitellogenin (VTG) in male fish

Vitellogenin production in several male freshwater fish species has now been reported in many places worldwide including Europe, North America, Australia and Japan. Similarly, vitellogenin production in male marine fish has also now been reported in many species and many countries worldwide, including: cod from the North Sea (Scott *et al.,* 2006) and UK estuaries (Kirby *et al.,* 2004); flounder from Denmark (*Platichthys flesus*); sole (*Pleuronectes yokohamae*) from Japan; sole from Puget Sound, USA; grey mullet (*Mugil cephalus*) from Japan; flounder from a Dutch harbour; a Dutch offshore spawning ground (for review see Matthiessen, 2003); and Mediterranean swordfish (*Xiphias gladius*) from the Strait of Messina near Sicily, where VTG induction was seen at very high levels (Fossi *et al.,* 2004).

Studies in UK freshwaters were the first to report the phenomenon of VTG production in male fish (Purdom *et al.,* 1994; Harries *et al.,* 1996), and similarly UK studies by Lye *et al.* (1997; 1998) were the first to report VTG induction and testicular abnormalities in a marine fish, the flounder (*Platichthys flesus).* In many UK fresh waters downstream of sewage treatment works, it seems that a large part of the estrogenic component is derived from the natural female hormones (estrone and estradiol-17b) and the contraceptive pill (ethinyl oestradiol) excreted in sewage (Jobling and Tyler, 2003). However, in some UK rivers, industrial chemicals, such as nonylphenol, have also been implicated as a causal factor in VTG production (Thorpe *et al.,* 2001; Lye *et al.,* 1999).

Intersex in fish

The presence of intersex or ovotestis (i.e. primary or secondary oocytes abnormally present in the testicular tissue) is now a frequently reported phenomenon in fish. This disrupted gonad development is almost certainly linked to endocrine disruption caused by exposure to estrogenic compounds. It can be induced experimentally through exposure at the larval stage, but not by exposure of the adult fish. In some very estrogenically contaminated UK estuaries (Mersey, Tyne, Clyde and Forth) up to a fifth of the male flounder and blenny (*Zoarces viviparous*) in some locations show ovotestes, whereas ovotestes has not been seen in flounder from a relatively uncontaminated reference estuary, the Alde (for review see Matthiessen, 2003). In comparison, intersex has been reported to varying degrees, in (up to 100% of freshwater roach (*R.rutilus. rutilus)* in some UK rivers (Jobling and Tyler, 2003), but it is not known whether this is due to species differences in response, higher exposures in the freshwater upstream, or the fact that breeding grounds for marine species are further offshore and therefore probably less contaminated.

Freshwater fish species in which abnormal intersex has been reported include: roach *(R.rutilus. rutilus)*; bream (*Abramis abramis*); chub; gudgeon (*Gobio gobio*); barbell (*Barbus plebejus*); perch (*Perca fluviatilis*); stickleback (*Gasterosteus aculeatus);*

and shovel-nosed sturgeon (*Scaphirhynchus platyorynchus)* (for review see Jobling and Tyler 2003).

Table 1. Effects reported in wildlife and humans, known or suggested to be linked to chemical contaminants

	Animals															
	Fish	Amphi-bian	Reptile alligator	Turtle	Birds	Rodent	Otter	Mink	Seal or sea lion	Whales or other cetacean	Polar Bears	Black/ brown Bears	Panther	Deer	Eland	Humans
Reduced reproduction	Y	Y	Y	Y	Y	Y	Y	Y	Y	Y	Y		Y	Y		Y
Intersex /testicular Abnormalities	Y	Y	Y	Y	Y	Y	Y			Y	Y	Y	Y	Y	Y	Y
Deformities of sex linkedstructure / reduced phallus	Y		Y	Y			Y	Y			Y			Y		Y
Thyroid Abnormalities	Y		Y	Y	Y					Y	Y		Y			Y
Immune system Effects	Y	Y	Y	Y	Y				Y	Y	Y		Y			Y
Bill, jaw, teeth, nail, or claw defects					Y			Y	Y							Y
Specific VTG in female	Y			Y	Y											
Specific other	Osmo-reg	Limb			Egg-shell		Eye CNS							antler		

Y = Effect reported and known or suggested to be linked to contaminants; VTG: vitellogenin; CNS: central nervous systems

Compared to the example of the roach in UK waters, there are numerous examples of milder cases of intersex in estuaries and coastal locations from elsewhere in Europe, and Japan. For example, flounder from the Seine estuary in France, flounder from the southern Baltic in Germany, swordfish from the Mediterranean, and flounder from Tokyo (for review see Matthiessen, 2003). Evidence of intersex has also been reported in catfish (*Clarias gariepinus*) in South Africa (Barnhoorn *et al.,* 2004). Similarly, numerous cases of intersex have been reported in North America. For example, lake whitefish (*Coregonus clupeaformis)* collected in 1996 from the polluted Sanit Lawrence river in Quebec exhibited neoplasms, lesions, and intersex (Michaelian *et al.,* 2002). In a polluted area of Lake Ontario in Canada, intersex gonads were seen in 83% of male white perch (*Marone Americana*) collected in 1999-2000, which was an increase on the previous year (Kavanagh *et al.,* 2004).

2.1.1. Deformities of sex-linked structures in fish

There are species differences in the response of fish to exposure to sex hormone disrupters. For example, sand gobies (*Pomatoschistus minutus*) from oestrogen-contaminated estuaries in the UK do not show either induction of vitellogenin or intersex, but instead male fish exhibit deformed and feminised urogenital papillae, which is the structure used by both sexes to deposit gametes (Matthiessen *et al.,* 2002). This phenomenon has been termed morphologically intermediate papilla syndrome (MIPS), and was found in males from the UK Tees, Mersey and Clyde estuaries. The causal agent(s) are not known with certainty, but in some UK estuaries and effluents the substances identified with steroidal activity included the natural steroids, 17B oestradiol, androsterone, nonylphenol, di-ethylhexylphthalate, cinnarizine (an anti-histamine drug), and cholesa-4,6-dien-3-one (a natural cholesterol degradation product) (Allen *et al.,* 2002).

Structural defects of the reproductive apparatus have also been noted in other fish species. In Florida USA, mosquitofish (*Gambusia holbrooki*) from the pesticide-polluted Lake Apopka were compared with those from less polluted lakes. Male fish from the polluted Lake Apopka had slightly shorter gonopodia and fewer sperm cells per milligram testis, when compared with the fish collected from Orange Lake and Lake Woodruff. The authors concluded that sexual characteristics of relevance to male reproductive capacity are altered in the Lake Apopka mosquitofish population, and that anti-androgenic chemicals were a possible cause of the effects (Toft *et al.,* 2003).

Elsewhere, studies in several areas including the EU, North America, and New Zealand, (Jobling and Tyler, 2003; Ellis *et al.,* 2003) have documented masculinization or reduced fecundity in female fish living in rivers below paper mill outfalls. For example, in the EU, female perch exposed to pulp mill effluent had reduced gonad weight and reduced fecundity (Karels *et al.*, 2001). In Florida, female mosquitofish had masculinized gonopodial and anal fin development and altered reproductive behaviour. Here, androgenic components in the effluent (Orlando *et al.*, 2002), some of which originate from the breakdown of wood products were considered to be the causal agent (Jobling and Tyler, 2003).

2.1.2. *Poor reproductive success / reduced hatching in fish*

In Lake Ontario, contaminants, probably dioxins (measured as 2,3,7,8-tetrachloro-p-dibenzodioxin (TCDD) equivalents), which are potent EDCs, were considered responsible for the loss of lake trout (*Salvelinus namaycush*) in the 1960s (Cook *et al.*, 2003). Moreover, even after the extinction of this population of lake trout, and the re-stocking, there was a lack of reproductive success, and reduced fry survival occurred after 1980 (Cook *et al.*, 2003).

Reduced spawning success or reduced hatching has been noted in several wild populations of marine fish, including: a DDT contaminated population of white croaker in California (*Genyonemus lineatus*); a flatfish species in the Puget Sound (in the USA); Baltic flounder; Baltic herring (*Clupea harengus*); and Baltic cod (*Gadus morhua*) (for review see Matthiessen, 2003).

2.1.3. *Thyroid disruption in fish*

In the Great Lakes, and particularly in Lake Erie, salmon were seen with enlarged thyroids, and indeed goitres and depressed thyroid hormone levels were induced in rodents fed with the contaminated fish. Epizootics of thyroid hyperplasia and hypertrophy (affecting 100% of the population) were reported in the introduced species of pacific salmon, including pink (*Oncorhynchus gorbuscha*), coho (*Oncorhynchus kisutch),* and Chinook salmon (*Oncorhynchus tshawytscha*) taken from every one of the Great Lakes (Leatherland, 1992). Nevertheless, forty years after first being identified, the precise identity of the causal agent(s) and the mechanisms of action were still elusive (see review Jobling and Tyler, 2003), but it is concluded to be related to pollution, rather than naturally occurring iodine deficiency. In the last few decades, there has been a temporal decline in the severity of the goitre seen in the Great Lake, presumed to be due to a reduction in the levels of some organochlorines. However, personal communication with Glen Fox (2005) confirms that there is still plenty of evidence of thyroid disruption in fish in the lower Great Lakes.

Thyroid disruption has also been noted in a marine fish species from Piles Creek, New Jersey. Here, mummichogs (*Fundulus hereroclitus)* had larger thyroid follicles and raised plasma thyroxine and exhibited poor prey capture and predator avoidance (see review, Matthiessen, 2003).

2.1.4. Immunosuppression in fish

Numerous studies show that fish living in polluted areas can exhibit compromised immune system function (Luebke *et al.,* 1997; Zelikoff *et al.,* 2000; Amado *et al.,* 2005; Reynaud and Deschaux, 2005; Kollner *et al.,* 2002; Maule *et al.,* 2005).

2.1.5. Altered osmoregulation in migrating fish

In Canada, nonylphenol ethoxylate surfactant contained in pesticide formulations has been linked with the decline in catches of Atlantic salmon (*Salmo salar*) and blueback herring (*Alosa aestivalis*). Nonylphenol appears to interfere with osmoregulation and the major hormonal changes that the fish must make during migration, when adapting to salt water from fresh water. However, it is not certain whether or not this is directly caused by endocrine disruption, although nonylphenol is known to have estrogenic properties, and gonadal steroids are known to exert an antagonistic action on the process (Fairchild *et al.,* 1999). Thyroid disruption may also play a role, as Iwata highlights that thyroid hormones are involved in a whole sequence of behaviours in migrating salmonids (Colborn, 2002). Similarly, controlled experiments by other workers on Arctic charr suggest that high doses of PCBs affect smoltification, with altered thyroid hormone levels noted during smoltification (Jorgensen *et al.,* 2004).

2.2. ***Birds***

In birds, estrogen is the differentiating hormone for both gonads, and for behaviour (Giesy *et al.,* 2003). This is in contrast to sexual differentiation in mammals, where it is androgen that causes the testes to develop, such that in the absence of androgen, the female is the default sex. In birds, in the absence of estrogen, both gonads develop into testes, whereas during normal female development, the left gonad develops into an ovary while the right gonad regresses (Fry, 1995). Such differences in early life development may make the response of birds to environmental endocrine disrupters, rather unique.

For birds to be exposed during the critical period of development, compounds must be passed from the female bird to her eggs. DDT is obviously readily transferred to the lipid-rich yolk, but it seems that other contaminants, including large molecules like deca brominated diphenylether (deca-BDE) can also find their way into eggs (EU RAR). Fish eating birds, and other birds of prey, will likely be most at risk.

2.2.1. *Abnormal VTG production in male birds*

In 2001, male herring gulls in a polluted area around the Great Lakes were found with elevated levels of vitellogenin in their blood. As in fish, this egg yolk protein is normally produced by breeding females (EC, 2003).

2.2.2. *Deformities of the reproductive tract, and ovo-testes in birds*

There appear to be few studies of the internal reproductive tract in birds. However, in 2001, a male herring gull (*Larus argentatus*), nesting in the lower Great Lakes (downstream of a polluted area) was found with a significantly feminized reproductive tract (EC, 2003).

Szczys and colleagues (2001) noted that off the coast of Massachusetts, the sex ratio of hatched Roseate Tern (*Sterna dougallii)* chicks was biased (55%) in favour of females. Common terns (*Sterna hirundo)* were subsequently studied as a surrogate, and in 1993/94, 60-90% of pipping male common tern embryos sampled exhibited ovarian cortical tissue in their testes (ovotestes). However, the examination of 21 day old common terns collected from Bird Island in 1995, suggested that the ovotestes may become fully regressed and therefore do not lead to permanent alterations in gonadal tissue that would be expected to impair reproduction. Indeed, it is speculated that ovotestes might occur naturally in some common terns at hatching, although the frequency with which it occurs might be enhanced by exposure to contaminants (Hart *et al.,* 2003).

2.2.3. *Embryonic mortality and reduced reproductive success in birds*

In studies undertaken in 2001, a low reproductive success was reported in gulls around the Great Lakes, with dead herring gull embryos found in higher numbers in the polluted areas (EC, 2003). Furthermore, in surveys during 2001-2004, reduced egg viability was still seen in herring gulls, and although the precise cause of this was not known, it was not due to infertility (Fox, 2005).

Bald eagles numbers are now recovering slowly in North America. However, those nesting near the Great Lakes have greater difficulty reproducing than those nesting further inland, presumably because their food supply remains contaminated. Furthermore, more than half the bald eagles that do manage to hatch along the shores of the Great Lakes die young (EC, 2001).

Reduced reproduction has also been noted in birds in the Arctic, and for example, bald eagles have less offspring on Adak Island, and this was associated with higher

DDE concentrations. Furthermore, organochlorine levels in many species of predator birds exceed those associated with effects on reproduction (AMAP, 2004).

2.2.4. *Altered thyroid function in birds.*

Adult herring gulls were found with enlarged thyroids (goitre) around the Great Lakes in 2001, and these gulls produced smaller amounts of hormone (EC, 2003). Organochlorine chemicals, particularly PCBs were among the chemicals suspected of causing such hormone disruption (McNabb and Fox, 2003). Recent work in 2001-2004 show that goitre and thyroid effects still persist in herring gulls in the Great Lakes area (Fox, 2005).

Similarly, in Bjornoya in the Arctic, high levels of HCB (hexachlorobenzene), p,p'-DDE, and summed PCBs were associated with reduced plasma thyroxine levels in male glaucous gulls (*Larus hyperboreus*) (AMAP, 2004).

2.2.5. *Immunosuppression in birds*

Immunosuppression has implications for survival, as it suggests increased susceptibility to infectious diseases, and reduced ability to grow, compete for food, and to withstand the rigors of weather and migration. Grassman has reported an association between exposure to polyhalogenated aryl hydrocarbons and decreased T cell immunity in the offspring of fish-eating birds (herring gulls and Caspian terns) at highly contaminated sites in the Great Lakes. Throughout the 1990s, associations were found between organochlorines and suppressed T cell function and enhanced antibody production in young Caspian terns (*Sterna caspia*) from a location at Lake Huron in the Great Lakes. Moreover, deficits in the functioning of the immune system were still found in studies of birds from the Great Lakes undertaken in 2001 (EC, 2003; Grasman and Fox, 2001).

Similarly, many birds in the Arctic have current levels of summed DDT, summed PCBs, and/or dioxin-like substances which exceed those associated with immunosuppressive effects (AMAP, 2004). For example, glaucous gulls from Bjornoya had invasive nematode density which was correlated with p,p-DDT, mirex and summed PCBs (AMAP, 2004).

To date, there have not been any reports associating measured immune suppression with an avian disease epidemic. It is, nevertheless, likely that immune suppression in birds resulting from chronic stresses (such as crowding, poor habitat and food, climate stress) and/or environmental contaminants causes chronic morbidity and mortality associated with multiple pathogens. Over time, this may significantly alter genetic diversity and species survival (Fairbrother *et al.,* 2004).

2.2.6. Eggshell thinning in birds

Eggshell thinning is a notorious pollution-related effect on bird reproduction, and is caused by DDE, the degradation product of DDT (see Giesy *et al.,* 2003). However, the precise mechanism of action is still not known. Effects have persisted for many years, and for example, peregrine falcons (*Falco peregrinus, tundrius* and *anatum* sub-species) breeding in the Canadian Arctic, still have eggs around 10% thinner than those produced prior to the introduction of DDT (Johnstone *et al.,* 1996; AMAP, 2004). This is presumably due to the chemical body burden being passed on to subsequent generations, coupled with the life-time exposure of each generation, rather than any congenital altered ability to form the eggshell.

Bird populations in the UK have not totally recovered in some areas, possibly due to many factors, including lack of food or nest disturbance. For example, after the ban of organochlorines, the numbers of peregrine falcon slowly recovered, and by the late 1990s had reached pre-crash levels over much of its former range, but in south-east and east of England the bird was slower to recover, and the range may actually be contracting again in northern Scotland (RSPB, 2001). In Europe, the peregrine population is currently stable or increasing over much of Europe, but numbers are declining in many Mediterranean and eastern European countries, and moreover, the northern populations in Fennoscandia are still depleted (RSPB, 2001).

2.2.7. Deformities of the bill and bone in birds

Crossed bill syndrome was part of a syndrome called GLEMEDS (Great Lakes embryo mortality, oedema and deformity syndrome) which was reported in the Great Lakes, and which correlated with planar chlorinated hydrocarbons (Gilbertson *et al.,* 1991). These dioxin-like effects in fish-eating colonial water birds are believed to be largely due to planar PCBs, except perhaps near dioxin (TCDD) point sources (Giesy, 2003).

Bill deformities are still found. For example, increasing numbers of beak deformities have been noted in recent years in the Arctic. In the Barents Sea, black-legged kittiwakes (*Rissa tridactyla*) have been seen with crossed bills and clump feet (AMAP, 2004). Indeed, over 30 species of birds have apparently been seen with curved beaks up to three times their usual size, and in many cases, this prevented the birds from eating or preening effectively. Handel of the US Geological Surveys'

Alaska Science Centre is reported as suggesting that PCBs and dioxins might be responsible (The Guardian, 2004).

Developmental deformities are still seen in the Bald eagle nestlings from around the shores of the polluted Great Lakes. These include beak and foot malformations, fused vertebrae and hip displacements. Exposure to PCBs and other pollutants, including metals, is confirmed, but the precise cause of these deformities is not determined (EC, 2001).

In the UK, bone deformities have been reported in herons (*Ardea cinerea).* Herons from a North Nottinghamshire heronry were reported dead and deformed during the survey period of 1996 – 2004. Many of the birds died for no apparent reason, but broken bones suggested that something was affecting calcium deposition in their bones. PCB and dioxin levels were higher in deformed nestlings and in nestlings found dead, compared to those apparently unaffected. However, after analysis of the 2004 data, it was concluded that in 2 sites, the deformities were associated with higher levels of PCBs, but not dioxins (Thompson *et al.,* 2005).

2.3. ***Reptiles***

Studies of long-lived species like turtles and alligators can provide a very useful indicator of the health of wetland ecosystems.

2.3.1. Effects in alligators

Guillette and others have reported population decline and numerous reproductive abnormalities in alligators (*Alligator mississippiensis*) from Lake Apopka in Florida, a lake which has been polluted with several organochlorine pesticides, including dicofol and DDT following a chemical spill in the 1980s (Woodward *et al.,* 1993; Guillette *et al.,* 1995; Guillette *et al.,* 2000; Guillette and Iguchi, 2003). However, effects have also been noted in alligators from Florida lakes polluted by diffuse sources. The following abnormalities in Florida alligators have been linked to EDCs: sex hormone disruption; smaller phallus; testicular abnormalities; reduced clutch size; fertilisation failure; embryo mortality; thyroid hormone disruption (Hewitt *et al.,* 2002); and alterations in thymus and spleen, which would be likely to affect immune function (Rooney *et al.,* 2003).

High embryo mortality in alligators and high exposure to organochlorine pesticides has been found in Florida in Lakes Apopka and Griffin, and Emeralda Marsh, as compared to less polluted sites at Lakes Woodruff and Orange (Sepulveda, 2004).

Furthermore, research by Tim Gross and co-workers has also noted that low rates of hatching were due to fertilization failure as well as early embryonic mortality (SBRP, 2003).

2.3.2. *Effects on turtles*

At a heavily polluted site on the Great Lakes in 2001, around 10% of the adult male snapping turtles (*Chelydra serpentina*) were found to be abnormally producing vitellogenin (VTG), indicating sex hormone disruption (EC, 2003).

Furthermore, studies in snapping turtles from the Great Lakes and the Saint Lawrence river in Canada have found differences in the physiology of adult turtles taken from highly contaminated sites, compared to those from less contaminated sites. At all sites, the precloacal length of male hatchlings was larger than that of females by an equal amount at any given body size. However, the precloacal length of both males and females from the polluted site increased with body size at a slower rate than males and females from the other two sites. These alterations in secondary sexual characteristics are believed to be initiated early in development, are linked to contaminant levels, and may result in permanent organizational changes in morphology (de Solla *et al.,* 2002). Precloacal length is also used as an estimator of penis length, and in a 2001 study, this was shorter in male adult turtles from the Detroit River and in juvenile males from two polluted sites, as compared to cleaner reference sites (EC, 2003).

Decreased hatching success has also been reported in snapping turtles in polluted sites around the Great Lakes compared to those from reference sites (EC, 2003). At a particularly polluted site, there were no signs of reproductive activity in the adult snapping turtles (EC, 2003). There is also a suspicion that deformities in Great Lakes hatchlings, which are found at higher rates than in cleaner reference locations, might be linked to chemicals

Turtles living in polluted lakes elsewhere are also affected. For example, in Lake Apopka in Florida, which is contaminated with several EDCs, many new-born red belly turtles (*Pseudemys nelsoni*) have been reported with genital disruption. Here, abnormal testes, including ovo-testes were found (Guillette *et al.,* 1995).

In addition, thyroid hormone disruption has also been noted around the Great Lakes. For example, young turtles with impaired thyroid function were found at all three polluted sites studied (EC, 2003).

Immune system impairment has also been reported. For example, snapping turtles from a polluted site on Lake Erie were found with lower globulin levels in the study period from 2001, and this is taken as a rough measure of antibody-mediated immune status (EC, March 2005, draft). Similarly, in North Carolina loggerhead sea turtles (*Caretta caretta*), sampled in the summers of 2000 and 2001, modulation of the immune system was reported, with a suggested association with mirex and dioxin-like PCBs (Keller *et al.,* 2004).

Green turtle fibropapillomatosis is a threat to the survival of the green turtle (*Chelonia mydas*), and environmental contaminants might be involved through plausible mechanisms such as co-carcinogenesis and contaminant-induced immune suppression, although this is not yet clear (Herbst and Klein, 1995).

2.4. ***Amphibians***

Many amphibian species worldwide are in decline. Even amphibians in apparently pristine areas have suffered. In California's snow-capped Sierra Nevada, populations of frogs and toads, including the yellow-legged frogs (*Rana boylii and Roma muscosa*) and the California red-legged frog (*Roma aurora*) have crashed, with some researchers suggesting that the high levels of pesticides carried there on the wind are responsible (Sparling *et al.,* 2001).

Moreover, in frogs, a disorder characterised by an extra or malformed limb(s) has emerged, with some scientists suggesting that this may be linked to chemicals, UV exposure, trematode infection, acid rain, viruses, nitrates, or a combination of these (Ankley *et al.,* 2004; EC, 2004). Some researchers have suggested that chemical contaminants with the ability to mimic or disrupt retinoids may be implicated (Gardiner *et al.,* 2003), and eye and CNS deformities, which have been found in some amphibians, might also be linked to pollutants.

It certainly seems that something in the environment, and particularly a chemical or chemicals, may be at least partly responsible. For example, researchers in Minnesota found that tadpoles of the northern leopard frog (*Rana pipiens*) reared in the presence of UV and extracts from sites where deformities were common, had higher deformity rates (Bridges *et al.,* 2004). Other controlled experiments have linked pesticide exposure with increased trematode infection and increased limb deformities in wood frogs (*Rana sylvatica*), with the suggested mechanism being pesticide-mediated suppressed immune system function (Kiesecker, 2002).

Altered immune system function has been found in amphibians in the wild, for example, in frogs from rice fields (Fenoglio *et al.,* 2005) and in toads taken from organochlorine polluted areas of Bermuda, where the study authors suggested that environmental pollutants may account for immunosuppression and possible decline in amphibian populations (Linzey *et al.,* 2003). Hayes *et al.* (1992) has also suggested that immunosuppression caused by exposure to pesticides may be an important factor in population decline, because when atrazine and eight other pesticides were mixed at relatively low levels (0.1ppb each compound) to replicate a Nebraska cornfield, over one third died. He concluded that the frogs developed an array of health problems, including meningitis, because the chemicals suppressed their immune systems. They also took longer to complete the metamorphosis from tadpole to frog, which would reduce their chances of survival. Metamorphosis is common to amphibians and may confer a particular susceptibility to thyroid hormone disruption.

Intersex features, linked to chemical exposure, have now also been seen in the wild in both frogs and toads, and these and other effects are outlined below.

Defects of the reproductive system, VTG production, and intersex in amphibians

Hayes and co-workers observed retarded gonadal development (gonadal dysgenesis) and testicular oogenesis (intersex or hermaphroditism) in leopard frogs (*Rana pipiens*) collected from atrazine-contaminated sites across the USA (Hayes *et al.,* 2003). They concluded that atrazine impacts amphibian populations in the wild, and showed that atrazine exposure in the laboratory (at 0.1 ppb) resulted in intersex characteristics in leopard frogs. Furthermore, Hayes and colleagues (2003) hypothesised that atrazine might induce aromatase, which converts testosterone to estrogen, and thereby could increase the production of endogenous estrogen. However, one group of researchers using mesocosms have subsequently suggested that oocytes in the testes of the African clawed frog (*Xenopus laevis)* may be a natural phenomenon (Jooste *et al.,* 2005), although many other amphibian experts have expressed doubts about this finding (Science News, June 9, 2005), and it may be that other endocrine disrupting chemicals contaminated the water in the mesocosms. The reports of atrazine's effects on amphibians has certainly been the subject of some controversy, and after reviewing Hayes' and other studies, a report drafted for the US EPA stated that 'the available data do not establish a concordance of information to indicate that atrazine will or will not cause adverse developmental effects in amphibians' (Steeger and Tietge, 2003).

However, Reeder *et al.* (1998, 2005) have suggested that chemical contaminants, including atrazine, contributed to the decline of cricket frogs (*Acris crepitans*) in Illinois. From studying museum collections, they considered that the proportion of intersex individuals peaked during the period 1946-1959.

Cane toads (*Bufo marinus*) in the wild are also exhibiting signs of sex hormone disruption. Gross and McCoy have studied populations in sugar cane fields in the Florida Everglades where high levels of triazine herbicides, such as atrazine, are used. All the cane toads living near the cane fields had female coloration, and about 30% of the male toads found at the contaminated sites were hermaphrodites (Renner, 2003). Levels of vitellogenin in the male toads resembled those of female toads, suggesting the role of increased levels of estrogen or an estrogen mimic.

2.5. ***Mammals***

Predator mammals in contaminated areas are at risk, because bioaccumulative contaminants can build up in the food chain, and even a remote area like the Arctic is under threat, because persistent organic pollutants are carried to the northern latitudes on air and ocean currents, in a process termed global re-distillation. Chemical contamination in many Arctic mammalian predator species is already at levels above those which cause effects on reproduction and immune system function in other mammalian species (AMAP, 2004). However, because of the difficulties of research on large but elusive or remote aquatic mammals, some of the best evidence of the effects of pollutants on reproduction and the immune system comes from semi-field studies on seals, minks and otters.

2.5.1. Feral rodents

A couple of studies on rodents living in areas with high organochlorine contamination show effects on reproduction and the testes (Damstra *et al.,* 2002). For example, significantly reduced testes weights have been reported in male white footed mice (*Peromyscus leucopus*) inhabiting PCB and cadmium contaminated land, and in addition, effects on reproduction were also noted, with numbers of juveniles and sub-adults reduced, compared to an unexposed population (Batty *et al.,* 1990).

2.5.2. Otters

Several decades ago, otters (*Lutra lutra*) completely disappeared in some UK and European rivers, due to contaminant induced reproductive problems (Mason and Macdonald, 2004). Monitoring in 1989-1991 suggested that, at least in some areas,

PCBs were still sufficiently high to exert detrimental effect on some UK otters (Mason and Macdonald, 1994). However, the pesticide dieldrin has now been implicated in this decline, and habitat destruction will also, of course, have had a negative impact (EA, 2003)

Captive bred otters were released in some river catchments in the UK and elsewhere (Fernandez-Moran *et al.,* 2002), and otters are now breeding again, although in some UK rivers the population growth is still slow (Mason and Macdonald, 2004). In other European countries, such as Denmark, the distribution range of the otter is still much reduced (Pertoldi *et al.,* 2001). Similarly, in southern Sweden, total PCB concentrations are still high and the indications of population improvement are weak (Roos *et al.,* 2001). Overall, in the EU, the otter population distribution is still reduced, and as well as PCBs, rodenticides are also a concern in some areas (Fournier-Chambrillon *et al.,* 2004). Similarly, the North American river otter presently occupies a greatly reduced range, and at least 17 states and one Canadian province have undertaken re-introduction programs (Kimber and Kollias, 2000).

Structural defects of the male reproductive tract have also been reported in male otters. Otters surveyed in the polluted Lower Columbia river in North America in the 1990s were reported to have abnormally small reproductive organs, and these reproductive tract disorders correlated with several environmental contaminants present in the river (NBS, 1996). Research by Henny and colleagues reported that the baculums and testicles of young males from the Lower Columbia River were shorter or smaller than in animals of the same age class from non- polluted areas. In the Portland Vancouver area, where the highest PCB and organochlorine levels were recorded, of the four animals collected, one otter even had no testicles. However, it was suggested that some of the effects on the young male river otters from the Lower Columbia River may be temporary, resulting from delayed development due to endocrine dysfunction (NBS, 1996).

Eye defects have also been noted. In a study of 88 otters found dead in south-west England between 1990 and 2000, some 26 otters had dysplastic changes in their retina. Moreover, those with dysplastic retina had significantly lower concentrations of vitamin A and higher concentrations of dieldrin than the otters with normal retinas (Williams *et al.,* 2004).

There also seem to be some problems in sea otters. For example, the Alaskan sea otter (*Emhyoher lutris kenyoni*) population in the Aleutian Islands (USA) has recently declined, as has the Southern sea otter around California (*Enhydra lutris nereis)* (Hanni, 2003). The Californian population of Southern sea otters began a pattern of slow decline in 1995, and is listed as a threatened species. Some have attributed this decline to high adult mortality rates, with infectious disease, including opportunistic pathogens, being the major cause of death. It was suggested that the immunological health of this population might be compromised (Schwartz *et al.*, 2005), because the otters that died of disease had higher levels of butyltin, PCB and DDT (Nakata *et al.*, 1998; Kannan *et al.*, 1998). Kannan and colleagues also noted that the TCDD equivalents of non- and mono-ortho PCBs in both the sea otters from the California coast and certain prey species were at or above the theoretical threshold for toxic effects (Kannan *et al.*, 2004).

2.5.3. Mink

The endangered European mink (*Mustela lutreola*) has suffered a rapid decline, and its distribution is still shrinking. In France, the range of the mink shrank by nearly 50% over the last 20 years (Fournier-Chambrillon *et al.*, 2004). Dioxins, PCBs, and rodenticide exposure via prey (Fournier-Chambrillon *et al.*, 2004), are all a concern, and it seems that mink are particularly susceptible to dioxins and structurally related PCBs.

In North America, in the 1970s, commercial mink farms reported reproductive failure in their mink which were fed fish from the Great Lakes, and it was subsequently shown that exposure to low levels of PCBs could impair reproduction. Data from 1982 and 1987 from surveys around the Great Lakes continued to indicate that wild mink populations were being affected by pollutants, particularly PCBs (Wren, 1991). Rather alarmingly, recently reported levels of PCBs in mink from western Lake Erie show increases from 1979 when they were last sampled, and moreover, many exceed the lowest observable effect level for reproductive impacts (EC, 2003; Fox, 2005).

The Housatonic River in Connecticut, downstream of an old General Electric Company plant, is particularly contaminated with PCBs. Mink fed fish from this river had offspring which had lower birth weights and higher infant mortality rates, compared to mink fed with Atlantic herring processed in the same way, such that it can reasonably be predicted that the wild population would likely be suffering adverse effects (Bursian *et al.*, 2003).

Structural defects have also been noted. In British Columbia in Canada, there was a significant negative correlation between total PCB concentrations and baculum length in juvenile mink, caught in the winters of 1994/5 and 1995/6, although a few individual animals with gross abnormalities of reproductive systems did not show high levels of contamination (Harding *et al.,* 1999). In addition, mink from the Kalamazoo River in the USA, which is affected by a PCB contaminated site, have been reported with jaw defects. This hyperplasia (excessive growth) of epithelium in the mandible and maxilla, is known to be caused by PCB 126 and TCDD (Beckett *et al.,* 2005).

2.5.4. Seals and sea lions: Reproductive problems, immunosuppression, adrenal and other effects in seals

In the 1970s, *harbour seals* (*Phoca vitulina*) in the polluted Dutch Wadden Sea (part of the North Sea) declined in number, with low reproduction being blamed on PCBs. A study showed that female harbour seals fed fish from the polluted Wadden Sea had half as many pups compared to seals fed fish from the less contaminated Atlantic (Reijnders, 1986), and this was linked to implantation failure associated with lower levels of estrogen (Reijnders, 1990). More recently, seals from the Baltic have also been reported to exhibit a compromised endocrine system, associated with high PCB and DDE/T levels (see review, Damstra *et al.,* 2002). Some of the disorders observed in the exposed seals included abortion in early pregnancy, uterine stenosis and occlusions, and sterility. Moreover, many seals from the Arctic, including some ringed (*Phoca hispida*) and northern fur seals (*Callorhinus ursinus*) are contaminated with summed PCB levels above the threshold for decreased reproduction in otter (AMAP, 2004).

In sea lions (*Zalophus californianus*), stillbirths and premature pupping were reported in the 1970s, and this was associated with high PCB and DDE levels (DeLong *et al.,* 1973). On San Miguel Island, some twenty percent of the California sea lion pups died due to premature birth. The p,p'-DDE levels in the premature parturient cows' blubber were 7.6 times greater than in the full-term animals, although it seems that infections may also have contributed (Gilmartin *et al.,* 1976). Moreover, sea lion populations have also suffered a decline in western Alaska, although the cause is not known, and it may be related to a decline in their prey (AMAP, 2004).

Immunosuppression in seals

Contaminant-induced immune suppression is a suggested contributory factor in the mass mortalities of seals due to infectious agents (Dietz *et al.,* 1989; Hall *et al.,* 1992; Van Loveren, 2000). The recurrence of numerous fatalities in the different seal populations of the North Sea (during the years 1988, 1989 and 2002), of the Baikal Lake and Caspian Sea (during the years 2000 and 2001) suggests that immune suppression is a factor in population declines and warrants more study (Bragulla, 2004).

This concern is particularly highlighted by an experiment which showed that female harbour seals fed fish from the polluted Wadden Sea had impaired natural killer cell activity and T-lymphocyte function (de Swart *et al.,* 1994), and delayed hypersensitivity (Ross *et al.,* 1995) compared with seals fed less contaminated fish from the Atlantic.

In 1999, it was noted that the reproductive problems, which had previously been most evident, were reduced in seals born after 1980, although there were indications of immune system effects becoming more prevalent (Bergman, 1999a). In a study of Baltic seals spanning the 20 year period 1977-1996, Bergman (1999a) reported increased prevalence of colonic ulcers in young seals. He suggested that the food consumed by the Baltic seals may contain "new" or increased amounts of hitherto unidentified toxic factors which affect their immune system.

In the Arctic, PCB levels have been associated with measured deficits in the immune system of seals, and deficits in the immune function of Steller sea lions (AMAP, 2004).

Adrenal and other effects in seals

The following effects have also been reported in seals: severe adrenocortical hyperplasia; osteoporosis; claw malformations; arteriosclerosis; and uterine cell tumours (Bergman and Olsson, 1985; Bergman, 1999a; 1999b).

The adrenal gland is involved in the regulation of stress, and environmental stress factors can lead to permanent strain, resulting in structural alterations of the adrenals that in turn are followed by hormonal imbalances (Bragulla *et al.,* 2004).

2.5.5. Whales and other cetaceans

Reduced reproduction, hermaphroditism and immunosuppression in cetaceans

Beluga whales (*Delphinapterus leucas*) in the Saint Lawrence estuary and orca whales (*Orcinus orca*) in the North Pacific are two very highly polluted wildlife populations (for review see Fossi and Marsili, 2003; Heiman *et al.,* 2000; Ross *et al.,* 2000). Neither is reproducing well. They have some of the highest PCB levels found in wildlife, higher than those associated with reduced reproduction in seals, although the effects of contaminants on whales is difficult to ascertain with certainty

(Trites and Barrett-Lennard, 2001). Orcas in the Pacific off British Columbia are very contaminated and compared to the beluga whales in the Saint Lawrence estuary, the fish-eating southern resident orcas are twice as contaminated with PCBs, and the seal-eating transient killer whales are four times as contaminated (Trites and Barrett-Lennard, 2001). Similarly, research by Wolkers and colleagues for WWF shows that orcas in the Norwegian Arctic are very highly contaminated (WWF, 2005).

In the Arctic, not only killer whales, but also some harbour porpoises (*Phocoena phocoena*) from Norway, long-finned pilot whales (*Globicephala melas*), narwhal (*Monodon monoceros*), and some minke whales (*Balaenoptera acutorostrata*) have summed PCB levels higher than those associated with decreased otter and mink reproduction (AMAP, 2004).

The beluga population in the Saint Lawrence estuary has not increased since hunting was banned in the 1970s. In a report published in 1995 by Douglas, the rate of pregnancy of beluga whales in the Saint Lawrence river estuary was only 3% compared with 35% in those from the Canadian Arctic (see Riedel *et al.,* 1997). De Guise (1995) has suggested that this could be linked with exposure to organochlorines.

To date, there have been few reports of congenital defect of the reproductive tract in whales, but few studies investigate internal organs. However, researchers have reported finding a true hermaphrodite (with 2 testicles and 2 ovaries) amongst the 129 belugas that were examined from the Saint Lawrence estuary in the period 1983-1999 (De Guise, 1995), and a pseudohermaphrodite (Michaelian *et al.,* 2003). Effects on sex hormones have been linked with pollutants, and for example, reduction in the testosterone levels in Dall's porpoises of the northwestern North Pacific has been associated with exposure to PCBs and DDE (Subramanian *et al.,* 1987).

In addition, many cetaceans are contaminated to such an extent that immune system effects are also likely, but research is needed to examine whether offspring are affected. For example, summed PCB levels in some orcas and long-finned pilot whales from the Arctic exceed the threshold for immunosuppression in harbour seals (AMAP, 2004), and similarly so do the levels of PCBs and dioxins in the blubber of some northern right whale dolphin (*Lissodelphis borealis*) and Pacific white-sided dolphin (*Lagenorhynchus obliquidens*) from the northern North Pacific, and dolphin and porpoise species from the seas around Japan, and the finless porpoise (*Neophocaena phocaenoides*) from Hong Kong (Minh *et al.,* 2000). Some

Mediterranean species are also very contaminated, and here the common dolphin has almost completely disappeared (Fossi and Marsili, 2003). Contaminant-induced immunosuppression may also be contributing to disease susceptibility in harbour porpoises from the Baltic Sea, as thymic atrophy was associated with pollutants (Beineke *et al.,* 2005). In Florida, decreased lymphocyte proliferation responses in wild bottlenose dolphins were associated with increased concentrations of contaminants, particularly PCBs and DDT related chemicals (Lahvis *et al.,* 1995).

When immunosuppression is widespread, outbreaks of infectious diseases may lead to mass mortality. For example, in the Mediterranean from 1990 to 1991, mass mortality among striped dolphins (*Stenella coeruleoalba*) occurred due to dolphin morbillivirus, and immunosuppression may have played a role (de Swart *et al.,* 1995b).

A high rate of cancers has been reported in Saint Lawrence belugas (De Guise *et al.,* 1994; Martineau *et al.,* 1994). Twenty eight tumours were found in 18 of the 45 animals subject to necropsy, and this was said to make up 37% of the total tumours (75) reported up until then in cetaceans. Immunosuppression was suspected, and indeed mice fed with a diet in which the fat content was replaced with blubber of the Saint Lawrence beluga exhibited immunosuppression (Fournier *et al.,* 2000).

In addition, thyroid abnormalities have been reported in whales, including goitre in pilot whales from Newfoundland, and adenomatous hyperplasia in beluga whales from the Saint Lawrence estuary and the Hudson Bay. Other lesions suggestive of endocrine disruption in the Saint Lawrence estuary belugas included degenerative changes in the adrenal gland (see Mikaelian *et al.,* 2003).

2.5.6. Polar bears

Many pollutants, including some PCBs, increased in polar bears in the period between 1967 and 1994 (Derocher *et al.,* 2003), and both pollutants and global warming may pose a threat to the long-term survival of this species.

Effects on reproduction in polar bears

The impacts of contaminants on the Svalbard polar bear population are inconclusive but there are suggestions of contaminant-related population level effects that could have resulted from reproductive impairment of females, lower survival rates of cubs, or increased mortality of reproductive females (Derocher *et al.,* 2003). Cubs of mothers with high levels of contaminants in their fat were found to be more likely to die during their first year than cubs of mothers with low levels. These cubs may be particularly vulnerable since polar bear milk is about 30 percent fats, and contaminants stored in the mother's fat are transferred to offspring during suckling. Certainly, the summed PCB levels in some Svalbard polar bears exceeded the levels known to be correlated with poor reproductive success in seals (AMAP, 2004).

Intersex / hermaphroditism and reduced testosterone in polar bears

In polluted areas, hermaphrodite polar bears may be more common, and that this condition could be due to excessive maternal androgen excretion caused by a tumour or endocrine disrupting pollutants (Wiig *et al.,* 1998). In 1996, two yearling Svalbard polar bears, which were believed to be female, were found with a normal vaginal opening and a 20 mm penis containing a baculum. Then on subsequent separate occasions, two other Svalbard bears were found to exhibit female pseudo-hermaphroditism as they had deformed genitals, manifest as clitoral hypertrophy (Wiig *et al.,* 1998). However, some of these later reported females may have been mis-diagnosed as pseudohermaphrodites, because a subsequent female was found with an enlarged clitoris, with no signs of any histological or structural changes which would be expected if hormone disruption was involved (Sonne *et al.,* 2005).

Normal sexual development and later reproductive function are dependent on testosterone, and organochlorine levels in adults have been linked with perturbation of testosterone levels. Both the sum of the pesticides (hexachlorocyclohexanes, hexachlorobenzene, chlordanes, p,p'-DDE) and the sum of the PCBs, were related to a reduction in plasma testosterone concentration. Researchers have therefore suggested that the continuous presence of high concentrations of organochlorines in male polar bears throughout their life could aggravate any reproductive toxicity that might have occurred during early development in the womb (Oskam *et al.,* 2003).

Thyroid disruption and neurotoxic effects in polar bears

Organochlorine chemicals, such as DDE and hexachlorobenzene, have been shown to be associated with altered thyroid hormone levels (Skaare *et al.,* 2001). Similarly, PCB levels have also been shown to affect five thyroid hormone variables in the female polar bears (total thryoxine (TT4), free thyroxin (F4), free triiodothyronine (FT3), TT3:FT3, TT4:TT3), whereas only two such variables were affected in males (FT3, FT4:FT3). This suggests that female polar bears could be more susceptible to thyroid disruption-related effects of PCBs than males (Braathen *et al.,* 2004). In Svalbard polar bears, the summed PCB levels also exceed the levels known to cause neurobehavioral effects in the offspring of monkeys and humans (AMAP, 2004).

Immune system suppression in polar bears

Associations were found between immunoglobulin G (IgG) levels and the organochlorine contaminants in the blood plasma of polar bears caught at Svalbard during 1991-1994. Immunoglobulin was negatively correlated with HCB, and with summed PCB levels, and with three individual PCB congeners (IUPAC numbers 99, 194, and 206). It was concluded that the significant negative organochlorine correlation with IgG levels may indicate an immunotoxic effect (Bernhoft *et al.,* 2000). This has been confirmed, in that high levels of organochlorines (OCs) have been shown to be associated with decreased ability to produce antibodies, in free-ranging polar bears. In 1998 and 1999, 26 and 30 polar bears from Svalbard in Norway, and Churchill in Canada, respectively, were recaptured around 36 days after immunization with inactivated influenza virus, reovirus, and herpes virus and tetanus toxoid. Blood was sampled at immunization and at recapture. The combination of PCBs (sum of 12 PCB congeners), and organochlorine pesticides (sum of 6), and biological factors accounted for 40-60% of the variation in the immunological parameters. This demonstrated that high levels of organochlorines may impair the polar bears' ability to produce certain antibodies and thus may produce impaired humoral immunity, which results in reduced resistance to infections (Lie *et al.,* 2004). A subsequent publication noted that organochlorine exposure also significantly influences specific lymphocyte proliferation responses and part of the cell-mediated immunity, which also is associated with impaired ability to produce antibodies (Lie *et al.,* 2005).

2.5.7. Black and brown bears

Unlike polar bears, where only pregnant female polar bears hole up in a den, brown and black bears hibernate in winter. This might make them uniquely sensitive to thyroid disrupting chemicals, which play an important role in hibernation (Tomasi *et al.,* 1998).

However, as yet it seems that only effects tentatively suggestive of sex hormone disruption have been reported. For example, Dunbar and colleagues found retained testes in 11 (16%) of 71 black bears (*Ursus americanus*) examined over a 3-year period in Florida (USA). Four of the 11 bears were older than one year and were therefore considered to be cryptorchid. The remaining seven bears may have had delayed testicular descent due to their apparent normal immature development. This is the first published report of the prevalence of cryptorchidism in black bears, and therefore it is difficult to draw conclusions (Dunbar *et al.,* 1996).

In bears in Alberta in Canada, a couple of decades ago there were some reported cases of masculinized females, but the cause of this pseudo-hermaphroditism was not known (Cattet, 1988). Nevertheless, given the frequency of the occurrence, environmental factors might have played a role. Black (*Ursus americanus*) and brown bears (*Ursus arctos*) are herbivorous, and so it was suggested that excessive maternal androgens or herbicides or plant derived alkaloids might have been involved. However, it is speculated here that atmospherically transported industrial pollutants deposited on foliage could perhaps also have played a role.

2.5.8. Florida panther

Facemire *et al.* (1995) reported that many of the small remaining population of Florida panther (*Felis concolor coryi*) have one or more of the following: increased number of abnormal sperm, low sperm density, and undescended testicles. These effects have been linked to the abnormally similar serum oestradiol levels found in male and females, suggesting that many males had been de-masculinized and feminized. Endocrine disrupting pollutants, taken up through the food chain, were the suggested cause. However, some researchers suggest it may largely be due to the genetics of the small population (Mansfield and Land, 2002). Other defects noted in these panthers, which Facemire and others argued were linked to pollutants, included thyroid dysfunction and adenomatous hyperplasia, altered immune system, and congenital heart defects. Moreover, both sterile males and an infertile female were reported (Facemire *et al.,* 1995).

2.5.9. Sitka black tail deer

Veeramachaneni and colleagues have reported that many sitka black tail deer *(Odocoileus hemionus Sitkenis)* on Kodiak Island, Alaska have undescended testicles, and antler dysgenesis, which are signs of defective androgen action. These researchers could not rule out a link with a recessive mutation in a founder animal, but considered that one or more environmental estrogenic chemicals were more likely to be the cause. These researchers noted experimental data illustrating that, while a variety of genetic mutations or chemicals can block normal testicular descent, not all of them cause germ cell transformation leading to CIS and development of tumours.

In the low lying Aliluik peninsula, two thirds of the deer examined (61 out of 94) had both testicle descended, and 70% of these (43) had abnormal antlers. The testes of some of the deer were subject to examination. Where both testes were undescended, there were many abnormalities, including no spermatogenesis and carcinoma in situ (CIS) cells (considered to be precursors of seminoma, a form of testicular cancer). Moreover, at least 2 scrotal testes were also found to contain CIS cells (Veeramachaneni *et al.,* 2005). Effects on reproduction at the level of the individual must therefore be apparent, as those with neither testicle descended are azoospermic (without sperm).

2.5.10. White tailed deer

From 1996 to 2000, accident-killed and injured white-tailed deer (*Odocoileus virginianus*) in Montana USA were collected and examined for genital abnormalities. Of the 254 male deer examined, approximately 33% were normal, but the remaining 67% showed varying degrees of genital developmental anomalies, specifically mis-positioned and undersized scrota and ectopic testes. The sex ratio was also skewed towards males, but this might be due to the males venturing out more onto trafficked roads. The authors discussed the possible role of endocrine disrupting pesticides, but could not give any firm conclusions as the cause of the abnormalities (Hoy *et al.,* 2002).

Earlier research by Marburger also suggested effects in white tailed deer due to defective androgenic action. He reported findings in 3 white tailed deer, which included effects like hypospadias (Veeramachaneni *et al.,* 2005).

2.5.11. Mule deer

Mule deer (*Odocoileus hemionus)* on a former plutonium production site along the Columbia River at the Hanford Site, Washington (USA) have also been found with abnormal testes and antlers. Some 27 of 116 adult males examined had unusually shaped, velvet-covered antlers and abnormally developed testicles (Tiller *et al.,* 1997). The severity of the testicular atrophy and apparent lack of other affected tissues led the researchers to suggest that radiation was unlikely to be responsible. Other possibilities include phytoestrogens in the diet or endocrine disrupting chemicals on the soil and vegetation or some genetic abnormality.

2.5.12. Eland

Eland are one of the genera of antelope from the *bovidae* species. Deformities reported in eland (*Tragelaphus oryx*) may represent the first evidence of wildlife being affected by EDCs in South Africa. Professor Bornman and colleagues collected testes and body fat from 11 eland. Focal white gritty areas were observed in the testes of all 11 eland, and spermatogenesis was generally impaired. It was suggested that the testicular lesions observed in eland could be associated with the relatively high body burden of nonylphenol, as the vacuolisation of Sertoli cells were similar to those observed in rats exposed to nonylphenol (Bornman *et al.*, 2005). Nonylphenol ethoxylate; which breaks down to nonylphenol, is used as an ingredient in many pesticides, as well as in leather or textile processing, metal working, and cleaning operations.

3. Summary of some effects found in both humans and wildlife

3.1. Testicular dysgenesis syndrome / Intersex

In humans, Skakkebaek *et al.* (2001) hypothesised that genital malformations in baby boys (including undescended testes and hypospadias), low sperm counts, and testicular cancer might all be symptoms of one underlying entity, testicular dysgenesis syndrome (TDS), originating from an *in-utero* event. Moreover, they considered that the rapid pace of the increase of these reproductive disorders argued for an environmental cause. Since anti-androgenic pollutants cause similar genital malformations in rodents, these chemicals are suspected to play a role in such de-masculinization of the male reproductive health (Gray *et al.*, 1999a; Key *et al.*, 1996). Estrogenically active compounds may also be involved, and in a small study of babies with ambiguous genitals, higher estrogenic activity was found in the newborns' serum (Paris *et al.*, 2006). Baskin *et al.* (2001) have reviewed studies suggesting that exposure to dioxins, or to chemicals used in agriculture, or leaking from hazardous waste sites, may increase the risk of hypospadias, a defect of the penis.

Certainly, many chemicals have been associated with deficits in sperm, and effects on reproduction have been noted. For example, some studies suggest that wives of pesticide workers (Sallmen *et al.*, 2003) or metal or solvent workers (Petrelli *et al.*, 2001) have a longer time to pregnancy. In addition, banana and pineapple plantation workers in several countries have been rendered infertile (azoospermic) from exposure to the pesticide, dibromochloropropane (Slutsky *et al.*, 1999).

However, it will be difficult to identify chemicals causing 'testicular dysgenesis syndrome' effects through epidemiological or field studies, unless there is a particular pollutant or group of pollutants that are largely responsible. This is suggested to be the case in South Africa, where some mothers have high p,p'-DDE levels due to spraying of DDT for malaria control, and a high number of babies with deformed and intersex genitals have been reported (Bornman *et al.*, 2005). Here, the authors conclude that the concordant high prevalence of urogenital birth defects and the DDE concentrations in cord blood in babies born in a DDT-sprayed area should be regarded as a matter of extreme international concern. In other locations, other pollutants may dominate. For example, in the USA, it may be that phthalates are responsible for a large part of any anti-androgenic activity, as work by Swan suggests that exposure to several phthalates may actually be impacting androgen related development in baby boys (Swan *et al.,* 2005). These workers found that almost all (11 out of 12) of boys with the highest combined exposure to certain phthalates had a short anogenital distance (AGD) (ie. below the 25th percentile for age and weight). Furthermore, 21% of the boys with short AGD had incomplete testicular descent, compared to 8% of other boys. AGD is a sensitive indicator of masculinization, and was also significantly correlated with the size (volume) of the penis. Similarly, work by Main and colleagues in EU countries, shows that perinatal exposure to phthalates, via breast milk, can alter the normal postnatal surge of reproductive hormones in newborn boys, which is a suggested sign of testicular dysgenesis (Main *et al.,* 2006).

If testicular dysgenesis syndrome is occurring in humans due to pollutants, then it should be found in wildlife in areas with high levels of pollutants, and this indeed does seem to be the case. This review concludes that symptoms comparable with testicular dysgenesis syndrome are now widespread in wildlife, and that these are likely to be caused by contaminants acting as steroid hormone disrupters. Intersex features have been reported in many species of wild fish, amphibians, reptiles, mammals, and to lesser extent, in birds.

3.2. Thyroid disruption

Thyroid hormones are responsible for normal metabolism and for orchestrating normal brain development, and thyroid disruption has now been seen in many wildlife species. Given the complexity and the long development time of the human brain, it might be reasonable to predict that this will be especially vulnerable to toxic insult.

In the Great Lakes, salmon and birds have been found with enlarged thyroids (goitre). Moreover, when these salmon were fed to rats, the rats themselves developed thyroid lesions (Leatherland, 1992), indicating a persistent chemical in the diet was the causal agent. Similarly, in humans living around the Great Lakes in Michigan in the late 1980s, Beierwaltes reported high rates of a thyroid disease, goitre (Fox, 2001).

In many developed nations, in order to assess and, if necessary, provide early treatment for thyroid hormone disorders, routine thyroid screening of newborns is undertaken. However, academic studies have found pollutant associated effects on thyroid hormones in babies in European countries. For example, in a Dutch mother and child cohort, which included over 200 mother and child pairs, higher PCDD, PCDF, and PCB levels in human milk, expressed as toxic equivalents, correlated significantly with lower plasma levels of maternal total triiodothyronine and total thyroxine (T4), and with higher plasma-levels of thyroid stimulating hormone (TSH) in the infants in the 2nd week and 3rd month after birth. Infants exposed to higher toxic equivalents levels had also lower plasma free thyroxine and total thyroxine levels in the 2nd week after birth (Koopman-Esseboom, 1994). Similarly in Taiwan, Wang *et al.* (2005) found significantly altered thyroid hormone levels associated with *in utero* exposure to non-ortho PCBs.

Given the role of thyroid hormones in orchestrating normal brain development, it should not be surprising that altered thyroid hormones have been accompanied by effects on brain function and behaviour. In some study populations, including children living around the Great Lakes and children in EU countries exposed in the womb to the high end of normal background levels, such effects have indeed been extensively reported (Jacobson *et al.,* 1985, 1990; Jacobson and Jacobson, 1993, 1996; Darvill *et al.,* 2000; Stewart *et al.,* 2003; Patandin *et al.,* 1999; Walkowiak *et al.,* 2001). However, although both effects on brain function and thyroid hormone have been reported, it is difficult to prove that the effects on brain function are actually a direct consequence of thyroid disruption and do not result from some other pollutant-related mechanism of action. Nevertheless, it is clear that these effects on brain function are a congenital defect resulting from exposure to chemical contaminants.

In Japan, it seems that between 1981 and 1995 the incidence of defective thyroid function (cretinism) in children tripled, despite the steady decline in the birth rate (Seo, 2002). Nagayama (1998) and also Seo, in a personal communication of 2003, have suggested that dioxin was the causal agent.

However, dioxins and PCBs are not the only chemicals that can affect thyroid hormones. Some brominated chemicals used as flame retardants, such as the polybrominated diphenyl ethers (PBDEs) have effects on the thyroid (for review see EU RARs; Fernie *et al.,* 2005), as do other industrial chemicals and pesticides (Zoeller, 2005; Brucker-Davis, 1998).

*3.3. **Immunotoxicity***

The immune system can also be adversely affected by developmental exposure to EDCs, and such changes can last for extended periods (Holladay *et al.,* 1991).

Immune system function can be altered by stress and the interaction between the reproductive and immune system is well known. Indeed, the immune system carries significant sexual dimorphism. Both reduced or hyperactive immune system function are a concern, the latter being associated with auto-immune diseases, which are more frequent in women.

Pollutants causing immunotoxicity in humans and wildlife could be a factor in a whole host of other diseases including infectious diseases and cancer. For example, dioxin exposure inhibits immunoglobulin secretion and decreases resistance to bacterial, viral, and parasitic infections in exposed animals (Baccarelli *et al.*, 2002). The potential for chemical onslaught to reduce the immune systems of wildlife and humans must also be viewed in the context of the threat of new infectious diseases.

In wildlife, reduced immune system function has been reported in many species and linked to pollutants. Furthermore, controlled experiments have found immune system deficits in seals fed on fish from polluted waters (de Swart *et al.*, 1995a), and as such fish were destined for the table of humans, it is not unreasonable to speculate that contaminant related immune suppression in humans is likely to be occurring. Indeed, numerous examples, both relating to accidental and background exposures, underline that this is indeed the case.

With regard to accidental exposure to high levels, there are several notable examples. For example, people exposed to dioxin contaminated waste oils, sprayed for dust control on a dirt road in a caravan park in Missouri in 1971, had altered immune systems (Damstra *et al.*, 2002). Similarly, immune system deficits were also observed in Taiwanese residents who consumed rice oil accidentally contaminated with furans and PCBs, and children born to exposed women had higher rates of bronchitis. Effects on the immune system may also persist for a long time. In Italy, nearly 20 years after the accident in Seveso, where people were exposed to high dioxin levels, plasma immunoglobulin levels were found to be lower in people with higher TCDD plasma levels (Baccarelli *et al.*, 2002).

Other chemical exposures are known to be linked to immune suppression. For example, diethylstilboestrol (DES), which was used as a pharmaceutical, has been shown to cause immune alterations in humans, via an endocrine disrupting mechanism (Blair, 1992). The developing immune system is more vulnerable that that of the adult, and therefore the chemically-related immunotoxic risks to newborns may be underestimated (Dietert and Piepenbring, 2005). Immune suppression has been reported in children exposed to DES in the womb, and data suggest that effects persist (Wingard and Turiel, 1988).

It is also clear that even 'normal' background levels of certain pollutants may cause effects. For example, in the Dutch cohort studies published by Weisglas-Kuperus and colleagues in the 1990s, which looked at children who were exposed to normal background levels of dioxins and PCBs, both pre and postnatal exposure to PCBs and dioxins was significantly associated with reduced monocyte and granulocyte counts at 3 months, but not at 18 months. However, other indications of immune deficiency persisted, such as more recurrent middle ear infection and lower antibody levels to measles (Weisglas-Kuperus *et al.,* 2000).

In the Arctic, an area where people tend to have higher pollutant levels, studies also suggest that prenatal organochlorine exposures are linked with deficits in immune function, giving rise to increased gastro-intestinal, respiratory and ear infections in infants (Dewailly *et al.,* 2000; van Oostdam *et al.,* 2005; Dallaire *et al.,* 2004).

3.4. Congenital defects of jaw, beak, claw, nails or teeth

Mink (*Mustela vison*) from the PCB contaminated Kalamazoo River in the USA, have been found with jaw deformities (Becket *et al.,* 2005). In birds in the Arctic, foot deformities and increased numbers of congenital beak deformities are evident, with PCBs and dioxins suggested to be the causal agents (Guardian, 2004). In addition, in seals, claw deformities have been reported. Perhaps similarly, in babies in Taiwan, where high level accidental exposure occurred, transplacental exposure to thermally degraded PCBs resulted in a generalized disorder of ectodermal tissue, manifest as abnormalities of the skin, nails, and teeth (Rogan *et al.,* 1988). It is speculated here that ectodermal tissue deformities in these species might originate from damage to similar tissue in the embryo during differentiation of the cells.

4. Conclusions

There is very good evidence that pollutant-related congenital defects are widespread in both humans and wildlife. An exhaustive review of all the literature has not been undertaken, but nevertheless it is clear that some of the most prevalent congenital effects reported in wildlife, which are associated with pollutants, are related to genital disruption (GD). Genital disruption includes an array of manifestations, such as, for example: ovotestes; obvious structural defects of the male reproductive tract such as cryptorchidism and smaller phallus; and intersex features or ambiguous genitals. GD is related to in-womb or *in-ovo* exposures to endocrine disrupting pollutants. Other overt congenital defects noted in wildlife include beak and jaw deformities. Pollutant-related thyroid disruption and immunotoxic effects have also been frequently reported in wildlife.

The effects noted in wildlife, show similarities across the species, as shown in table 1. Such wildlife effects should raise concerns for contaminant-induced genital disruption, immune suppression and thyroid disruption in human infants, and indeed these, and effects on brain function, have all been reported and associated with pollutants.

The potential for transgenerational effects also heightens the concern, and underlines the need for a more precautionary approach to regulation. More research is also needed to fully understand these phenomena. Furthermore, resources are needed to ensure better monitoring of contaminants and effects in wildlife, in locations throughout the globe.

Most importantly, current findings merit urgent action to drastically reduce exposure to chemicals of very high concern, such as endocrine disrupting chemicals, and persistent and bioaccumulating chemicals. Eliminating exposure, and using safer alternatives is the best way forward, particularly because of the potential for additive effects and because there may be no threshold for effects, for some hormonally active chemicals (Sheahan, 2005). Such action is imperative in order to protect wildlife populations and also ensure that our own offspring are not 'doomed from the womb'.

References

Allen, Y., Balaam, J., Bamber, S., Bates, H., Best, G., Bignell, J., Brown, E., Craft, J., Davies, I.M., Depledge, M., Dyer, R., Feist, S., Hurst, M., Hutchinson, T., Jones, G., Jones, M., Katsiadaki, I., Kirby, M., Leah, R., Matthiessen, P., Megginson, C., Moffat, C.F., Moore, A., Pirie, D., Robertson, F., Robinson, C.D., Scott, A.P., Simpson, M., Smith, A., Stagg, R.M., Struthers, S., Thain, J., Thomas, K., Tolhurst, L., Waldock, M. and Walker, P. (2002) *Endocrine disruption in the marine environment (EDMAR),* ISBN 0 907545 16 5, UK Department for Environment, Food and Rural Affairs, London.

Amado, L.L., Rosa, C.E., Leite, A.M., Moraes, L., Pires, W.V., Pinho, G.L., Martins, C.M., Robaldo, R.B., Nery, L.E., Monserrat, J.M., Bianchini, A., Martinez, P.E. and Geracitano, L.A. (2005) Biomarkers in croakers Micropogonias furnieri (Teleostei: Sciaenidae) from polluted and non-polluted areas from the Patos Lagoon estuary (Southern Brazil): Evidences of genotoxic and immunological effects, *Mar. Pollut. Bull.* E pub.

AMAP (Arctic Monitoring and Assessment Programme) (2004) *AMAP Assessment 2002: Persistent organic pollutants in the Arctic,* AMAP, Oslo, Norway. (online) www.amap.no [last accessed 16/03/06]

Ankley, G.T., Degitz, S.J., Diamond, S.A. and Tietge, J.E. (2004) Assessment of environmental stressors potentially responsible for malformations in North American anuran amphibians, *Ecotoxicol. Environ. Safety* **58**, 7-16.

Anway, M.D., Cupp, A.S., Uzumcu, M. and Skinner, M.K. (2005) Epigenetic transgenerational actions of endocrine disrupters and male fertility, *Science* **308**, 1466-1469.

Baccarelli, A., Mocarelli, P., Patterson, D.G. Jr., Bonzini, M., Pesatori, A.C. and Caporaso, N.L. (2002) Immunologic effects of dioxin: new results from Seveso and comparison with other studies, *Environ. Health Perspect.* **110**, 1169-1175.

Barnhoorn, I.E., Bornman, M.S., Pieterse, G.M. and van Vuren, J.H. (2004) Histological evidence of intersex in feral sharptooth catfish (*Clarias gariepinus*) from an estrogen-polluted water source in Gauteng, South Africa, *Environ. Toxicol.* **19**, 603-608.

Baskin, L.S., Himes, K. and Colborn, T. (2001) Hypospadias and endocrine disruption: is there a connection? *Environ. Health Perspect.* **109**, 1175-1183.

Batty, J., Leavitt, R.A., Biondo, N. and Polin, D. (1990) An ecotoxicological study of a population of the white footed mouse (*Peromyscus leucopus*) inhabiting a polychlorinated biphenyls-contaminated area, *Arch. Environ. Contam. Toxicol.* **19**, 283-290.

Beckett, K.J., Millsap, S.D., Blankenship, A.L., Zwiernik, M.J., Giesy, J.P. and Bursian, S.J. (2005) Squamous epithelial lesion of the mandibles and maxillae of wild mink (*Mustela vison*) naturally exposed to polychlorinated biphenyls, *Environ. Toxicol. Chem.* **24**, 674-677.

Beineke, A., Siebert, U., McLachlan, M., Bruhn, R., Thron, K., Failing, K., Müller, G. and Baumgärtner, W. (2005) investigations of the potential influence of environmental contaminants on the thymus and spleen of harbor porpoises (Phocoena phocoena), *Environ. Sci. Technol.* **39** 3933 -3938.

Bergman, A. (1999a) Health condition of the Baltic grey seal (*Halichoerus grypus*) during two decades. Gynaecological health improvement but increased prevalence of colonic ulcers. *APMIS* **107**, 270-282.

Bergman, A. (1999b) Prevalence of lesions associated with a disease complex in the Baltic grey seal (*Halichoerus grypus*) during 1977-1996. In: T.J.O'Shea, R.R. Reeves, and A.K.Long (eds.), *Marine Mammals and Persistent Ocean Contaminants*: Proceedings of the Marine Mammal Commission workshop. Keystone, Colorado, 12-15 October 1998, pp139-143.

Bergman, A. and Olsson, M. (1985) Pathology of Baltic gray seal and ringed seal females with special reference to adrenocortical hyperplasia: Is environmental pollution the cause of a widely distributed disease syndrome? *Finn. Game Res.* **44**, 47–62.

Bernhoft, A., Skaare, J.U., Wiig, O., Derocher, A.E. and Larsen, H.J. (2000) Possible immunotoxic effects of organochlorines in polar bears (*Ursus maritimus*) at Svalbard, *J. Toxicol. Environ. Health* **59**, 561-574.

Blair, P.B. (1992) Immunologic studies of women exposed *in utero* to diethylstilboestrol, in T.Colborn and C.Clement *(eds.)*, *Chemically Induced Alterations in Sexual and Functional Development*: The Wildlife/Human Connection, Princeton Scientific Publishing, Princeton, NJ, pp289-293.

Bornman, M.S., Barnhoorn, I.E.J., Dreyer, L., Veeramachaneni, D.N.R. and De Jager, C. (2004) *Testicular degeneration coincident with fat residues of nonylphenol in the common eland (Tragelaphus oryx): a possible link to endocrine disruption?* [Abstract]. In: proceedings of the CREDO cluster workshop on ecological relevance of chemically induced endocrine disruption in wildlife, 5-7 July 2004, University of Exeter, Exeter, UK.

Bornman, M.S., Delport, R., Becker, P., Risenga, S. and de Jager, C. (2005) *Urogenital birth defects in newborns from a high-risk malaria area in Limpopo province*, South Africa. Conference abstract from September 2005, Sandon, Johannesburg.

Bowers, W.J., Nakai, J.S., Chu, I., Wade, M.G., Moir, D., Yagminas, A., Gill, S., Pulido, O. and Meuller, R. (2004) Early developmental neurotoxicity of a PCB/organochlorine mixture in rodents after gestational and lactational exposure, *Toxicol. Sci.* **77**, 51-62.

Braathen, M., Derocher, A.E., Wiig, O., Sormo, E.G., Lie, E., Skaare, J.U. and Jenssen, B.M. (2004). Relationships between PCBs and thyroid hormones and retinol in female and male polar bears, *Environ. Health Perspect.* **112**, 826-833.

Bragulla, H., Hirschberg, R.M., Schlotfeldt, U., Stede, M. and Budras, K.D. (2004) On the structure of the adrenal gland of the common seal (Phoca vitulina vitulina), *Anat. Histol. Embryol.* **33**, 263-272.

Brian, J.V., Harris, C.A., Scholze, M., Backhaus, T., Booy,P., Lamoree, M., Pojana, G., Jonkers, N., Runnalls, T., Bonfà, A., Marcomini, A. and Sumpter, J.P. (2005) accurate prediction of the response of freshwater fish to a mixture of estrogenic chemicals, *Environ. Health Perspect.* **113**, 721-728.

Bridges, C., Little, E., Gardiner, D., Petty, J. and Huckins, J. (2004) Assessing the toxicity and teratogenicity of pond water in north-central Minnesota to amphibians, *Environ. Sci. Pollut. Res. Int.* **11**, 233-239.

Brouwers, M.M., W.F.J. Feitz, W.F.J., Roelofs, L.A.J., Kiemeney, L.A.L.M., de Gier, R.P.E. and Roeleveld, N. (2005) Hypospadias: a transgenerational effect of diethylstilboestrol? *Hum. Reprod. (*Online Nov 17), *Hum. Reprod.* **21**, 666-669.

Brucker-Davis, F. (1998) Effects of environmental synthetic chemicals on thyroid function. *Thyroid* **8**, 827-856.

Bursian, S.J., Richard, J.A., Yamini, B. and Tillitt, D.E. (2003) Dietary exposure of mink to fish from the Housatonic River: Effects on reproduction and survival. US Environmental Protection Agency, (online) http://www.epa.gov/region1/ge/thesite/restofriver/reports/final era/ SupportingInformation%20and%20Studies%20for%20the%20HousatonicRiverProject/Dietary %20Exposure%20of%20Mink.pdf [last accessed 16/03/06].

Catalani, A., Casolini, P., Scaccianoce, S., Patacchioli, F.R., Spinozzi, P. and Angelucci, L. (2000) Maternal corticosterone during lactation permanently affects brain corticosteroid receptors, stress response and behaviour in rat progeny, *Neuroscience* **100**, 319-325.

Cattet, M. (1988) Abnormal sex differentiation in black bears (*Ursus americanus*) and brown bears (*Ursos arctos*), *J. Mamm.* **69,** 849-852.

Colborn, T. (2002) Clues from wildlife to create an assay for thyroid system disruption, *Environ. Health Perspet.* **110(**Supp3), 363-367.

Cook, P.M., Robbins, J.A., Endicott, D.D., Lodge, K.B., Guiney, P.D., Walker, M.K., Zabel, E.W. and Peterson, R.E. (2003) Effects of aryl hydrocarbon receptor-mediated early life stage toxicity on lake trout populations in Lake Ontario during the 20th century, *Environ. Sci. Technol.* **37**, 3864-3877.

Crofton, K.M., Craft, E.S., Hedge, J.M., Gennings, C., Simmons, J.E., Carchman, R.A., Hans Carter, W. and DeVito, M.J. (2005) Thyroid hormone disrupting chemicals: Evidence for dose dependent additivity or synergism, *Environ. Health Perspect.* **113**, 1549-1554.

Dallaire, F., Dewailly, E., Muckle, G., Vézina, C., Jacobson, S.W., Jacobson, J.L. and Ayotte, P. (2004) Acute infections and environmental exposure to organochlorines in Inuit Infants from Nunavik, *Environ. Health Perspect.* **112**, 1359-1364.

Damstra, T., Barlow, S., Bergman, A., Kavlock, R. and Van Der Kraak, G. (eds) (2002) *Global assessment of the state-of-the science of endocrine disrupters.* WHO/PCS/EDC/02.2. World Health Organization/International Programme on Chemical Safety, Geneva.

Darvill, T., Lonky, E., Reihman, J., Stewart, P. and Pagano, J. (2000) Prenatal exposure to PCBs and infant performance on the fagan test of infant intelligence, *Neurotoxicology* **21**, 1029-1038.

De Guise, S., Martineau, D., Beland, P. and Fournier, M. (1995) Possible mechanisms of action of environmental contaminants on Saint Lawrence beluga *whales (Delphinapterus leucas*), *Environ. Health Perspect.* **103**, 73-77.

De Solla, S.R., Bishop, C.A. and Brooks, R.J. (2002) Sexually dimorphic morphology of hatchling snapping turtles (*Chelydra serpentina*) from contaminated and reference sites in the Great Lakes and Saint Lawrence River basin, North America, *Environ. Toxicol. Chem.* **21**, 922-929.

De Swart, R.L., de Ross, P.S., Vedder, L.J., Timmerman, H.H. van Loveren, H., Vos, L.G., Reijnders, P.J.H. and Osterhaus, A.D.M.E (1994). Impairment of immune functions in harbour seals (*Phoca vitulina*) feeding on fish from polluted coastal waters. *Ambio* **23**, 155-159.

De Swart, R.L., Harder, T.C., Ross, P.S., Vos, H.W. and Osterhaus, A.D. (1995b) Morbilliviruses and morbillivirus diseases of marine mammals, *Infect. Agents. Dis.* **4**, 125-30.

De Swart, R.L., Ross, P.S., Timmerman, H.H., Vos, H.W., Reijnders, P.J., Vos, J.G. and Osterhaus, A.D. (1995a) Impaired cellular immune response in harbour seals (*Phoca vitulina*) feeding on environmentally contaminated herring, *Clin. Exp. Immunol.* **101**, 480-486.

DeLong, R., Gilmartin, W.G. and Simpson, J.G. (1973) Premature births in California sea lions: association with high organochlorine pollutant residue levels, *Science* **181**, 1168-1170.

Derocher, A.E., Wolkers, H., Colborn, T., Schlabach, M., Larsen, T.S. and Wiig, O. (2003) Contaminants in Svalbard polar bear samples archived since 1967 and possible population level effects, *Sci. Total Environ.* **301**, 163-74.

Dewailly, E., Ayotte, P., Bruneau, S., Gingras, S., Belles-Isles, M. and Roy, R. (2000) Susceptibility to infections and immune status in Inuit infants exposed to organochlorines, *Environ. Health Perspect.* **108**, 205-211.

Dietz, R., Heide-Jorgensen, M.P. and Harkonen, T. (1989) Mass deaths of harbour seals (*Phoca vitulina*) in Europe, *Ambio* **18**, 258-264.

Dunbar, M.R., Cunningham, M.W., Wooding, J.B. and Roth, R.P. (1996) Cryptorchidism and delayed testicular descent in Florida black bears, *J. Wildl. Dis.* **32**, 661-664.

EA (Environment Agency of England and Wales) (2003) *Fourth otter survey of England 2000-2002*, UK. (online) http://www.environmentagency.gov.uk/commondata/acrobat/otter_introduction.pdf [last accessed 15/03/06].

EC (Environment Canada) (2001) *Bald Eagle populations in the Great Lakes region*, Great Lakes Fact Sheet, Catalogue no CW69-17/1-2001E, ISBN 0-662-298098-X, Ontario,

Canada, (online) http://www.on.ec.gc.ca/wildlife/publications-e.html [last accessed 15/03/06].

EC (Environment Canada) (2003) *Great Lakes fact sheet*, Catalogue no CW66-223/2003E, ISBN 0-662-34076-0, Ontario, Canada. (online) http://www.on.ec.gc.ca/wildlife/publications-e.html [last accessed 15/03/06].

Ellis, R.J., van den Heuvel, M.R., Bandelj, E., Smith, M.A., McCarthy, L.H., Stuthridge, T.R. and Dietrich, D.R. (2003) *In vivo* and *in vitro* assessment of the androgenic potential of a pulp and paper mill effluent, *Environ. Toxicol. Chem.* **22**, 1448-1456.

EU RAR (*EU Risk Assessment Reports*), (online) http://ecb.jrc.it/existing-chemicals/ [last accessed 15/03/06].

Facemire, C.F., Gross, T.S. and Guillette, L.J. Jr. (1995) Reproductive impairment in the Florida panther: nature or nurture? *Environ. Health Perspect.* **103 (**Supp 4), 79-86.

Fairbrother, A., Smits, J. and Grasman, K. (2004) Avian immunotoxicology, *J. Toxicol. Environ. Health B. Crit. Rev.* **7**, 105-137.

Fairchild, W.L., Swansburg, E.O., Arsenault, J.T. and Brown, S.B. (1999) Does an association between pesticide use and subsequent declines in catch of Atlantic salmon (*Salmo salar*) represent a case of endocrine disruption? *Environ. Health Perspect.* **107**, 349-358.

Fenoglio, C., Boncompagni, E., Fasola, M., Gandini, C., Comizzoli, S., Milanesi, G. and Barni, S. (2005) Effects of environmental pollution on the liver parenchymal cells and Kupffer-melanomacrophagic cells of the frog (*Rana esculenta*), *Ecotoxicol. Environ. Safety.* **60**, 259-268.

Fernandez-Moran, J., Saavedra, D. and Manteca-Vilanova, X. (2002) Reintroduction of the Eurasian otter (*Lutra lutra*) in northeastern Spain: trapping, handling, and medical management, *J. Zoo Wild. Med.* **33**, 222-227.

Fernie, K.J., Shutt, J.L., Mayne, G., Hoffman, D., Letcher, R.J., Drouillard, K.G. and Ritchie, I.J. (2005) Exposure to polybrominated diphenyl ethers (PBDEs): changes in thyroid, vitamin a, glutathione homeostasis, and oxidative stress in american kestrels (Falco sparverius), *Toxicol. Sci.* **88**, 375-383.

Fossi, M.C. and Marsili, L. (2003) Effects of endocrine disrupters in aquatic mammals, in Special topic issue on the implications of endocrine active substances for humans and wildlife, *Pure. Appl. Chem.* **75**, 2235-2247.

Fossi, M.C., Casini, S., Marsili, L., Ancora, S., Mori, G., Neri, G., Romeo, T. and Ausili, A. (2004) Evaluation of ecotoxicological effects of endocrine disrupters during a four-year survey of the Mediterranean population of swordfish (*Xiphias gladius*). *Mar. Environ. Res.* **58**, 425-429.

Foster, P. M., Turner, K. J. and Barlow, N. J. (2000) Antiandrogenic effects of a phthalate combination on *in utero* male reproductive development in the Sprague-Dawley rat: additivity of response? *Toxicologisty* **66**, 233.

Fournier, M., Degas, V., Colborn, T., Omara, F.O., Denizeau, F., Potworowski, E.F. and Brousseau, P. (2000) Immunosuppression in mice fed on diets containing beluga whale blubber from the Saint Lawrence estuary and the Arctic populations, *Toxicol. Lett.* **15**, 112-113, 311-317.

Fournier-Chambrillon, C., Berny, P.J., Coiffier, O., Barbedienne, P., Dasse, B., Delas, G, Galineau, H., Mazet, A., Pouzenc, P., Rosoux, R. and Fournier, P. (2004). Evidence of secondary poisoning of free-ranging riparian mustelids by anticoagulant rodenticides in France: implications for conservation of European mink (*Mustela lutreola*). *J. Wild. Dis.* **40**, 688-695.

Fox, G.A. (2001) Wildlife as sentinels of human health effects in the Great Lakes-Saint Lawrence basin, *Environ. Health Perspect.* **109** (Supp 6), 853-861.

Fry, D.M. (1995) Reproductive effects in birds exposed to pesticides and industrial chemicals, *Environ. Health Perspect.* **103**, 165-171.

Gardiner, D., Ndayibagira, A., Grun, F. and Blumberg, B. (2003) Deformed frogs and environmental retinoids, in special topic issue on the implications of endocrine active substances for humans and wildlife, *Pure Appl. Chem.* **75**, 2263-2273.

Giesy, J.P., Feyk, L.A., Jones, P.D., Kannan, K. and Sanderson, T. (2003) Review of the effects of endocrine disrupting chemicals in birds, in special topic issue on the implications of endocrine active substances for humans and wildlife, *Pure. Appl. Chem.* **75**, 2287-2303.

Gillis A. M. (1995). *Smithsonian National Zoological Park website*, (online) http://nationalzoo.si.edu/Publications/ZooGoer/1995/4/cautionarytales.cfm [last accessed 15/03/06].

Gilman, A.P., Beland, P., Colborn, T., Fox, G., Giesy, J., Hesse, J., Kubiak, T. and Pierkarz, D. (1991) Environmental and wildlife toxicology of exposure to toxic chemicals, in R.W.Flint and J.Vena J (eds.), *Human health risks from chemical exposure*: The Great Lakes Ecosystem, Lewis Publishers Inc, Chelsea, Michigan, pp 61-91.

Gilmartin, W.G., Delong, R.L., Smith, A.W., Sweeney, J.C., De Lappe, B.W., Risebrough, R.W., Griner, L.A., Dailey, M.D. and Peakall, D.B. (1976) Premature parturition in the California sea lion, *J. Wild. Dis.* **12,** 104-115.

Grasman, K.A. and Fox, G.A. (2001) Associations between altered immune function and organochlorine contamination in young Caspian terns (*Sterna caspia*) from Lake Huron, 1997-1999, *Ecotoxicology* **10**, 101-114.

Gray, L.E. Jr., Ostby, J., Furr, J., Wolf, C.J., Lambright, C., Parks, L., Veeramachaneni, D.N., Wilson, V., Price M., Hotchkiss, A., Orlando, E. and Guillette, L. (2001) Effects of environmental antiandrogens on reproductive development in experimental animals, *Hum. Reprod. Update* **7**, 248-264.

Gray, L.E. Jr., Wolf, C., Lambright, C., Mann, P., Price, M., Cooper, R.L. and Ostby, J. (1999a) Administration of potentially antiandrogenic pesticides (procymidone, linuron, iprodione, chlozolinate, p,p'-DDE, and ketoconazole) and toxic substances (dibutyl- and diethylhexyl phthalate, PCB 169, and ethane dimethane sulphonate) during sexual differentiation produces diverse profiles of reproductive malformations in the male rat, *Toxicol. Ind. Health* **15**, 94-118.

Gray, L.E., Ostby, J., Wilson, V., Lambright, C., Bobseine, K., Hartig, P., Hotchkiss, A., Wolf, C., Furr, J., Price, M., Parks, L., Cooper, R.L., Stoker, T.E., Laws, S.C., Degitz, S.J., Jensen, K.M., Kahl, M.D., Korte, J.J., Makynen, E.A., Tietge, J.E. and Ankley, G.T. (2002) Xenoendocrine disrupters-tiered screening and testing. Filling key data gaps, *Toxicology* **181-182**, 371-382.

Gray, M.A., Teather, K.L. and Metcalfe, C.D. (1999b) Reproductive success and behaviour of Japanese medaka, *Oryzias Latipes* exposed to 4-tert-octylphenol, *Environ. Toxicol. Chem.* **18**, 2587-2594.

Guardian Unlimited (2004) *Beak deformities in Alaska spread,* Thursday April 15, (online) http://www.guardian.co.uk/life/dispatch/story/0,12978,1191738,00.html [last accessed 15/03/06].

Guillette, L.J. Jr., Crain, D.A., Galle, A., Gunderson, M., Kools, S., Milnes, M.R., Orlando, E.F., Rooney, A.A. and Woodward, A.R. (2000) Alligators and endocrine disrupting contaminants: a current perspective, *Am. Zool.* **40**, 438-452.

Guillette, L.J. Jr., Crain, D.A., Rooney, A.A. and Pickford, D.B. (1995) Organization versus ctivation: The role of endocrine disrupting contaminants (EDCs) during embryonic development in wildlife, *Environ. Health Perspect.* **103**(suppl7), 157-164.

Guillette, L.J.Jr. and Iguchi, T. (2003) Contaminant induced endocrine and reproductive alterations in reptiles, in special topic issue on the implications of endocrine active substances for humans and wildlife, *Pure. Appl. Chem.* **75**, 2275-2286.

Hall, J.A., Law, R.J., Wells, D.E., Harwood, J., Ross, H.M., and Kennedy, S., Allchin, C.R., Campbell, L.A., Pomeroy, P.P. (1992) Organochlorine levels in common seals (*Phoca vitulina*) which were victims and survivors of the 1988 phocine distemper epizootic, *Sci. Total Environ.* **115**, 145-162.

Hanni, K.D., Mazet, J.A., Gulland, F.M., Estes, J., Staedler, M., Murray, M.J., Miller, M. and Jessup, D.A. (2003) Clinical pathology and assessment of pathogen exposure in southern and Alaskan sea otters, *J. Wild. Dis.* **39**, 837-850.

Harding, L.E., Harris, M.L., Stephen, C.R. and Elliott, J.E. (1999) Reproductive and morphological condition of wild mink (*Mustela vison*) and river otters *(Lutra canadensis*) in relation to chlorinated hydrocarbon contamination, *Environ. Health Perspect.* **107**, 141-147.

Harries, J.E., Sheahan, D.A., Jobling, S., Matthiessen, P., Neall, P., Routledge, E., Rycroft, R.., Sumpter, J.P. and Tylor, T (1996) A survey of estrogenic activity in United Kingdom in land waters*, Environ. Toxicol. Chem.* **15**, 1993-2002.

Hart, C.A., Nisbet, I.C., Kennedy, S.W. and Hahn, M.E. (2003) Gonadal feminization and halogenated environmental contaminants in common terns (*Sterna hirundo*): evidence that ovotestes in male embryos do not persist to the prefledgling stage, *Ecotoxicology* **12**, 125-140.

Hauser, R., Williams, P., Altshul, L. and Calafat, A.M. (2005) Evidence of interaction between polychlorinated biphenyls and phthalates in relation to human sperm motility. *Environ. Health Perspect.* **113**, 425-430.

Hayes, T.B., Case, P., Chui, S., Chung, D., Haefele, C., Haston, K., Lee, M., Mai, V.P., Marjuoa, Y., Parker, J. and Tsui, M. (1992) Organochlorine levels in common seals (*Phoca vitulina*) which were victims and survivors of the 1988 phocine distemper epizootic, *Sci. Total Environ.* **115,** 145-162.

Hayes, T.B., Case, P., Chui, S., Chung, D., Haefele, C., Haston, K., Lee, M., Mai, V.P., Marjuoa, Y., Parker, J. and Tsui, M. (2006) Pesticide mixtures, Endocrine disruption, and amphibian declines: Are we underestimating the impact? *Environ. Health Perspect* doi:10.1289/ehp.8051. (online) http://www.ehponline.org/docs/2006/8051/abstract.pdf. [last accessed 15/03/06].

Hayes, T.B., Haston, K., Tsui, M., Hoang, A., Haeffele, C. and Vonk, A. (2003) Atrazine-induced hermaphroditism at 0.1 ppb in American leopard frogs (*Rana pipiens*): laboratory and field evidence, *Environ. Health Perspect.* **111**, 568-575.

Health Canada (1997) *State of knowledge report on environmental contaminants and human health in the Great Lakes Basin*, ISBN 0-662-26-169-0, Ottowa.

Heiman, M., Brown, M., Middaugh, J., Berner, J., Cochran, P., Davis, M., Marcy, S., Hild, C., Johnson, P., Hohn, J., Miller, P., Wang, B., Wright, B. and Bradley. M. (2000) *Contaminants in Alaska: is America's Arctic at risk*? A white paper published by the Department of the Interior and the State of Alaska.

Herbst, L.H. and Klein, P.A. (1995) Green turtle fibropapillomatosis: challenges to assessing the role of environmental cofactors, *Environ. Health Perspect.* **103**, 27-30.

Hewitt, E.A., Crain, D.A., Gunderson, M.P. and Guillette, L.J. Jr.. (2002) Thyroid status in juvenile alligators (*Alligator mississippiensis*) from contaminated and reference sites on Lake Okeechobee, Florida, USA, *Chemosphere* **47**, 1129-1135.

Holladay, S.D., Lindstrom, P., Blaylock, B.L., Comment, C.E., Germolec, D.R., Heindell, J.J. and Luster, M.I. (1991) Perinatal thymocyte antigen expression and postnatal immune development altered by gestational exposure to tetrachlorodibenzo-p-dioxin (TCDD), *Teratology* **44**, 385-393.

Hoy, J.A., Hoy, R., Seba, D. and Kerstetter, T.H. (2002) Genital abnormalities in white-tailed deer (*Odocoileus virginianus*) in west-central Montana: pesticide exposure as a possible cause, *J. Environ. Biol.* **23**, 189-197.

Iwata, M. (1995) Downstream migratory behaviour of salmonids and its relationship with cortisol and thyroid hormones: a review, *Aquaculture* **135**, 131-139.

Jacobson, J. L. and Jacobson, S. W. (1993) A 4-year follow up study of children born to consumers of Lake Michigan fish, *J. Great Lakes Res.* **19**, 776-783.

Jacobson, J. L. and Jacobson, S. W. (1996) Intellectual impairment in children exposed to polychlorinated biphenyls *in utero*, *N. Engl. J. Med.* **335**, 783-789.

Jacobson, J. L., Jacobson, S. W. and Humphrey, H. E. (1990) Effects of *in utero* exposure to polychlorinated biphenyls and related contaminants on cognitive functioning in young children, *J. Pediatr.* **116**, 38-45.

Jacobson, S. W., Fein, G. G., Jacobson, J. L., Schwartz, P. M. and Dowler, J. (1985) The effect of intrauterine PCB exposure on visual recognition memory, *Child. Dev.* **56,** 853-860.

Jobling S. and Tyler, C.R (2003). Endocrine disruption in wild freshwater fish, in Special topic issue on the implications of endocrine active substances for humans and wildlife, *Pure. Appl. Chem.* **75**, 2219-2234.

Johnstone, R.M., Court, G.S., Fesser, A.C., Bradley, D.M., Oliphant, L.W. and MacNeil, J.D. (1996) Long-term trends and sources of organochlorine contamination in Canadian tundra Peregrine Falcons, (*Falco peregrinus tundrius)*, *Environ. Pollut.* **93**, 109-120.

Jooste, A.M, Du Preez, L.H., Carr, J.A., Giesy, J.P., Gross, T.S., Kendall, R.J., Smith, E.E., Van der Kraak, G.L. and Solomon, K.R. (2005) Gonadal development of larval male *Xenopus laevis* exposed to atrazine in outdoor microcosms, *Environ. Sci. Technol.* **15**, 5255-5261.

Jorgensen, E.H., Aas-Hansen, O, Maule, A.G., Strand, J.E.T. and Vijayan, M.M. (2004) PCB impairs smoltification and seawater performance in anadromous Arctic charr (*Salvelinus alpinus*), *Comp. Biochem. Physiol.* Part C **138**, 203-212.

Kannan, K., Guruge, K.S., Thomas, N.J., Tanabe, S. and Geisy, J. (1998) Butyltin residues in southern sea otters (*Enhydra lutris nereis*) found dead along California coastal waters, *Environ. Sci. Technol.* **32,** 1169-1175.

Kannan, K., Kajiwara, N., Watanabe, M., Nakata, H., Thomas, N.J., Stephenson, M., Jessup, D.A. and Tanabe, S. (2004) Profiles of polychlorinated biphenyl congeners, organochlorine pesticides, and butyltins in southern sea otters and their prey, *Environ. Toxicol.* **23**, 49-56.

Karels, A., Markkula, E. and Oikari, A. (2001) Reproductive, biochemical, physiological, and population responses in perch (*Perca fluviatilis L.*) and roach (*Rutilus rutilus L.*) downstream of two elemental chlorine-free pulp and paper mills, *Environ. Toxicol. Chem.* **20**, 1517-1527.

Kavanagh, R.J., Balch, G.C., Kiparissis, Y., Niimi, A.J., Sherry, J., Tinson, C. and Metcalfe, C.D. (2004) Endocrine Disruption and Altered Gonadal Development in White Perch (Morone americana) from the Lower Great Lakes Region, *Environ. Health Perspect.* **112**, 898-902.

Keller, J.M., Kucklick, J.R., Stamper, M.A., Harms, C.A. and McClellan-Green, P.D. (2004) Associations between organochlorine concentrations and clinical health parameters in loggerhead sea turtles from North Carolina, USA, *Environ. Health Perspect.* **112,** 1074-1079.

Key, T.J., Bull, D., Ansell, P., Brett, A.R., Clark, G.M., Moore, J.W., Chilvers, C.E. and Pike, M.C. (1996) A case-control study of cryptorchidism and maternal hormone concentrations in early pregnancy, *Br. J. Cancer* **73**, 698-701.

Kiesecker, J.M. (2002) Synergism between trematode infection and pesticide exposure: a link to amphibian limb deformities in nature? *PNAS* **99**, 9900-9904.

Kimber, K.R. and Kollias, G.V. (2000) Infectious and parasitic diseases and contaminant-related problems of North American river otters (*Lontra canadensis*): a review, *J. Zoo Wild. Med.* **31**, 452-472.

Kirby, M.F., Allen, Y.T., Dyer, R.A., Feist, S.W., Katsiadaki, I., Matthiessen, P., Scott, A.P., Smith, A., Stentiford, G.D., Thain, J.E., Thomas, K.V., Tolhurst, L. and Waldock, M.J. (2004) Surveys of plasma vitellogenin and intersex in male flounder (*Platichthys flesus*) as measures of endocrine disruption by estrogenic contamination in United Kingdom estuaries: temporal trends, 1996 to 2001, *Environ. Toxicol. Chem.* **23**, 748-758.

Kollner, B., Wasserrab, B., Kotterba, G. and Fischer, U. (2002) Evaluation of immune functions of rainbow trout (*Oncorhynchus mykiss*) - how can environmental influences be detected? *Toxicol. Lett.* **10**, 83-95.

Koopman-Esseboom, C., Morse, D.C., Weisglas-Kuperus, N., Lutkeschipholt, I.J., Van der Paauw, C.G., Tuinstra, L.G., Brouwer. A. and Sauer, P.J. (1994) Effects of dioxins and polychlorinated biphenyls on thyroid hormone status of pregnant women and their infants, *Pediatr. Res.* **36**, 468-473.

Lahvis, G.P., Wells, R.S., Kuehl, D.W., Stewart, J.L., Rhinehart, H.L. and Via, C.S. (1995). Decreased lymphocyte responses in free-ranging dolphins *(Tursiops truncates*) are associated with increased concentrations of PCBs and DDT in peripheral blood, *Environ. Health Perspect.* **103**, 67-72.

Leatherland, J.F. (2002) Endocrine and reproductive function in Great Lakes salmon, in T.Colborn and C.Clement (eds.), *Chemically induced alterations in sexual and functional development: the wildlife/human connection*, Princeton Scientific Publishing, Princeton, NJ, pp129-145.

Lie, E., Larsen, H.J., Larsen ,S., Johansen, G.M., Derocher, A.E., Lunn, N.J., Norstrom, R.J., Wiig, O. and Skaare, J.U. (2004) Does high organochlorine (OC) exposure impair the resistance to infection in polar bears (*Ursus maritimus*)? Part I: Effect of OCs on the humoral immunity, *J. Toxicol. Environ. Health A* **67**, 555-582.

Lie, E., Larsen, H.J., Larsen, S., Johansen, G.M., Derocher, A.E., Lunn, N.J., Norstrom, R.J., Wiig, O. and Skaare, J.U. (2005) Does high organochlorine (OC) exposure impair the resistance to infection in polar bears (*Ursus maritimus*)? Part II: Possible effect of OCs on mitogen- and antigen-induced lymphocyte proliferation. *J. Toxicol. Environ. Health A* **68**, 457-484.

Linzey, D., Burroughs, J., Hudson, L., Marini, M., Robertson, J., Bacon, J., Nagarkatti, M. and Nagarkatti, P. (2003) Role of environmental pollutants on immune functions, parasitic infections and limb malformations in marine toads and whistling frogs from Bermuda. *Int. J. Environ. Health Res.* **13**, 125-148.

Luebke, R.W., Hodson, P.V., Faisal, M., Ross, P.S., Grasman, K.A. and Zelikoff, J. (1997) Aquatic pollution-induced immunotoxicity in wildlife species, *Fund. Appl. Toxicol.* **3**, 1-15.

Lye, C.M., Frid, C.L.J. and Gill, M.E. (1998) Seasonal reproductive health of flounder. Platichthys flesus exposed to sewage effluent, *Mar. Ecol. Progr. Ser.* **170**, 249-260.

Lye, C.M., Frid, C.L.J., Gill, M.E. and McCormick, D. (1997) Abnormalities in the reproductive health of flounder *Platichthys* flesus exposed to effluent from a sewage treatment works, *Mar. Pollut. Bull.* **34**, 34-41.

Lye, C.M., Frid, C.L.J., Gill, M.E., Cooper, D.W. and Jones, D.M. (1999) Estrogenic alkylphenols in fish tissues, sediments, and waters from the U.K. Tyne and estuaries, *Environ. Sci. Technol.* **33**, 1009-1014.

Lyons, B.P., Pascoe, C.K. and McFadzen, I.R. (2002) Phototoxicity of pyrene and benzo[a]pyrene to embryo-larval stages of the Pacific oyster (*Crassostrea gigas), Mar. Environ. Res.* **54**, 627-631.

Main, K.M., Mortensen, G.K., Kaleva, M.M., Boisen, K.A., Damgaard, I.N., Chellakooty, M., Schmidt, I.M., Suomi, A.M., Virtanen, H.E., Petersen, D.V., Andersson, A.M., Toppari, J. and Skakkebaek, N.E. (2006) Human breast milk contamination with phthalates and alterations of endogenous reproductive hormones in infants three months of age, *Environ. Health Perspect.* **114,** 270-276.

Mansfield, K.G. and Land, E.D. (2002) Cryptorchidism in Florida panthers: prevalence, features, and influence of genetic restoration, *J. Wild. Dis*. **38**, 693-698.

Marburger, R.G., Robinson, R.M. and Thomas, J.W. (1967) Genital hypoplasia of white-tailed deer, *J. Mammal.* **48,** 674-676.

Martineau, D., De Guise, S., Fournier, M., Shugart, L., Girard, C. Legace, A. and Beland, P. (1994) Pathology and toxicology of beluga whales from the Saint Lawrence estuary, Quebec Canada. Past, present and future, *Sci. Total Environ.* **154,** 201-215

Mason, C.F. and Macdonald, S.M. (1994) PCBs and organochlorine pesticide residues in otters (*Lutra lutra*) and in otter spraints from SW England and their likely impact on populations, *Sci. Total Environ.* **144**, 305-312.

Mason, C.F. and Macdonald, S.M. (2004) Growth in otter (*Lutra lutra*) populations in the UK as shown by long-term monitoring, *Ambio* **33**, 148-152.

Matthiessen, P. (2003) Endocrine disruption in marine fish, in special topic issue on the implications of endocrine active substances for humans and wildlife, *Pure. Appl. Chem.* **75**, 2249-2261.

Matthiessen, P., Allen, Y., Bamber, S., Craft, J., Hurst, M., Hutchinson, T., Feist, S., Katsiadaki, I., Kirby, M., Robinson, C., Scott, S., Thain, J. and Thomas, K (2002) The impact of oestrogenic and androgenic contamination on marine organisms in the United Kingdom-summary of the EDMAR (Endocrine Disruption in the Marine Environment) programme, *Mar. Environ. Res.* **54**, 645-649.

Maule, A.G., Jorgensen, E.H., Vijayan, M.M. and Killie, J.E. (2005) Aroclor 1254 exposure reduces disease resistance and innate immune responses in fasted Arctic charr, *Environ. Toxicol. Chem.* **24**, 117-124.

McNabb, F.M. and Fox, G.A. (2003) Avian thyroid development in chemically contaminated environments: is there evidence of alterations in thyroid function and development? *Evol Dev.* **5,** 76-82.

Mikaelian, I., Labelle, P., Kopal, M., De Guise S. and Martineau, D. (2003) adenomatous hyperplasia of the thyroid gland in beluga whales (*Delphinapterus leucas*) from the Saint Lawrence Estuary and Hudson Bay, Quebec, Canada *Vet. Pathol.* **40**, 698-703.

Minh, T.B., Nakata, H., Watanabe, M., Tanabe, S., Miyazaki, N., Jefferson, T.A., Prudente, M. and Subramanian, A.(2000) Isomer-specific accumulation and toxic assessment of polychlorinated biphenyls, including coplanar congeners, in cetaceans from the North Pacific and Asian coastal waters, *Arch. Environ. Contam. Toxicol.* **39**, 398-410.

Moore A. and Lower, N. (2001) The impact of two pesticides on olfactory-mediated endocrine function in mature male Atlantic salmon (*Salmo salar L.*) parr, *Comp. Biochem. Physiol. B Biochem. Mol. Biol.* **129,** 269-276.

Nagayama, J. L. (1998) *endocrine disrupters and dioxin (in Japanese)*. Kagakudojin, Japan.

Nakata, H., Kannan, K., Jing, L., Thomas, N., Tanabe, S. and Geisy, J.P. (1998) Accumulation pattern of organochlorine pesticides in southern sea *otters (Enhydra lutris nereis*) found stranded along coastal California, US, *Environ. Pollut.* **103**, 45-53.

National Biological Service (NBS) (1996). *Press Release: male otters from Columbia river have abnormally small reproductive organs*, February 14.

Newbold, R.R., Hanson, R.B., Jefferson, W.N., Bullock, B.C., Haseman, J. and McLachlan, J.A. (1998). Increased tumors but uncompromised fertility in the female descendants of mice exposed developmentally to diethylstilboestrol, *Carcinogenesis* **19**, 1655-1663.

Newbold, R.R., Hanson, R.B., Jefferson, W.N., Bullock, B.C., Haseman, J. and McLachlan, J.A. (2000) Proliferative lesions and reproductive tract tumors in male descendants of mice exposed developmentally to diethylstilboestrol, *Carcinogenesis* **21**, 1355-1363.

O'Regan, D., Welberg, L.L., Holmes, M.C. and Seckl, J.R. (2001). Glucocorticoid programming of pituitary-adrenal function: mechanisms and physiological consequences, *Semin. Neonatol.* **6**, 319-329.

Orlando, E.F., Davis, W.P. and Guillette, L.J. Jr. (2002) Aromatase activity in the ovary and brain of the eastern mosquitofish (*Gambusia holbrooki)* exposed to paper mill effluent, *Environ. Health Perspect.* **110**, 429-433.

Oskam, I.C., Ropstad, E., Dahl, E., Lie, E., Derocher, A.E., Wiig, O., Larsen, S., Wiger, R. and Skaare, J.U. (2003) Organochlorines affect the major androgenic hormone, testosterone, in male polar bears (*Ursus maritimus*) at Svalbard, *J. Toxicol. Environ. Health A* **66**, 2119-2139.

Paris, F., Jeandela, C., Servant, N. and Sultana, C. (2006) Increased serum estrogenic bioactivity in three male newborns with ambiguous genitalia: A potential consequence of prenatal exposure to environmental endocrine disrupters, *Environ. Res.* **100**, 39-43.

Patandin, S., Lanting, C. I., Mulder, P. G. H., Boersma, E.R., Sauer, P. J. J. and Weisglas-Kuperus, N. (1999) Effects of environmental exposure to polychlorinated biphenyls and dioxins on cognitive abilities in Dutch children at 42 months of age, *J. Pediatr.* **134**, 33-41.

Pertoldi, C., Hansen, M..M, Loeschcke, V., Madsen, A.B., Jacobsen, L. and Baagoe, H. (2001) Genetic consequences of population decline in the European otter (*Lutra lutra*): an assessment of microsatellite DNA variation in Danish otters from 1883 to 1993, *Proc. Biol. Sci.* **268**, 1775-1781.

Petrelli, G., Lauria, L. and Figa-Talamanca, I. (2001) Occupational exposure and male fertility. Results of an Italian multicenter study in an exposed population, *Med. Lav.* **92,** 307-313.

Pottinger, T.G. (2003) Interactions of endocrine-disrupting chemicals with stress responses in wildlife, in special topic issue on the implications of endocrine active substances for humans and wildlife, *Pure. Appl. Chem.* **75**, 2321-2333.

Purdom, C.E., Hardiman, P.A., Bye, V.J., Eno, N.C., Tyler, C.R. and Sumpter, J.P. (1994) Estrogenic effects of effluents from sewage treatment works, *Chem. Ecol.* **8**, 275-285.

Reeder, A.L., Foley, G.L., Nichols, D.K., Hansen, L.G., Wikoff, B., Faeh, S., Eisold, J., Wheeler, M.B., Warner, R., Murphy, J.E. and Beasley, V.R. (1998) Forms and prevalence of intersexuality and effects of environmental contaminants on sexuality in cricket frogs (*Acris crepitans*), *Environ. Health Perspect.* **106**, 261-266.

Reeder, A.L., Ruiz, M.O., Pessier, A., Brown, L.E., Levengood, J.M., Phillips, C.A., Wheeler, M.B., Warner, R.E. and Beasley, V.R. (2005) Intersexuality and the cricket frog decline: historic and geographic trends, *Environ. Health Perspect.* **113**, 261-265.

Reijnders, P.J.H. (1986) Reproductive failure in common seals feeding on fish from polluted coastal waters, *Nature* **324**, 456-457.

Reijnders, P.J.H. (1990) Progesterone and oestradiol-17 beta concentration profiles throughout the reproductive cycle in harbour seals (*Phoca vitulina*), *J. Reprod. Fert.* **90**, 403-409.

Renner, R. (2003) More evidence that herbicides feminize amphibians, Science News, January 14, *Environ. Sci. Technol,* (online) http://pubs.acs.org/subscribe/journals/esthag-w/2003/jan/science/rr_frogs.html [last accessed 15/03/06]

Reynaud, S. and Deschaux, P. (2005) The effects of polycyclic aromatic hydrocarbons on the immune system of fish: A review. *Aquat. Toxicol.* Dec 24; [Epub ahead of print].

Rice, D.C. (2000) Parallels between attention deficit hyperactivity disorder and behavioral deficits produced by neurotoxic exposure in monkeys, *Environ. Health Perspect.* **108,** 405-408.

Riedel, D., Tremblay, N. and Tompkins, E. (1997) *State of knowledge report on environmental contaminants and human health in the Great Lakes Basin.* ISBN 0-662-26-169-0. Health Canada, Canada.

Rogan, W.J., Gladen, B.C., Hung, K.L., Koong, S.L., Shih, L.Y., Taylor, J.S., Wu, Y.C., Yang, D., Ragan, N.B. and Hsu, C.C. (1988) Congenital poisoning by polychlorinated biphenyls and their contaminants in Taiwan, *Science* **241**, 334-336.

Rooney, A.A., Bermudez, D.S. and Guillette, L.J. Jr. (2003) Altered histology of the thymus and spleen in contaminant-exposed juvenile American alligators, *J. Morphol.* **256**, 349-359.

Roos, A., Greyerz, E., Olsson, M. and Sandegren, F. (2001) The otter (*Lutra lutra*) in Sweden-population trends in relation to sigma DDT and total PCB concentrations during 1968-99, *Environ. Pollut.* **111**, 457-469.

Ross, P.S., De Swart, R.L., Reijnders, P.J., Van Loveren, H., Vos, J.G. and Osterhaus, A.D. (1995) Contaminant-related suppression of delayed-type hypersensitivity and antibody responses in harbor seals fed herring from the Baltic Sea, *Environ. Health Perspect.* **103**, 162-167.

Ross, P.S., Ellis, G.M., Ikonomou, M.G., Barrett-Lennard, L.G. and Addison, R.F. (2000) High PCB concentrations in free-ranging pacific killer whales, *Orcinus orca*: effects of age, sex, and dietary preferences, *Marine Pollut. Bull.* **40**, 504-515.

RSPB (2001) *Royal society for the protection of birds*, Fact Sheet (online) (http://www.rspb.org.uk/birds/peregrine/population_trends.asp) [last accessed 15/03/06].

Sallmen, M., Liesivuori, J., Taskinen, H., Lindbohm, M.L., Anttila, A., Aalto, L. and Hemminki, K. (2003) Time to pregnancy among the wives of Finnish greenhouse workers, *Scand J. Work Environ. Health* **29**, 85-93.

SBR (2003) *Research brief 106: The effects of chlorinated hydrocarbons on wildlife.* 1st Oct, Doumenting research by Dr Tim Gross. (online) http://list.niehs.nih.gov/pipermail/sbrp-brief/2003-October/000043.html [last accessed 15/03/06].

Schwartz, J., Aldridge, B., Blanchard, M., Mohr, F.C. and Stott, J. (2005) The development of methods for immunophenotypic and lymphocyte function analyzes for assessment of Southern sea otter (*Enhydra lutris nereis)* health, *Vet. Immunol. Immunopathol.* **104**, 1-14.

Scott, A.P., Katsiadaki, I., Witthames, P.R., Hylland, K., Davies, I.M., McIntosh, A.D. and Thain, J. (2006) Vitellogenin in the blood plasma of male cod (*Gadus morhua*): A sign of oestrogenic endocrine disruption in the open sea? *Mar. Environ. Res.* **61**, 149-170.

Seo, H. (2002) *Endocrine disrupters and thyroid function. Program and Abstracts*, International Symposium on Environmental Endocrine Disprupters, November 29th, Hiroshima, Japan.

Sepulveda, M.S., Wiebe, J.J., Honeyfield, D.C., Rauschenberger, H.R., Hinterkopf, J.P., Johnson, W.E. and Gross, T.S. (2004) Organochlorine pesticides and thiamine in eggs of largemouth bass and American alligators and their relationship with early life-stage mortality, *J. Wild. Dis.* **40**, 782-786.

Sheahan D.M. (2006) No-threshold dose–response curves for nongenotoxic chemicals: Findings and applications for risk assessment, *Environ. Res.* **100**, 93-99.

Skaare, J.U., Bernhoft, A., Wiig, O., Norum, K.R., Haug, E., Eide, D.M. and Derocher, A.E. (2001) Relationships between plasma levels of organochlorines, retinol and thyroid hormones from polar bears (*Ursus maritimus*) at Svalbard, *J. Toxicol. Environ. Health Part A.* **62**, 227-241.

Skakkebaek, N.E., Rajpert-De Meyts, E. and Main, K.M. (2001) Testicular dysgenesis syndrome: an increasingly common developmental disorder with environmental aspects, *Hum. Reprod.* **16**, 972-978.

Slutsky, M., Levin, J.L. and Levy, B.S. (1999) Azoospermia and oligospermia among a large cohort of DBCP applicators in 12 countries, *Int. J. Occup. Environ. Health* **5**, 116-122.

Sonne, C., Leifsson, P.S., Dietz, R., Born, E.W., Letcher, R.J., Kirkegaard, M., Muir, D.C.G., Andersen, L.W., Riget, F.F. and Hyldstrup, L. (2005) Enlarged clitoris in wild polar bears (*Ursus maritimus*) can be misdiagnosed as pseudohermaphroditism, *Sci. Total Environ.* **337**, 45-58.

Sparling, D.W., Fellers, G. and McConnell, L. (2001) Pesticides are involved with population declines of amphibians in the California Sierra Nevadas, *Sci. World J.* **1**, 200-201.

Steeger, T. and Tietge, J. (2003) *White Paper on the potential developmental effects of atrazine on amphibians*, Office of Prevention, Pesticides and Toxic Substances, US EPA, Washington.

Stewart, P., Reihman, J., Lonky, E., Darvill, T. and Pagano, J. (2003) Cognitive development in preschool children prenatally exposed to PCBs and MeHg, *Neurotoxicol. Teratol.* **25**, 11-22.

Stewart, P., Reihman, J., Lonky, E., Darvill,T. and Pagano, J. (2000) Prenatal PCB exposure and neonatal behavioural assessment scale (NBAS) performance, *Neurotoxicol. Teratol.* **22,** 21-29.

Subramanian, A., Tanabe, S., Tatsukawa, R., Saito, S. and Miyazaki, N. (1987) Reduction in the testosterone levels by PCBs and DDE in Dall's porpoises of the northwestern North Pacific, *Mar. Pollut. Bull.* **18**, 643-646.

Swan, S.H., Main, K.M., Liu, F., Stewart, S.L., Kruse, R.L., Calafat, A.M., Mao, C.S., Redmon, J.B., Ternand, C.L., Sullivan,. S. and Teague, J.L. (2005) Decrease in anogenital distance among male infants with prenatal phthalate exposure, *Environ. Health Perspect.* **113**, 1056-1061.

Szczys, P., Nisbet, I.C.T., Hatch, J.J. and Kesseli, R.V. (2001) Sex ratio bias at hatching and fledging in the Roseate tern, *Abstract for CONDOR* **103**.

Thompson, H.M., Rose, M., Fernandes, A., White, S., Blackburn, A. and Ashton, D. (2005) *Causes of nestling deformities in a UK heronry* – possible sources of pollutants. Poster presentation at SETAC Europe 15th Annual Meeting, 22-26 May, Lille, France.

Thorpe, K.L., Hutchinson, T.H., Hetheridge, M.J., Scholze, M., Sumpter, J.P. and Tyler, C.R. (2001) Assessing the biological potency of binary mixtures of environmental estrogens using vitellogenin induction in juvenile rainbow trout *(Oncorhynchus mykiss*), *Environ. Sci. Technol.* **35**, 2476-2481.

Tiller, B.L., Dagle, G.E. and Cadwell, L.L. (1997) Testicular atrophy in a mule deer population, *J. Wild. Dis.* **33,** 420-429.

Toft, G., Edwards, T.M., Baatrup, E. and Guillette, L.J. Jr. (2003) Disturbed sexual characteristics in male mosquitofish (*Gambusia holbrooki*) from a lake contaminated with endocrine disrupters, *Environ. Health Perspect.* **111**, 695-701.

Tomasi, T.E., Hellgren, E.C. and Tucker, T.J. (1998) Thyroid hormone concentrations in black bears (*Ursus americanus*): hibernation and pregnancy effects, *Gen. Comp. Endocrinol.* **109**, 192-199.

Trites, A.W. and Barrett-Lennard L.G. (2001) *COSEWIC status report addendum on killer whales (orcinus orca)*, Marrine Mammal Research Unit, Vancouver.

Van Loveren, H., Ross, P.S., Osterhaus, A.D. and Vos, J.D. (2000) Contaminant-induced immunosuppression and mass mortalities among harbor seals, *Toxicol. Lett.* **15**, 319-324.

Van Oostdam, J., Donaldson, S.G., Feeley, M., Arnold, D., Ayotte, P., Bondy, G., Chan, L., Dewaily, E., Furgal, C.M., Kuhnlein, H., Loring, E., Muckle, G., Myles, E., Receveur, O., Tracy, B., Gill, U. and Kalhok, S. (2005) Human health implications of environmental contaminants in Arctic Canada: A review, *Sci. Total Environ.* **351-352**, 165-246.

Veeramachaneni, D.N.R., Amann, R.P. and Jacobson, J.J. (2005) Testis and antler dysgenesis in Sitka black tailed deer on Kodiak Island, Alaska. Sequela of environmental endocrine disruption, *Environ. Health Perspect,* (online) http://www.ehponline.org/docs/2005/8052/abstract.html [last accessed 15/03/06].

Walkowiak, J., Wiener, J. A., Fastabend, A., Heinzow, B., Kramer, U., Schmidt, E., Steingruber, H.J., Wundram, S. and Winneke, G. (2001) Environmental exposure to polychlorinated biphenyls and quality of the home environment: effects on psychodevelopment in early childhood, *Lancet* **358**, 1602-1607.

Wang, S.L., Su, P.H., Jong, S.B., Guo, Y.L., Chou, W.L. and Papke, O. (2005) *In utero* exposure to dioxins and polychlorinated biphenyls and its relations to thyroid function and growth hormone in newborns, *Environ. Health Perspect.* **113**, 1645-1650.

Weisglas-Kuperus, N., Patandin, S., Berbers, G.A., Sas, T.C., Mulder, P.G., Sauer, P.J. and Hooijkaas, H. (2000) Immunologic effects of background exposure to polychlorinated biphenyls and dioxins in Dutch preschool children, *Environ. Health Perspect.* **108**, 1203-1207.

Welberg, L.A., Seckl, J.R. and Holmes, M.C. (2001) Prenatal glucocorticoid programming of brain corticosteroid receptors and corticotrophin-releasing hormone: possible implications for behaviour, *Neuroscience* **104**, 71-79.

Wiig, O., Derocher, A.E., Cronin, M.M. and Skaare, J.U. (1998) Female pseudohermaphrodite polar bears at Svalbard, *J. Wild. Dis.* **34**, 792-796.

Williams, D.L., Simpson, V.R. and Flindall, A. (2004) Retinal dysplasia in wild otters (*Lutra lutra*), *Vet. Rec.* **155**, 52-6.

Wingard, D.L. and Turiel, J. (1988) Long term effects of exposure to diethylstilboestrol, *West J. Med.* **149**, 551-554.

Woodward, A.R., Jennings, M.L., Percival, H.F. and Moore, C.T. (1993) Low clutch viability of American alligators on Lake Apopka, FL, *Science* **56**, 52-63.

Wren, C.D. (1991) Cause-effect linkages between chemicals and populations of mink (*Mustela vison*) and otter (*Lutra canadensis*) in the Great Lakes basin, *J. Toxicol. Environ. Health* **33**, 549-85.

WWF UK (World Widelife Fund for Nature United Kingdom) (2005) *Killing killer whales with toxics*, (online) www.panda.org/news_facts/newsroom/features/index.cfm?u–NewsID=53540 [last accessed 15/03/06].

Zelikoff, J.T., Raymond, A., Carlson, E., Li, Y., Beaman, J.R. and Anderson, M. (2000) Biomarkers of immunotoxicity in fish: from the lab to the ocean, *Toxicol. Lett.* **112-113,** 325-331.

Zoeller, R.T. (2005) Environmental chemicals as thyroid hormone analogues: new studies indicate that thyroid hormone receptors are targets of industrial chemicals? *Mol. Cell. Endocrinol.* **242**, 10-15.

EPIDEMIOLOGICAL METHODS

A. ROSANO[1*] AND E. ROBERT-GNANSIA[2]
[1] *Italian Institute of Social Medicine*
Via P.S. Mancini 29
Rome
ITALY
[2] *Institut Européen des Génomutations*
rue Edmond Locard 86
Lyon
FRANCE
(author for correspondence, Email: a.rosano@iims.it)*

Summary

This chapter discusses the principles of study design and related methodological issues in the epidemiology of congenital anomalies, with specific regard to environmental factors. We present the major types of experimental and observational designs used in environmental epidemiology, namely the basic designs involving the individual as the unit of analysis and the ecological designs, which involve groups or geographical areas as units of analysis. We also include a brief discussion on study designs in genetic epidemiology. Examples of how the various study designs are applied in order to investigate the environmental risk factors for congenital anomalies are described with comments. We present methodological issues related to outcome, including methods and the timing for ascertaining cases with malformations, diagnostic criteria, and problems in grouping malformations for purposes of analysis. Other methodological issues include controlling for confounders, inadequate measurement, duration, timing of exposure, and exposure-response relationships.

1. Introduction

Generally speaking, an environmental risk factor is any non-genetic factor that modifies (increasing or, occasionally, decreasing) the risk of occurrence of a disease

P. Nicolopoulou-Stamati et al. (eds.), Congenital Diseases and the Environment, 89–130.

(Dolk, 2004). Researchers have become very interested in the relationship between the environment and malformations, bearing in mind that the developing foetus is more sensitive to all sorts of insults than adults are, and that these insults may result in a life-long handicap. Congenital anomalies (CAs) include structural defects, chromosomal and monogenic syndromes, and inborn errors in the metabolism.

Most studies have focused on structural birth defects (such as neural tube defects, cleft lip and palate, congenital heart disease, dysgenesis of the kidneys, limb anomalies and conditions such as prematurity) which are easily identifiable. Researches on functional abnormalities, like attention deficit disorder and mild mental retardation are rarely carried out, since it is difficult to obtain epidemiologic measures.

The cause of most CAs is unknown. Nutritional, infectious, and other environmental factors such as radiation, pharmaceuticals, and toxic chemicals have been investigated as potential risk factors. Most CAs is likely to result from multiple factors, such as an interaction between one or more genes and one or more environmental factors.

What epidemiology does is to look first at the presence or absence of birth defects and then carry out studies to correlate their occurrence to the presence or absence of genetic and/or environmental exposures. Identification of risk factors for CAs can be achieved using experimental or observational *ad hoc* studies, whereas temporal surveillance or spatial surveillance may be useful for generating hypotheses. Temporal surveillance is aimed at identifying significant changes over time, in the occurrence of CAs. It is conducted through the comparison of the current prevalence rate with the expected rate (baseline rate). Several international surveillance systems are currently operating such as surveillance in the world:

- the International Clearinghouse for Birth Defects Surveillance and Research (ICBDSR) (see www.icbd.org), at worldwide level,

- the "European Surveillance of Congenital anomalies" network (EUROCAT) in Europe (see www.eurocat.ulster.ac.uk) the National Birth Defect Prevention Network in the United States (see www.nbdpn.org), and

- the "Estudio Latinoamericano de Colaborativo Malformaciones Congénitas" ECLAMC, in South America (see eclamc.ioc.fiocruz.br).

Spatial surveillance is conducted by analysing the spatial distribution of the prevalence of CAs. When more cases than expected (i.e. a cluster) occur simultaneously in a certain area, an environmental factor (EF) is sought, in order to explain the excess of cases. *Ad hoc* studies are conducted to investigate specific hypotheses on associations between EFs and CAs which may arise from population surveillance, local observations (case reports from a hospital, observations made by the public or the news media), or treating doctors. The studies can be performed by means of analytical or ecological studies. The analytical studies, by using information at individual level, test the relationship between the occurrence of a CA, or a group of CAs, and variables identifying specific risk factors. Ecological studies use information at aggregate level in conjunction with indicators of environmental pollution (Mastroiacovo and Rosano, 2001).

CAs are rare events, thus research studies must access cases from large populations. Defects that occur only a few times in every 10,000 births present obvious challenges, even for the largest databases. Basic information for investigating the relationship between EFs and CAs include retrospective interviews on nutrition and environmental factors affecting either the mother or the father, blood and urine samples from subjects and parents, and a DNA bank. Most of these factors may play a role at the time of conception, or even before. Conditions of data collection regarding environmental factors should be included in the data, and information should be collected as close to conception as possible, and ideally with a long term follow-up of children, in order to recognise a CA not detectable in the neonatal period. The short delay between exposure and outcome must be taken into account when designing epidemiologic studies for testing relationships between EFs and CAs. The appropriateness of the study design will depend on the extent of the consideration actually given to the peculiar characteristics of the CAs.

In the following paragraphs, we will analyse the principal sources of data available; we then provide an overview of the different types and the characteristics of study designs, with some examples of investigations conducted. After a discussion of the gene-environment issue, we will deal with the main problems that can be encountered in the study of environmental risk factors for CAs.

2. Sources of data, coding and classification

2.1. Source of data for congenital anomalies

Registries are the principal source of data on CAs. They can retrieve information from a closed population limited in size, where all births can be monitored. In this case, the registries are said to be population-based. In reality, no population is closed because population movements always occur, but when a population represents a country, it is more closed than if it represents an area within a country (Källén, 1988). Other types of registries operate by sampling the population from a number of hospitals and they are said to be hospital-based. The above-mentioned ECLAMC is an example of such a program, which takes samples from hospitals in most countries of South America. Using hospital-based data makes it possible to cover a large geographical area but may create selection bias, because high risk mothers nowadays are referred to certain hospitals for specialist services in case of a prenatal diagnosis of severe malformation. A hospital-based program may nevertheless be the only possibility of surveillance, especially in developing countries where registries are pure research programs and rely heavily on an interested person/team.

CA registries are based on retrospective data collection, and often have a limited follow-up (rarely over one year after the birth), which hampers the tracking of functional birth defects or other CAs which are not likely to be diagnosed early in life. A few registries exist for evaluating this type of event, with long-term follow-up (e.g. the British Columbia Health Status Registry). They include children registered for mental retardation or other functional congenital diseases diagnosed when they are in school or later.

Prospective biomonitoring studies with long follow up are time-consuming and expensive but have undoubted advantages. Large numbers of mother/infant pairs are enrolled prospectively, and biological samples as well as epidemiological data are collected in early pregnancy. Such studies are being conducted in the United States and Scandinavian countries, but they started recently, and no results have been published yet, based on these data. With a prospective approach, it is possible to quantify a number of potentially contributing variables, including environmental factors, diet, lifestyle, infectious diseases and sociological variables.

Ad hoc studies, such as case-control studies, are used to test specific hypotheses concerning teratogenic exposures. Data for *ad hoc* studies can be retrieved from CA registries, but most information on malformed infants is collected *ad hoc* to address new or more elaborate hypotheses.

2.2. *Coding and classification of congenital anomalies*

The primary goal of coding is to represent infants with CAs accurately, completely, and concisely. Coding procedures need to accommodate the expected objectives: for example, research activities may require different coding procedures from those that focus on linking infants to services. There are several challenges in coding CAs, including the need to distinguish infants with multiple defects and syndromes from those with isolated defects, and the need for strategies for coding suspected defects for which confirmation is not available. Selection of a coding system is central to the use of collected data. A modification of the International Classification of Diseases and Related Health Problems (ICD) is the most used system, but there are other systems too. The important goal in coding CAs is to be able to retrieve as many different types as possible, even rare ones. Because of the importance of coding, a process for evaluating coding quality is beneficial. A validation process may be adopted, in which a sample of cases is recoded to ensure that defects are being coded accurately and consistently (Rasmussen and Moore, 2001). In any case, it is essential to have access, when needed, to written report forms with detailed descriptions of the malformations, when necessary supplemented with copies of relevant medical documents such as pictures, X- rays or autopsy reports.

Isolated and associated CAs may have distinct pathogeneses. For example, alcohol use during pregnancy is a known cause of birth defects, and may lead to a newborn phenotype called fetal alcohol syndrome, including typical craniofacial anomalies. Maternal consumption of alcohol is more linked with isolated than associated forms of facial clefts (Munger *et al.,* 1996). A specific gene-environment interaction may explain this: the TGFα, TGFβ3, and MSX1 genes are suspected of playing a role in the cause of facial clefts when combined with exposure to alcohol (Romitti *et al.,* 1999), and the most likely candidate, the MSX1 gene, is associated with isolated forms of facial clefts (Jezewski *et al.,* 2003). Therefore, classifying CAs as isolated or associated can be fundamental for identifying EFs as possible causal factors.

CAs are a heterogeneous group of conditions, from the point of view of etiology and pathogenesis, and it is unlikely that an exposure would affect all types of congenital defects. It is therefore useless to analyse this outcome as the total rate, and divisions into categories must be made. A common way is to use organ system subdivisions, but the group of « limb malformations » for instance is wide-ranging, and a change in a specific association with one type (e.g. radial ray defect) may be hidden by large numbers of more common defects. Another problem is that the pathogenesis of specific defects is often unclear, or completely unknown, and it is therefore not possible to set up a strict categorization of CAs that mirrors pathogenesis (Källén,

2006). With maternal cocaine use during pregnancy, for example, defects involving vascular disruption seem to be implicated. However, a biological basis for positing subgroups of interest is often lacking (Hatch and Duncan, 1993). The decision should be made *ad hoc* for each study, bearing in mind that if the subdivision is carried too far, the number of cases in each subgroup becomes so low that the analysis is meaningless.

2.3. Accuracy of diagnosis and ascertainment of congenital anomalies

Accuracy of diagnosis and ascertainment are vital for etiologic studies of CAs. Overall case ascertainment probably never reaches 100%, and its level depends on the methods of data collection. Use of multiple sources of information (hospital discharge records, maternity records, fetal ultrasound screening, laboratory records - cytogenetic, molecular, pathology-specialised departments and birth certificates) is always advisable. Under-ascertainment of some anomalies may occur, if sources of information stop in the early neonatal period, as diagnoses may be made later than this. Services treating children later than the postneonatal period may be used for confirmation of diagnosis.

The methodological problems, in case ascertainment and accuracy of information, belong to three aspects:

1) The nature of the CA: some malformations are easily diagnosed before birth or in the first week of life, but others will display symptoms later in life, sometimes even years after birth, while others are likely to induce early fetal death. Then the quality of ascertainment and the quality of data will depend on the source of data for each type of malformation;

2) Use of medical records, in the absence of standard examination protocols: different doctors may have different diagnostic criteria, the autopsy rate may vary across hospitals, as well as the rate of cytogenetic investigations, and the quality of ultrasound examinations will influence the rate of prenatal diagnosis;

3) The characteristics of the health information system are variable and variations in reporting and coding CAs are significant (Dolk, 1996).

2.4. Source of data, coding and assessment of environmental exposures

Methodological issues relevant to studies concerning environmental exposure pertain to the definition of exposure: what it is, who is exposed (mother, father, or fetus) and when in the periconceptional period it occurs, the amount (dose) experienced by a study subject, how it is measured, from where information about it was obtained.

Environmental exposures are often ascertained at aggregate level. Data may come from resident information such as population-based registries and environmental data such as sample databases of water conditions or air quality records. Census data such as geographic databases may provide accurate locations of population and information is sometimes available on land use patterns or air pollution distribution.

Information on exposure at individual level can be obtained by direct interview, or from hospital records. For many exposures, it is difficult to obtain accurate and unbiased exposure information from study subjects. Therefore, researchers have become very interested in biological markers, when they are able to provide a quantitative estimate of exposure levels.

EFs concern many areas, such as illnesses, toxic agents, drugs and chemical elements. Various coding and classification systems are used, such as the Anatomical Therapeutic Chemical (ATC) classification for drugs or the ICD for illnesses and the Globally Harmonized System (GHS) for the classification and labelling of hazardous chemicals (UN, 2003). The main issue is to identify teratogenic substances, which then makes it possible classify the EFs in terms of potential teratogenicity. Certain environmental factors are known as teratogens, such as specific organic solvents and ionising radiations, but the effects may vary greatly in severity, according to the dose level and timing of exposure.

Most substances, including drugs to which pregnant women are exposed, can cross the placenta and enter the tissues of the developing embryo and fetus, with a consequent risk of teratogenic effects. Fetal exposure will reflect the exposure pattern of the mother: dietary habits and lifestyle, alcohol consumption, smoking, medical treatments and occupational exposure. Persistent substances accumulated in the body of the mother (from previous exposures) may be redistributed, thus leading to exposure of the fetus. The embryo-fetal exposure to chemical substances depends on the amount the mother is, and has been (if the substance is accumulated in the body), exposed to and on the extent to which the substance crosses the placenta. For example, organic mercury and lead compounds cross the placenta and reach the fetus, while cadmium will to some extent be withheld in the placenta. During embryogenesis, the placental transfer of weak acids is favoured, while during late gestation, the pH of the fetal compartment changes and favours the transfer of weakly alkaline substances (Nielsen, 2001).

The reproductive toxicity of substances has been studied using mostly *animal models*. In the past, there was a tendency to consider only malformations or embryo-fetal death as relevant endpoints in teratological studies. Today it is assumed that all of the four manifestations of developmental toxicity (death, structural abnormalities, growth alterations and functional deficits) are of concern (OECD, 1989). As a consequence, the name of the test has been changed to "prenatal developmental toxicity test" and the exposure period is extended to the day before birth (OECD, 1999). To assess the developmental toxicity of a chemical, it is therefore important to include information on other developmental effects such as minor anomalies, variations, fetal death and growth. In addition, malformations of organs developing after the period of major organogenesis, e.g. the sex organs and the brain, may at present not be detected in the teratology study. An example is the suspected endocrine disrupter dibutyl phthalate, where exposure during the period of male sexual differentiation resulted in major disturbances in the morphological and functional development of the male reproductive system (Mylchreest *et al.,* 1999). Teratological studies are suitable for the demonstration of intra-uterine death after implantation (resorptions). In studies where animals are treated before implantation, pre-implantation loss may also be assessed. Fetal weight can be assessed in a teratological study, but it is important to take into account variations due to different litter sizes or sex distribution in control vs. exposed groups in the analysis.

In *human teratology,* investigations conducted in exposed populations are constrained by the difficulty of identifying exposed individuals accurately (Kline *et al.,* 1989). Parental exposure to a possible teratogen can be established on at least two different levels. When the putative exposure of interest is dispersed in the environment) rather than due to individual activity (e.g. consumption of a particular substance), documentation at the individual level, through the use of assays measuring biomarkers, is the first and preferable option. Alternatively, the exposure can be established at the ecological level. Defining individuals at risk will require documentation of i) the presence of the suspected teratogen in the ambient environment and ii) the location of the allegedly exposed individual at the time(s) of (peak) exposure. With only ecological-level information regarding exposure, there is a limit to the clarity that can be achieved regarding the biological activity, in a given individual, of the ambient toxin (Hindin *et al.,* 2005).

It must be considered that very early miscarriage is another reproductive endpoint affected by teratogenic substances. The observed drop in the birth rate in many European countries 7 – 9 months after the Chernobyl disaster presumably relates to an excess of miscarriages induced by that event – obscuring the underlying teratogenic impact (Hoffmann, 2001).

3. Study designs

Study design refers to the methodology that is used to investigate exposure-disease relationships. As shown in figure 1, there are three general design strategies for conducting population research using information at individual level (Morgenstern and Thomas, 1993):

a) experiments in which the investigators randomly assign (randomise) subjects to two or more treatment (exposure) groups,
b) quasi-experiments in which the investigators make the assignments to treatment groups non-randomly, and
c) observational studies in which the investigators simply observe exposure (treatment) status in subjects without assignment (Kleinbaum *et al.*, 1982).

Case reports may also give clues for detecting risk factors. Each approach has its strengths and limitations, which are discussed in the following paragraphs.

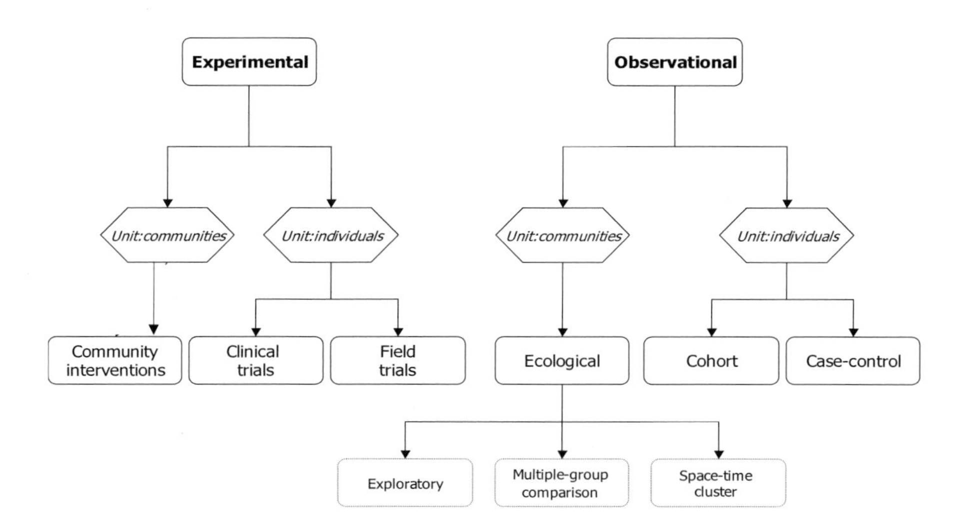

Figure 1. Study designs

In the experimental design, there are usually two exposure groups. One group is assigned to receive the new experimental intervention and the other (control) group is assigned to receive no intervention, placebo, or other available intervention. Experiments can be conducted on affected subjects (clinical trial) or on subjects who have not got diseases (field trial). Because experiments are best suited ethically and practically to the study of health benefits, not hazards, experiments in environmental epidemiology would usually be limited to the study of preventive interventions (field trial). As an example, we can consider the study of MRC (MRC Vitamin Study Research Group, 1991) carried out to determine whether supplementation with folic acid or a mixture of seven other vitamins (A, D, B1, B2, B6, C and nicotinamide) around the time of conception could prevent neural tube defects (NTD). A total of 1,817 women at high risk of having a pregnancy with a NTD, because of a previous affected pregnancy, were allocated at random to one of four groups (folic acid, other vitamins, both, or neither). The results showed a 72% reduced recurrence risk of NTD with folic acid. This is one of the very few epidemiological studies on EFs with an individual randomization of the exposure.

A quasi-experiment may be done in a similar way to an experiment by comparing two or more non-randomised groups, or by comparing one or more groups over time, before and after the intervention is initiated in at least one group (community interventions). With the latter approach, the composition of each group may change over time so that the subjects observed before the intervention are not the same subjects observed after the intervention. As an example, to verify whether the decreasing neural tube defects birth prevalence rates in Chile were due to folic acid fortification, introduced in 2000, or to pre-existing decreasing trends, birth prevalence rates of spina bifida were calculated for pre-fortification and fortification periods (before 2000 *vs.* 2001-2002). A 51% decrease in the birth prevalence rates of this anomaly was observed during the fortified period. Because no historical trend for prevalence of spina bifida at birth was observed before food fortification with folic acid, the decrease may be attributable to the intervention (Lopez-Camelo *et al.*, 2005). Because subjects were not individually randomised in this study, it is possible that pregnant women after intervention differ from those who became pregnant before the intervention with regard to other risk factors for neural tube defects, such as other diet characteristics.

Unlike experiments and quasi-experiments, observational studies are used to estimate the effects of exposures hypothesised to be harmful: fixed attributes (e.g. race and genotype), or behaviours/exposures over which the investigator has little or no control (e.g. body mass index, chronic or acute diseases), and other exposures for which manipulation or randomization would be unethical or infeasible.

Observational studies are mainly conducted with secondary or retrospective data and/or without following-up the subjects.

The formal testing of hypotheses regarding environmental risk factors for CAs is usually carried out by conducting observational studies. They provide weaker empirical evidence than experimental studies because of the potential for large confounding biases. The investigators describe the health status of a population or characteristics of a number of patients. A description is usually made with regard to time and person. As far as timing is concerned, the study population is *longitudinal*, involving the detection of incident events during a follow-up period. In the longitudinal studies, the case collection can be done prospectively (from a certain date onward) or retrospectively (from a certain date backward). The sampling of persons may involve the complete selection of the entire population from which study cases are identified (*cohort*), or the incomplete sampling of a fraction (<100%) of the non-cases in the population from which study cases are identified (*case-control*) (Greenland and Morgenstern, 1988). When observational studies are conducted at group level, the appropriate study design is the *ecological study*, in which aggregate data on risk factors and disease prevalence from different population groups is compared, in order to identify associations.

3.1. ***Cohort or follow-up study***

A cohort or follow-up study is prospective (or retrospective), analytical, observational, based on data and usually primary, from a follow-up period of a group in which some subjects have been, are, or will be, exposed to the element under investigation, to determine the association between that exposure (e.g. chemical exposure or nutritional status), and an outcome. The question is whether a greater proportion of pregnancies with a certain risk factor/exposure have a diagnosed birth defect than without that risk factor/exposure.

For example, the use of depleted uranium in recent military conflicts was suspected of being a risk factor for CAs (Arfsten *et al.,* 2001). A large-scale cohort study of reproductive outcome is the appropriate study design to investigate such a factor. After the first Gulf war, a cohort study was conducted in the UK to assess whether the offspring of UK veterans of the first Gulf war were at increased risk of fetal death or congenital malformation. The cohort consisted of all UK armed forces personnel deployed to the Gulf area at any time between August 1990 and June 1991. The comparison cohort comprised demographically similar Armed Service personnel, who were in service on 1st January 1991 and were appropriately fit, but were not deployed to the Gulf. The similarity between cohorts is an important aspect. Considering that the main focus of the investigation is the exposure effect, in cohort studies the groups to be compared should be as similar as possible, with the only substantial differences coming from the exposure under investigation. The study revealed evidence for an increased risk among veterans' offspring of malformations of the genital system, urinary system (renal and urinary tract), and defects of the digestive system, musculo-skeletal system, and non-chromosomal (non-syndrome) anomalies (Doyle *et al.,* 2004).

Cohort studies are suitable for analysing the recurrence risk of CAs, that is the probability of a mother with a previous affected pregnancy having another malformed infant. The risk of having a similar or a different anomaly is compared between mothers who, in previous pregnancies, had a malformed infant and a reference cohort of mothers who have had no observed malformations in previous pregnancies. The recurrence risk is high for many anomalies, such as cleft lip and palate, neural tube defects, certain congenital heart defects, hypospadias, talipes equinovarus and congenital dislocation of the hip (Brent, 2001). Social status at the time of birth seems to be a stronger predictor of the recurrence of similar anomalies, which points toward long-lasting environmental exposure related to life-style factors or occupational exposure (Basso *et al.,* 1999).

Cohort studies are susceptible to bias by differential loss to follow-up, the lack of control over risk assignment and thus to confounder symmetry. The major practical limitation of a cohort design is its inefficiency for studying rare outcome events like CAs. Because exposure status and other covariates must be observed at the start of follow-up in the entire study population, it is likely that most observed newborns will not be affected by a CA. Comparing a small number of cases with a large number of non-cases is statistically and economically inefficient, unless the attributable risk is high (Morgenstern and Thomas, 1993).

A large cohort of pregnant women with long-term follow-up of the offspring was put together in Denmark for collecting information about diseases, demography, social conditions and biological data in order to investigate different outcomes, including CAs. Only about 35% of eligible women participated in the Danish birth cohort (Basso and Olsen, 2005). This could cause bias if participants were at a different level of risk compared to the non-participants. A low participation rate exposes the cohort study to a high risk of biases and limitations in terms of generalisability, that is the extent to which the results of a study provide a correct basis for other circumstances.

3.2. Case-control studies

Case-control studies are the studies most commonly used to investigate the relationship between environmental factors and CAs. In this approach, the exposure rate among cases (infants with a CA) is compared with that among controls (usually healthy infants). This methodology therefore starts with the identification of cases and controls and can then study many different maternal exposures. Cases can be ascertained in different ways.

When a case-control study is conducted using a representative sample of all cases in a precisely identified population, e.g. by using from a population-based registry of CAs, and controls are sampled directly from the background population that gave rise to the cases, the study is said to be population-based. In this case, it is possible to estimate rates of the disease in exposed and not exposed subjects and to derive relative and attributable risks from these estimates. This scenario is rarely realised. Most of the time, cases and controls are identified from treating hospitals. A classical example is the large Baltimore-Washington Infant Study 1981-1989 (Ferencz *et al.,* 1997) in which live-born infants with congenital cardiac defects were identified from different sources in the Baltimore and Washington D.C. areas.

Alternatively, cases and controls are gathered from the hospitals where the cases are diagnosed. This option offers advantages in the study of CAs: the severe ones are usually diagnosed at delivery or before birth, when quite accurate information on clinical aspects as well as environmental exposures can be obtained from the parents of the index case and the controls. Hospital-based studies are more feasible, but home-born babies are missed out by this approach. It must also be considered that in the hospital-based studies the source population does not correspond to the population of a defined geographical area, but only to the people attending the hospital.

When controls are not easily or correctly retrievable, a proportional case control study can be carried out, in which selected controls have diseases different from the disease under study (e.g. minor malformations) (Kleinbaum *et al.,* 1982). By definition, therefore, this is not a population-based design, since controls may not be representative of the base population from which the study cases came.

In many instances, the identification of cases has been made from existing registers of congenital malformations. Such registers were often started for monitoring purposes: to rapidly identify a possible new epidemic similar to that caused by thalidomide - their greatest value has up to now been as a data source for scientific studies. In all such instances, the ascertainment of a case is reasonably independent of exposure.

As an example, the MADRE project (malformation and drug exposure), implemented by the ICBDSR, collects data from CA registries in different countries about malformed infants with a positive history of first trimester drug exposure in the mother. To analyse the association between a specific drug (D-1) and a specific CA (CA-1), the dataset is analysed in a case-control fashion: cases are defined as infants presenting a specific CA-1 and control infants are those with other CAs, different from CA-1. Similarly, exposure is defined as the use by the mother of drug D-1 during the first trimester of pregnancy: infants exposed to drugs other than the drug D-1 are considered as unexposed.

One problem with this approach is that the exposure effect could be underestimated, because of the inclusion among controls of infants with malformations possibly associated with the drug under study (Robert *et al.,* 1994). On the other hand, using a dataset containing only affected infants reduces the recall bias, caused by the higher probability of women who have given birth to malformed children to recall the drugs they took during pregnancy, compared to women who have given birth to healthy children (Koren *et al.,* 1998). Some authors argue that exposure misclassification due to recall bias appears unlikely to produce a sufficient bias to explain a strong association, unless the exposure is highly prevalent, whilst it is easy to underestimate a positive association owing to selection bias arising from the use of malformed controls (Swan *et al.,* 1992; Khoury *et al.,* 1994a).

Other methods have been suggested for assessing and adjusting for recall bias: one method is to ask respondents to identify the exposures which they believe are relevant risk factors for the disease, at the end of the interview. Another method is to use a validity scale in which the respondents are asked about their prior exposure to factors which are known not to be associated with the disease. In situations in which there is insufficient information for identifying factors that have no causal importance, it was suggested that the construction of the validity scale be based on exposure information for which it is possible to verify the reports of the individual respondents independently, for example by using medical records (Raphael, 1987).

Matching is a procedure for selecting a reference series that is identical, or nearly so, to the index series, with regard to the distribution of one or more potentially confounding factors. In case-control studies, matching is a popular strategy for reducing confounding and improving efficiency, even though matching alone cannot be assumed to control confounding by matching factors (Rothman and Greenland,

1998). If a factor is previously known to be related to a CA (e.g. maternal age for Down's syndrome) the sampling of controls may be set up so that vulnerable categories (e.g. mothers over 35 years old) are over-represented in the case series. Controls can be recruited with a probability proportional to the expected frequency of the selected known factor among cases. This procedure is known as "randomised recruitment" and imposes a systematic distortion on the otherwise random selection of controls that can be corrected in a simple way in the analysis (Weinberg and Wacholder, 1990). Advantages of randomized recruitment include:

1) the distribution of a matching factor in the general population does not need to be known in advance,
2) controls can be recruited simultaneously with cases,
3) the effect of the matching or first-stage screening factors can be assessed.

Randomized recruitment was developed to reduce costs and improve statistical efficiency in studies investigating environmental exposure and rare diseases (Weinberg and Sandler, 1991).

The major advantage of the case-control design is its efficiency for studying rare diseases. Thus, given a fixed sample size, case-control sampling in a study of a CA enhances the precision and power for estimating and testing the environmental exposure effect. In addition, some case-control studies, particularly proportional morbidity designs, tend to be relatively inexpensive and feasible because they can be based on readily available data sources.

3.3. Ecological designs

In ecological studies, aggregate data on risk factors and disease prevalence from different population groups are compared, in order to identify associations. Exposure and disease are usually measured in groups, typically persons living in a geographical area such as a census tract, region, or other administrative area. Because all data are aggregate at group level, relationships at individual level cannot be empirically determined but are instead inferred from the group level. Given a dichotomous exposure, for example, we would not know the numbers of exposed and unexposed cases in each group. Thus, we cannot estimate the exposure effect directly by comparing the disease rate for exposed and unexposed populations.

Ecological designs are therefore incomplete, in the sense that they lack certain information ordinarily contained in the basic designs. In environmental epidemiology, an inability or lack of resources for accurately measuring environmental exposures at an individual level is quite frequent. Thus, the widespread use of ecological designs in environmental epidemiology reflects a fundamental problem of exposure measurement. However, ecological studies represent an inexpensive design option for linking available data sets or record systems, even when exposures are measured at the individual level. The appeal of this alternative is that aggregate summaries of many exposures, including socio-demographic and other census variables, are often available for the same regions that are used to summarise morbidity and mortality data. The major pitfall of an ecological study is the ecological fallacy, which arises when a conclusion regarding associations at the individual level is based on an analysis of group-level data, while groups of individuals may differ greatly from the larger group.

Bias may also occur when the exposure effect varies across groups. This condition may result from extraneous risk factors being differentially distributed across groups or from a mis-specification of the model form used to analyse the data. If every group were completely exposed or unexposed, there would be no ecological bias attributable to confounding or effect modification by group. Indeed, if all covariates were measured at individual level, such a study would no longer be an ecological design. Thus, to reduce ecological bias, one should select areas that minimise the within-area exposure variation and maximise the between-area variation (Greenland, 1992). One strategy for achieving these goals is to choose the smallest unit of analysis for which required data are available (e.g. census tracts or blocks).

The value of ecological studies depends on the nature of the exposure. This design is most suitable for community exposure (e.g. air, water, soil, food) and has been used to investigate the association between components of drinking water and occurrence of neural tube defects (Macdonell *et al.,* 2000) or herbicides in the soil and oral clefts (Nelson *et al.,* 1979). In order to assess a community's exposures to contaminants in the environment, the level or concentration of contaminants in the environmental media must be measured and evaluated.

There is empirical evidence that different directions and magnitudes of association exist between ecological and case-control analyses of the same data (Richardson *et al.,* 1987). Ecological studies are often based on very crude data and, as a result, tend to have substantial limitations, in addition to any biases caused by the use of aggregated data or unrecognized confounders, operating at the level of the geographical unit employed. (Darby *et al.,* 2001).

However, the possibility of bias does not demonstrate the presence of bias, and a conflict between ecological and individual-level estimates does not by itself demonstrate that the ecological estimates are the more biased ones. This is because

(1) the two types of estimates are subject to overlapping but distinct sets of biases, and it may happen that the individual level estimates are more biased, and

(2) the effects measured by the two types of estimates are overlapping but distinct, with ecological estimates incorporating a contextual component that is frequently absent from the individual estimates, due to contextual (population) restrictions on individual level studies.

Indeed, the contextual component may be viewed as both a key strength and weakness of ecological studies, for it is often of the greatest substantive importance even though it is especially vulnerable to confounding (Greenland, 2001).

Under certain assumptions, an aggregate level analysis may provide valid estimates for the corresponding individual level parameters of interest. For a unifactorial level exposure-disease relationship, aggregate and individual level estimations of the slope may be theoretically expected to provide similar results (Cohen, 1990).

Ecological studies may be classified into different design types that differ in several ways, including methods of subject selection and methods of analysis (Morgenstern, 1982). The most common designs are: 1) exploratory studies, 2) multiple-group comparison studies, and 3) space-time cluster studies.

3.3.1. Exploratory studies

In exploratory ecological studies, we compare the rate of disease among many contiguous regions during the same period, or we compare the rate over time in one region. The purpose is to search for spatial or temporal patterns that might suggest an environmental etiology or more specific etiologic hypotheses. No exposures are measured and, generally, no formal data analysis is used. The simplest type of exploratory study of spatial patterns is a graphical comparison of relative rates across all regions (i.e. a mapping study), possibly accompanied by a statistical test for the null hypothesis of no geographic clustering (Ohno *et al.,* 1979). In mapping studies, however, a simple comparison of estimated rates across regions is often complicated by two statistical issues. First, regions with smaller numbers of observed cases show greater variability in the estimated rate; thus, the most extreme rates tend to be estimated for those regions with the fewest cases. Second, nearby regions tend to have more similar rates than do distant regions (i.e. positive autocorrelation). A statistical method for dealing with both complications involves Bayes' empirical estimation of rates using an auto-regressive spatial model (Cressie, 1993).

In exploratory studies of spatial patterns, regions are characterized in terms of general ecological indicators, such as degree of urbanization (urban versus rural), degree of industrialization (agricultural versus non-agricultural) and population density. Such a design is suited for testing the general worry that a polluted environment may cause birth defects and other adverse perinatal outcomes. The analysis of these data usually involves comparisons of regions or countries grouped by one or more ecological indicators. The priority is to obtain an exposure variable, e.g. along an urban/rural gradient, based on the mother's residence at the child's birth. The problems which are likely to be encountered do not only encompass the great variability in urbanization, but also problems related to how urbanization is measured in different countries (ICBDMS, 2003).

An exploratory study of temporal patterns is generally done by comparing disease rates for a geographically defined population over a long period. One can assess time trends in birth prevalence of congenital malformations from various sources. The main sources are the reports based on birth defects registry data. Significant and real time trends are recognized only for a few malformations in different parts of the world. Significant trends, which may have some relationship with environmental factors, have been described for neural tube defects (Rosano *et al.,* 1999), hypospadias (Toppari *et al.,* 1996) and gastroschisis (Di Tanna *et al.,* 2002).

The decreasing trend in NTDs in all western countries has partly resulted from an increase in the detection and terminations of NTD-affected pregnancies, but other changes may explain this decrease, especially improved maternal nutrition (Rankin *et al.,* 2000). The increasing prevalence at birth of hypospadias has been related to environmental factors such as environmental oestrogen-like compounds. This was confirmed in some studies (Paulozzi *et al.,* 1997) and not in others (Aho *et al.,* 2000), and the question remains whether changes in completeness of registration account for a substantial proportion of the reported increases in the prevalence of hypospadias. Possible explanations for the increasing trends of gastroschisis in different parts of the world are: low body mass index during pregnancy (Torfs *et al.,* 1996), smoking and cocaine use, aspirin use, and living near landfill sites (Elliot *et al.,* 2001). However, none of them seem to be sufficiently convincing.

3.3.2. Multiple-group comparison study

In the multiple-group comparison study, we observe the association between the average exposure level and the disease rate among several groups. This means that a specific environmental factor is identified as a potential causal agent and is measured in various groups, along with disease rates. Investigations on water contaminants are often conducted using a multiple group comparison study. In 1981, in a small area in the Tucson Valley, a groundwater source was found to be contaminated with trichloroethylene, known as a possible causal factor for congenital heart defects (Johnson *et al.,* 1998). Water contamination with trichloroethylene was investigated by comparing the birth prevalence of congenital heart disease in three areas (Yuma, Arizona; Sierra Vista, Arizona; and Baltimore, Washington). The three areas differed for levels of water contaminants, with the highest levels found in Yuma. There was a significantly higher birth prevalence of congenital heart disease in Yuma compared with both the Baltimore and Sierra Vista areas (Mayberry *et al.,* 1990).

Another well known multiple-group comparison study was conducted in Seveso, an Italian area close to Milan that was the scene of a chemical plant accident in 1976 that spread dioxin and other chemicals over an area with a population of 37,000. A comparison was made between the frequency and kinds of birth defects in children born in three Seveso areas (one being the most contaminated, the second less contaminated, and another one still less contaminated) and those in children born in the surrounding, uncontaminated area. The data collected failed to demonstrate any increased risk of birth defects associated with dioxin contamination (Mastroiacovo *et al.,* 1988).

3.3.3. Space-time cluster study

Space-time clustering refers to the interaction between place and time of disease occurrence, in the sense that cases that occur closely in space also occur closely in time. Evidence of space-time clustering may suggest the effects of point-source exposures. The analytic search for space-time clusters requires special statistical techniques that may or may not incorporate information on the base population and covariates (Wallenstein *et al.,* 1989; Roberson, 1990).

Space-time cluster analyses may be used when members of a community perceive a cluster or excess number of cases of one or more diseases in their area. This activity is often motivated by the suspicion that the apparent cluster is caused by a specific environmental exposure, such as chemical waste, pesticides or electromagnetic fields.

When an investigation begins, the first step is to verify the diagnoses of all reported cases and identify any additional cases in the cluster area, which must be defined. In addition to space-time cluster analyses, the investigators will probably want to compare the disease rate in the cluster area population with the rate in another population thought to be unexposed (information usually obtained from birth defects registries), and they may conduct a population-based case-control study to identify the risk factors for the disease.

Reports in the media of a "cluster" of congenital anomalies are relatively frequent. A community may become aware of an aggregation of cases in their area, and seek the nearest reason such as a waste site or power line. This approach to the problem has been likened to the "Texan sharpshooter" who draws his gun and fires at the barn door, and only afterwards goes and draws the target in the middle of the densest cluster of bullet holes. Since random "clusters" are expected to occur and there are usually relatively few cases for investigation, some argue that the likelihood of finding a common causal factor is so low that it may often be better not to investigate but instead to "clean up the mess" of the suspected contaminant without demanding causal proof (Rothman, 1990).

Others have tried to derive guidelines for deciding which clusters are worth investigating, often containing some reference to the number of cases and the nominal "significance" of the difference between the observed and expected number of cases, but also increasingly containing some appraisal of the type of concerns expressed locally (Williams *et al.,* 2002). To respond quickly and with an optimal use of resources to the management of a cluster response, a detailed protocol should be available. A series of steps should be taken before a full-scale epidemiological study is undertaken to investigate the etiology of a potential disease-exposure relationship (e.g. a case control study or an investigation of similar sites) (Quataert *et al.,* 1999). Multi-site studies can be a useful line of investigation in response to local clusters, for example for investigating all communities with municipal incinerators, rather than just the one where a cluster of congenital anomalies was observed (Dolk, 1999).

Studies of associations between living in the vicinity of potential point sources of pollution and CAs have so far been inconclusive. This lack of consistency in the reported outcomes may be interpreted in several ways. It may imply that any associations are complex, or weak, and are therefore difficult to detect through traditional epidemiological studies. It may also be due, at least in part, to the relatively poor measures of exposure and the limited statistical power of many previous studies.

3.4. Case reports

Finally, we cite *case reports* as a potential source of hypotheses. A case report is simply a description of a patient's diagnosis and disease progression, often published in medical literature by physicians who recognize a pattern or something unusual. Case reports may be useful when unusual defects suddenly show up in a cluster of children and are recognized by astute parents or clinicians. Investigation of the use of the thalidomide drug during pregnancy and the resultant severe arm and leg defects in children exposed prenatally is an example of an instance when case reports were helpful. Early suspicions of harmful effects were ignored in some countries, but case reports ultimately led to case-control studies that confirmed the link, tragically only after a large number of children had been damaged (Lenz, 1961; McBride, 1961). For a variety of reasons, however, investigations of case reports of clusters of defects may fail to find a cause, though they may generate hypotheses that warrant further study.

4. Genetic epidemiology

As previously described, epidemiological investigation is an important tool for detecting environmental risk factors for CAs. In the past, this research effort primarily focused on environmental exposures as possible etiologic factors. However, with the recent advances in molecular genetics and a more thorough understanding of molecular biology, the area of epidemiological investigation has widened significantly. Investigators now have an enhanced ability to examine the contributions of both maternal and fetal genotypes to disease risk. Examination of the interplay between genetic predispositions/susceptibilities and environmental exposures is a growing area of study, with potentially major implications with regard to understanding birth defect etiology. Genetic variants confer increased susceptibility to a variety of environmental factors (including chemical, infectious, physical, social, psychological, behavioural, and nutritional factors), thus increasing the risk of carriers for many diseases including birth defects (Khoury, 2000).

Genetic epidemiology is a term used to refer to ''the study of the role of genetic factors and their interaction with environmental factors in the occurrence of disease in human populations'' (Khoury *et al.,* 1993). In the field of CAs, researchers have focused their attention on the genes involved in folate metabolism (Botto and Yang, 2000) and the genes involved in the detoxification of xenobiotics (Van Rooij *et al.,* 2002; Shaw *et al.,* 2003). The rapid identification of genes critical for normal development also opens up opportunities for exploring the potential role of environmental factors in altering the structure and function of these major genes (Kelada *et al.,* 2003).

4.1. ***Study designs in genetic epidemiology***

Three basic types of information might shed light on the genetic component of such interactions:

- a classification of the subjects' genotypes at a major locus for disease susceptibility,
- some observable host characteristic (phenotype) that is genetically determined and linked with the genotype that was responsible for sensitivity, and
- family history as a surrogate for genetic (or shared environmental) influences (Morgenstern and Thomas, 1993).

The choice of the study design will depend upon which of these is sought.

The most powerful approaches are applicable if information on genotypes is available. Their feasibility will increase as more and more genes are identified and assays for them become available. If the genotype is observable, it can be considered simply as another risk factor and any of the basic design and analysis strategies used in epidemiology are applicable. Identifying host characteristics that interact with environmental exposures can be done in essentially the same way. Before trying to identify a specific gene that is related to sensitivity to environmental exposures, one should assess whether there is any evidence that such sensitivity has a genetic basis. This also requires the collection of family data, but unlike the standard analyses aimed at examining the main effect of genetics, it would also be advisable to examine interactions between family history and environmental exposures. The two major design issues to be addressed in such studies are the method of ascertainment of families and the information to be collected on family members, which is usually very scarce in epidemiological studies.

Concerns about selecting appropriate control subjects for case-control studies have led to the development of several non-traditional approaches in the study of genetic risk factors (Khoury and Flanders, 1996). These approaches involve the use of an internal control group rather than an external one.

The case-only design has been promoted as an efficient and valid approach to screening for gene-environment interaction. With this design, the association between an exposure and a genotype is only investigated among case subjects. Each case is coded as positive or negative for a genetic factor and an environmental factor. The case-only odds ratio is derived from the cross-categorization of the study sample on genetic factors and environmental exposure status. This innovative design offers an opportunity to study interaction that is more statistically efficient than the analogous case-control study and is not subject to the common biases arising from control selection. However, the validity of the case-only study hinges on one assumption - that the genetic and environmental factors of interest are independent of one another. This assumption may seem reasonable for a wide variety of genes and exposures, but there are some genes whose presence may be associated with a higher or lower likelihood of the exposure, on the basis of some biological mechanisms (Khoury and Flanders, 1996).

Family-based association studies such as case-parent triads were originally proposed to minimise the population stratification bias. Using controls that are family members of the cases ensures that controls and cases are matched with regard to genetic ancestry, since both groups come from the same families (Khoury, 1994b) Family-based sampling also provides an excellent control for a host of important environmental factors. Among family-based designs, the best known is the case-parent triad design which uses an 'index case' (an offspring affected with the disease of interest) and his/her parents. The genomic DNA for all triad members must be available in order to determine their allelic status. For example, the increased risk of non-syndromic oral clefts is well-known in the children of women who drink alcohol during the first trimester of pregnancy, and it has been shown that the mutated allele of the ADH1C enzyme carried by the child seems to have a protective effect against the risk of oral clefts (Chevrier, 2004). Other family-based designs compare cases with siblings, cousins, or other family members.

Another innovative approach for analysing gene-environment interaction is "Mendelian randomization", which is the random assortment of genes from parents to offspring that occurs during gamete formation and conception. We have seen that conventional genetic epidemiology investigates the association between genetic and phenotype variations within a population, in order to elucidate the genetic basis of the phenotype or to characterize gene function. In such studies, genetic variation is assessed by using markers, often single nucleotide polymorphisms, and the informative markers are those that show sufficient variation within a population and are of high enough prevalence to allow meaningful comparisons to be made. Yet it is also possible to exploit the random assignment of genes as a means of reducing confounding when examining exposure–disease associations: this is Mendelian randomization in an epidemiological context. The underlying idea is that: "if polymorphisms produce phenotypic differences that mirror the biological effects of modifiable environmental exposures which in turn alter disease risk, then the different polymorphisms should themselves be related to disease risk to the extent predicted by their influence on the phenotype." (Davey Smith and Ebrahim, 2003).

The power of Mendelian randomization lies in its ability to avoid the often substantial confounding seen in conventional observational epidemiology. However, confounding can occur in several ways that need to be considered. This approach cannot be used without a reliable association between genotype and disease: biases may arise from differences in the distribution of confounders or the inadequate power of the studies collected for the Mendelian randomisation (Little and Khoury, 2003). The lack of a gene-disease association in an epidemiological study would not exclude the gene-environment interaction in a subgroup (Khoury *et al.,* 1987). The potential for polymorphisms to have more than one specific phenotypic effect (*pleiotropy*) may induce confounding in the genotype-intermediate phenotype–disease association. Another problem is developmental compensation. If a person has developed and grown up in an environment in which one factor is perturbed because of a particular genetic variant, he/she may be rendered resistant to its influence through permanent changes in tissue structure and function that counterbalance its effects (Davey Smith and Ebrahim, 2005). A complete overview of the uses and limitations of Mendelian randomisation is summarised in table 2.

An illustration of this is that some data showed that the risk for spina bifida associated with the homozygosity for the MTHFR C677T gene may depend on the nutritional status of the mother (e.g. intake of vitamins). This would suggest that the intra-uterine environment may play a substantive role in increasing the risk of NTD, rather than the genotype of the offspring (Botto and Yang, 2000). By using a Mendelian randomization approach, it would be possible to provide strong evidence of the beneficial effect of folic acid supplementation, from the evidence of epidemiological studies demonstrating the relationship between MTHFR C677T gene and NTD, even before data became available from controlled clinical trials, avoiding the biased recall of diet and supplement use (Davey Smith and Ebrahim, 2003). However, the association between NTD and MTHFR gene may be confounded by the effect of other folate-related genes (e.g. cystathionine-beta-synthase and methionine synthase reductase) (Botto and Yang, 2000) or by the "pleiotropic" effect of the MTHFR gene controlling on both homocysteine and folate availability (Thomas and Conti, 2004), not revealed in the Mendelian randomization study.

Table 2. Use and limitations of Mendelian randomisation in observational studies (from Davey and Ebrahim, British Medical Journal, 2005, 330, p. 1078. Reproduced with permission from BMJ Publishing Group.)

Uses
Confounding: genetic variants will not generally be liable to confounding by behavioural, socio-economic, and physiological factors.
Reverse causation: genetic variants will not be influenced by the onset of disease or by the tendency for individuals with disease to report exposure history differentially.
Selection biases: genetic variants will not generally be influenced by factors determining how participants are selected for a study, either as a case or as a control.
Attenuation by errors (regression dilution bias): genetic variants will indicate differences in exposure level across a lifetime and associations will not be attenuated by random imprecision in measurement of the exposure.
Limitations
Cannot be used without a reliable association between genotype and disease.
Confounding of genotype by linkage disequilibrium between the genetic variant of interest and another genetic variant that influences the outcome.
Genetic variants with multiple effects may lead to misleading conclusions.
Developmental compensation, which consists in the developmental buffering against the effect of a polymorphism during fetal development.
Inadequate biological understanding of the functions of genetic variants.

5. Problems in environmental epidemiology when studying its effects on reproduction

There are several problems in environmental epidemiology that tend to limit causal inference and that should therefore influence design decisions. A summary of the main problems is presented in table 3. In the following paragraphs, we discuss in detail questions about the living status of cases, the genetic susceptibility, the rarity of the diseases, the low-level exposures, the identification, quantification and timing of the exposure and finally the different sources of more common biases.

5.1. Living status of cases

In the field of birth defects, investigators sometimes restrict the analysis to liveborn infants, which has long been recognized as a potential source of bias (Khoury *et al.*,

1989; Hook and Regal, 1991). In order to compare prevalence rates between populations, in relation to possible underlying environmental causes, it is necessary to calculate a "total" or "adjusted" prevalence rate, including stillbirths and pregnancy terminations. However, the inclusion of terminations, especially if they occur relatively early in pregnancy and relate to congenital anomalies with a high spontaneous fetal death rate (like Down's syndrome), can artificially inflate prevalence rates compared to those based on populations without terminations, since they include affected fetuses which would otherwise have been lost as unrecorded spontaneous abortions.

Table 3. Problematic issues in the investigation of the relationship between congenital anomalies and environmental factors

Miscarriage and fetal death are reproductive endpoints, hard to ascertain, which may result from exposure to teratogenic substances.
Not all people are equally susceptible to birth defects. Genetic and nutritional factors may combine with other environmental factors to increase the risk of the occurrence of a congenital anomaly.
Studies are often unable to identify and/or estimate accurately exposures that occurred months or years previously.
The ability of an agent to cause birth defects often depends on when exposure to that agent occurs. If researchers do not study exposures at that particular point in time, they will probably miss its ability to cause birth defects.
Most birth defects are rare, making it hard to design studies powerful enough to detect them reliably.
Interactions between multiple factors (environmental, genetic, health) make it difficult to pinpoint the contribution of each factor.

It should be also considered that EFs are likely to cause a CA in fetuses which are spontaneously aborted in the first weeks of pregnancy, and for this reason not detected. Reported EFs in miscarriages are maternal infections, exposure to anaesthetic gases, workplace toxicants (such as lead, ethylene glycol, carbon disulfide, cytostatic drugs, polyurethane, heavy metals, organic solvents, petrochemicals and mercury) and hyperhomocysteinemia (Garcia-Enguidanosa *et al.,* 2002).

5.2. Genetic susceptibility

Not all people are equally susceptible to birth defects. Genetic and nutritional factors may combine with other environmental factors to increase the risk of the occurrence of a CA. This combination of factors makes it difficult to conduct epidemiological studies in populations where the distribution of risk factors is not identified accurately. For example, a relationship has been shown between alleles of various genes and maternal cigarette smoking, with regard to the risk of orofacial clefts. Positive associations are seen between maternal cigarette smoking and polymorphisms of the transforming growth factor alpha (TGFα) gene taq1 (Shaw *et al.,* 1996), of the endothelial nitric oxide synthase (NOS3) gene (Shaw *et al.,* 2005) and of the glutathione S-transferases GSTT1 and GSTM1 (Lammer *et al.,* 2005).

5.3. Rare diseases, low-level exposures, and small effects

When studying CAs and environmental hazards, statistical significance is difficult to obtain because of the infrequent occurrence of cases, because of the low prevalence or levels of environmental exposures in the general population and because of the weakness of potential associations. A critical consequence is the substantial loss of the precision and power with which effects are estimated and tested. In addition, it is difficult for the investigator to separate the effect of the exposure under investigation from the distorting effects of confounding factors. Causal inference can be then seriously compromised.

Sample size is often a problem in studies of CAs and therefore the practice of broadly grouping CAs is often used. However, this is of dubious biological validity, as already discussed in paragraph 2.3. Proven teratogenic agents are known to increase the risk for specific malformations. One would not expect that all types of malformations would be increased by specific environmental exposures during pregnancy. The determination of sample size requires consideration of these biological issues (Shaw and Croen, 1993).

In a situation when there is relatively little variability of exposure in the background population, we expect an imprecise estimation of the exposure effect. Although such inefficiency is usually quite apparent in cohort studies, it may not be so apparent in case-control studies, especially when the investigator does not know the exposure distribution in the background population. For example, if environmental exposure levels are high throughout a given area, a comparison of cases and controls would result in an unstable estimate of effect and low power (Morgenstern and Thomas, 1993).

Small effects and sample size problems may be resolved by the use of the meta-analysis technique. Meta-analysis is a quantitative statistical technique combining the results of individual studies in order to obtain an overall estimate of the effect size of a given intervention or exposure. Unlike traditional epidemiologic methods, meta-analysis uses the summary statistics from individual studies as the data points. A key assumption of this analysis is that each study provides a differing estimate of the underlying relationship within the population. Meta-analysis permits the extraction of information that would not otherwise be apparent from the results of individual studies because of small effects or a lack of statistical power.

However, one should also consider the phenomenon of publication bias, which is the tendency on the part of investigators, reviewers and editors to submit or accept manuscripts for publication based on the direction or strength of the study findings. It is usually easier to get a report based on small numbers published if a risk increase is indicated. By adding a number of small studies into a meta-analysis, this can result in an exaggeration of the risk estimate (Källén, 2005).

5.4. *Identifying, quantifying, and timing exposures*

Identifying, quantifying, and timing exposures during fetal development are major challenges when investigating the role of environmental factors in causing CAs. A large body of scientific research shows that not only the magnitude of exposure but also its timing is extremely important determinants of risk, because of the specific sequencing of developmental events. To detect the exact timing between exposure and disease, the investigator must either conduct an expensive prospective study or rely on retrospective measurements of the exposure information.

Exposure may vary substantially across the different stages of pregnancy. The crucial role of the timing of exposure during pregnancy indicates that any misclassification may affect studies with a single exposure for the entire pregnancy. An accurate assessment of exposure requires an examination of the appropriate time window. For outcomes in which the relevant time window is short or unknown and the exposure is likely to vary, recording information on a month-by-month basis is to be preferred, if feasible (Hertz-Picciotto *et al.,* 1996).

If the timing of potentially harmful exposures is imprecisely known or completely unknown, a link between birth defects and environmental factors may be missed. For example, children exposed to the drug thalidomide during the third to sixth week of gestation often suffered severe limb deformities, while children exposed later had either no or different health effects. Bias could also arise from the timing of the exposure ascertainment. Most epidemiological studies of environmental risk factors are based on maternal residence at the time of delivery. Such an assessment would be invalid, however, in instances where the mother had moved prior to delivery. In fact, it is well understood that the effects of environmental teratogens occur early in embryogenesis (Khoury *et al.,* 1988), so assessing the influence of environmental exposures must be related temporally to conception.

5.5. Sources of errors

For assessing the validity of epidemiological studies, one has to consider several possible sources of errors in the estimation of effect: selection bias, information bias, confounding and the problem of multiple testing. The various types of bias are not entirely separate concepts. The amount of confounding, for example, can depend on how subjects are selected for the study or the analysis. When either disease status or exposure status influences the selection of subjects to a different extent in the groups to be compared, a selection bias may occur. The common types of bias in epidemiological studies are summarised in table 4.

Table 4. Common types of bias in epidemiological studies

Type of bias	Typical cause of bias
Selection	Can occur when there are differential selection rates of subjects by disease and/or risk factor characteristics; e.g. bias related to hospital-based studies.
Detection	Can occur when persons with a risk factor are more likely to have a disease detected because of more intense medical follow-up.
Information	Can occur when the nature or quality of measurement or data collection distorts the effect estimate.
Recall	Can occur in retrospective studies when persons with disease tend to report past events and exposures differently from persons without disease.
Misclassification	Can occur when there is an error in classifying subjects by disease or risk factor that tends to distort associations between disease and risk factors.
Confounding	Can occur when an association between a risk factor and disease can be explained by a factor associated with both disease and risk factor.

Selection bias is due to systematic differences in characteristics between those who are selected for study and those who are not. The selection bias induces a non-comparability between exposed and unexposed subjects, even if they were comparable before the selection. Depending on the pattern of interaction between the teratogenic factor and the CA in affecting the measure of occurrence, selection bias may lead to overestimation or underestimation of the true association measure (relative risks, odds ratio).

The "healthy worker effect" is the most common selection bias in occupational studies and occurs because relatively healthy individuals are more likely to gain employment and to remain employed. For examples the studies on the risk of CAs among the children of US Persian Gulf War veterans failed to find an excess risk in the deployed veterans, in contrast with the studies on UK veterans (Doyle *et al.,* 2004). The US study compared the rates of birth defects in the offspring of veterans included in two samples, one active duty personnel deployed to the war zone (exposed group) and the other sample randomly selected among the personnel on active duty, but not deployed to the war zone (not exposed group). (Cowan *et al.,* 1997) The CAs were identified solely among veterans remaining on active duty. Nothing is known about the CAs for either male or female veterans who had separated from active duty, which is the main source of the suspected selection bias from unequal follow-up due to the above-mentioned "healthy worker effect" (Haley, 1998).

Voluntary reporting may also produce selection bias, since some physicians may preferentially report certain types of CAs over others or investigate more attentively newborns from areas where some environmental exposures have been claimed.

Selection bias may also affect etiologic inferences in case-control studies conducted in spontaneous abortions. Both teratogens and defects may be associated with an increased risk of prenatal mortality. Teratogens could lead to reproductive loss, which may depend on the dose and timing of exposure and abnormal fetuses tend to be preferentially lost during the intra-uterine lifetime (Warkany, 1978). Unless the effects of exposure and the presence of abnormality are related in a particular mathematical way (namely, multiplicative effect), the usual epidemiologic measures of association determined in case-control studies of CAs are bound to be biased (Khoury *et al.,* 1989).

Information bias means that the nature or quality of measurement or data collection distorts the effect estimate. The primary source of information bias is error in measuring one or more variables. In fact, a major challenge in environmental epidemiology is to measure accurately each individual's exposure to hypothesised risk factors (i.e. the biologically relevant dose) (Hatch and Thomas, 1993). This task is made very difficult by the lack of information about environmental sources of emission, the complex pattern of most long-term exposures, the individual's ignorance of previous opportunities for exposure, the lack of good biological indicators of exposure level and the lack of sufficient resources for collecting individual exposure data on large populations. The consequences of exposure mismeasurement are a probable bias in the estimation of effect and a possible loss of the precision and power with which effects are estimated and tested (Shy *et al.*, 1978; Fleiss, 1986).

Information bias resulting from exposure misclassification usually tends to obscure a true exposure-related risk. For example:

- in environmental studies of proximity to pollution sources, when the dispersion pattern of relevant exposures is not known and the migration of residents between organogenesis and birth is not taken into account, or

- in studies of drinking water exposures which allocate exposure according to the residents' water source, without information on other sources of drinking water or the fluctuation of water contamination over time in relation to the organogenetic period.

Biological markers of individual exposure (such as cotinine or arsenic in urine, or PCB levels in blood serum) can improve exposure assessment, but costs usually dictate that they must be used within the framework of a case-control study. Case-control studies with exposure data or biospecimens collected retrospectively are particularly vulnerable to poor measurement of exposure, especially if collected long after the exposure occurs (Chatterjee and Wacholder, 2002).

Confounding refers to a lack of comparability between exposure groups (e.g. exposed versus unexposed) such that disease risk would be different, even if the exposure were absent or the same in both populations (Greenland and Morgenstern, 1988). A confounder is then a factor that is associated with the exposure and, independently of that exposure, is a risk factor for the disease. When studying CAs and environmental factors, many confounders may bias the detection of the true relationship between the occurrence of a CA and a potential risk factor. For example, when studying the potential preventive effect of neural-tube defects through the periconceptional use of folic acid, some confounding factors must be accounted for. Women before and during pregnancy may be treated with other drugs as well as multivitamins. The proportion of treated women may differ between women who did or did not take folic acid in the periconceptional period. Given that the use of other drugs may interfere with the potential preventive effect of folic acid, medicine use must be considered as a potential confounder and adjusted for the use of appropriate statistical methods.

Confounding may arise from maternal or paternal exposure. Even though maternal exposures are more frequently investigated, paternal environmental exposures have been studied, with sometimes positive associations with CAs, e.g. smoking and hypospadias (Pierik *et al.,* 2004). Several plausible biological mechanisms could mediate the effects of paternal environmental exposure: in fact, exposures to pesticides and solvents around the time of fertilization have been found to play a role after fertilization, causing male-mediated adverse reproductive outcomes (Davis *et al.,* 1992). However, teratogenic effects due to paternal exposures are more likely to be due to secondary exposures of the embryo, examples being passive smoking or "carry-home" exposures. Remote paternal exposure to chemicals and radiation does not appear to increase the risk of CAs in future offspring (Friedman, 2003). Confounding depends on which parent is considered as the unit of analysis. Different mechanisms and timings of exposure have to be considered: long-term exposures are more likely to concern the fathers, whereas acute or periconceptional exposures are more frequently related to the mothers.

Environmental exposures are usually geographically delineated (e.g. background radiation, air pollution and water quality), and there is a geographical unevenness in the genetic constitution of the population. It is therefore possible to observe a co-variation between environmental factors and a malformation because of the coincidental association between the environment and a gene pool.

When local factors are to be studied, like air pollution conditions, water supply, and food habits, the co-variation with socio-economic levels may be relevant and constitute an important confounder. An example of this is the observation that lead in drinking water correlated with mental retardation in children (Beattie *et al.,* 1975), but lead in drinking water means old water pipes, old houses, and low social class.

Some putative confounders are easy to correct for, e.g. year of birth, maternal age, and parity. Other confounders are often not known and when searched for retrospectively, the recall bias problem exists as much as for exposures. Ideally, confounders should also be collected prospectively - which of course means that they must be known in advance. Some factors are very important for studies of, for instance, gestational duration, birth weight, and neurodevelopment. Examples are maternal smoking, race or ethnicity, use of alcohol or other non-medical substances, maternal socio-economic level and education, previous pregnancy outcome, and length of a period of possible infertility. For congenital malformations, the confounding of such factors is weaker but may exist (Källén, 2006).

Finally, it is worthwhile to mention the multiple testing problem. The relationship between EFs and CAs, as well as gene-environmental interactions, may not be easy to analyse, due to the large number of variables and measurements involved in the analysis and their complex structure. Once the data sets related to these studies are constructed, they often consist of a large number of possible associations that researchers wish to test from many angles. A major drawback of multiple testing is the increased probability of getting "false-positive" results, that is to say statistically significant associations, interactions, etc, that are not real or associations that either cannot be replicated or cannot be corroborated by other similar studies. Thus, if a large number of exposures and a large number of different malformations are tested, an association can only be regarded as a "signal" to be confirmed or rejected by independent studies. In all studies, irrespective of methodology, which are not made because of a suspected association between a specific exposure and a specific outcome, the possibility of random findings due to multiple testing has to be considered.

6. Conclusions

The epidemiologic methods for investigating the relationship between CAs and EFs include all the standard study designs such as randomized control trials, cohort, case-control, ecological studies and community intervention. Challenges specific to environmental epidemiology have given rise to new designs, such as the two-stage randomized recruitment.

Epidemiologic studies are most likely to be productive if they are based on clearly stated hypotheses. These can be developed from the results of descriptive studies. Basically, these try to demonstrate an association between specific CAs and exposure to specific environmental agents.

The exposure assessment of EFs is a challenging issue. For many suspected teratogens, such as those dispersed in the environment, researchers have had to be satisfied with crude surrogate exposure information, without being able to assess the dose that reaches either the germ cells of the parents or the developing embryo or fetus. It has been suggested that potential long-lasting effects of EFs should be considered in addition to acute effects. When facing the challenges of the epidemiologic study of birth defects and environmental exposure to harmful agents, some authors have contemplated waiting until necessary developments in epidemiologic methods and advances in other fields have taken place (Bracken, 1997). A more practical and useful approach should now be to ask what type of studies we need and to implement well designed epidemiologic studies; that is what is addressed in this chapter.

Several study design and analysis issues need to be considered in evaluating epidemiological research and for planning projects. Improved methods for exposure data collection and assessment using questionnaires and biospecimens remain a major challenge, in order to better assess genetic factors and interaction with environmental exposures. To-day, epidemiologists need to bring the tools of molecular genetics into their research, and they cannot proceed without laboratory scientists. This is bringing the epidemiology back to the laboratory, as happened when this specialty was only the study of infectious diseases. The present challenge is to deal with the complexity and diversity of scientific methods that must be coordinated to do this work.

It is rare that single epidemiological studies give a final answer on hazards associated with specific exposure. Findings are often inconsistent from one study to another, but in some instances a consistent pattern emerges, and then provides arguments for an increased risk of birth defects with exposures to certain kinds of EFs.

Weak associations may reflect underlying biological mechanisms and could be the result of a combination of factors such as unmeasured confounding, exposure misclassification, outcome misclassification, gene-environment interactions and differential survival rates among cases. These issues can be addressed by using biological markers of exposure and susceptibility, careful dysmorphological evaluation of affected infants, subgroup analyses for etiologic heterogeneity, searches for biologic interactions and the use of prospectively collected data.

An association between EFs and CAs may be considered as real:

- when the association is strong and specific,
- when there is some consistency of findings across studies, and

- when there is a biological plausibility, based for example on the timing of exposures, on what is known about the pathogenesis of a CA, and on the existence of similar results in experimental work (see table 5 for details).

There are relatively few environmental pollution exposures for which strong conclusions can be drawn, with respect to their potential to cause congenital anomalies. Transmitting to the society and to the public the interpretation of the available studies is challenging, as well as making decisions in terms of public health and clinical practice. A precautionary approach should normally be adopted at both community and individual level.

Table 5. Criteria for evaluating the causal nature of an association in teratology (Extracted from Scialli, 1992)

Criteria	Question addressed
Strength of the association	What is the statistical likelihood that the association did not occur by chance alone?
Consistency of the association	Is the same association seen in several reports or studies? In different populations?
Specificity of the association	How often does the environmental exposure occur without causing the effect, and how often does the effect occur separately from environmental exposure?
Appropriate timing	Does the association make chronological sense? Was the EF taken at the critical time in development to affect the target organ?
Dose-response relationship	Does the likelihood and magnitude of response increase with dose of the EF?
Biological plausibility	Does the association make biological sense? Does the suspected agent pass through the placenta?

References

Aho, M., Koivisto, A.M., Tammela, T.L. and Auvinen, A. (2000) Is the incidence of hypospadias increasing? Analysis of finish hospital discharge data 1970-1994, *Environ. Health Perspect.* **108**, 463-465.

Arfsten, D.P., Still, K.R. and Ritchie, G.D. (2001) A review of the effects of uranium and depleted uranium exposure on reproduction and fetal development, *Toxicol. Ind. Health.* **17**, 180-189.

Basso, O. and Olsen, J. (2005) Subfecundity and neonatal mortality: longitudinal study within the Danish national birth cohort, *Br. Med. J.* **330**, 393–394.

Basso, O., Olsen, J. and Christensen, K. (1999) Recurrence risk of congenital anomalies--the impact of paternal, social, and environmental factors: a population-based study in Denmark, *Am. J. Epidemiol.* **150**, 598-604.

Beattie, A.D., Moore, M.R., Goldberg, A., Finlayson, M.J., Graham, J.F., Mackie, E.M., Main, J.C., McLaren, D.A., Murdoch, K.M. and Steward, G.T. (1975) Role of chronic low-level lead exposure in the aetiology of mental retardation, *Lancet* **1**, 549-551.

Botto, L.D. and Yang, Q. (2000) 5,10-Methylenetetrahydrofolate reductase gene variants and congenital anomalies: a huge review, *Am. J. Epidemiol.* **151**, 862–877.

Bracken, M.B. (1997) Musings on the edge of epidemiology, *Epidemiology* **8**, 337-339.

Brent, R.L. (2001) Addressing environmentally caused human birth defects, *Pediatr. Rev*. **22**, 153-165.

Chatterjee, N. and Wacholder, S. (2002) validation studies: bias, efficiency, and exposure assessment epidemiology, *Epidemiology* **13**, 503-506.

Chevrier, C., Perret, C., Bahuau, M., Nelva, A., Herman, C., Francannet, C., Robert-Gnansia E. and Cordier, S. (2004) Interaction between the ADH1C polymorphism and maternal alcohol intake in the risk of nonsyndromic oral clefts: An evaluation of the contribution of child and maternal genotypes, *Birth Defects Res. A Clin. Mol. Teratol.* **73**, 114-122.

Cohen, B.L. (1990) Ecological versus case-control studies for testing a linear-no threshold dose-response relationship, *Int. J. Epidemiol.* **19**, 680-684.

Cowan, D.N., DeFraites, R.F., Gray, G.C., Goldenbaum, M.B. and Wishik, S.M. (1997) The risk of birth defects among children of Persian Gulf War veterans, *N. Engl. J. Med.* **336**, 1650-1656.

Cressie, N. (1993) *Statistics for Spatial Data*, Revised Edition, Wiley, New York, USA.

Darby, S., Deo, H., Doll, R. and Whitley, E. (2001) A parallel analysis of individual and ecological data on residential radon and lung cancer in South -West England, *J. Royal Stat. Soc. Series A* **164**, 193-203.

Davey Smith, G. and Ebrahim, S. (2003) ‘Mendelian randomization’: can genetic epidemiology contribute to understanding environmental determinants of disease? *Int. J. Epidemiol.* **32**, 1–22.

Davey Smith, G. and Ebrahim, S. (2005) What can Mendelian randomisation tell us about modifiable behavioural and environmental exposures? *Br. Med. J.* **330**, 1076-1079.

Davis, D.L., Friedler, G., Mattison, D. and Morris, R. (1992) Male-mediated teratogenesis and other reproductive effects: biologic and epidemiologic findings and a plea for clinical research, *Reprod. Toxicol.* **6**, 289–292.

Di Tanna, G.L., Rosano, A. and Mastroiacovo, P. (2002) Retrospective time trends analysis of gastroschisis: evidence of an epidemic? *Br. Med. J.* **325**, 1389-1390.

Dolk, H. and De Wals, P. (1996) Congenital anomalies in “*Geographical and Environmental Epidemiology*” (Eds) Elliot P., Cuzick J., English D and Stern R. Oxford University Press, Oxford, UK.

Dolk, H. (1999) The role of the assessment of spatial variation and clustering in the environmental surveillance of congenital anomalies, *Eur. J. Epidemiol.* **15**, 839-845.

Dolk, H. (2004) Epidemiologic approaches to identifying environmental causes of birth defects, *Am. J. Med. Genet. C. Semin. Med. Genet.* **125**, 4–11.

Doyle, P., Maconochie, N., Davies, G., Maconochie, I., Pelerin, M., Prior, S. and Lewis, S. (2004) Miscarriage, stillbirth and congenital malformation in the offspring of UK veterans of the first Gulf war, *Int. J. Epidemiol.* **33**, 74-86.

Elliott, P., Briggs, D., Morris, S., de Hoogh, C., Hurt, C., Jensen, T.K., Maitland, I., Richardson, S., Wakefield, J. and Jarup, L. (2001) Risk of adverse birth outcomes in populations living near landfill sites, *Br. Med. J.* **323**, 363-368.

Ferencz, C., Loffredo, C.A., Correa-Villasenor A. and Wilson, P.D. (1997) Genetic and environmental risk factors of major cardiovascular malformations, the baltimore-Washington infant study, (1981-1989), *Perspectives in Pediatric Cardiology*, vol.**5**. Armonk, New York Futura Publishing Co. Inc.

Fleiss, J.L. (1986) Statistical factors in early detection of health effects, in D.W. Underhill and E.P. Radford (eds), *New and sensitive indicators of health impacts of environmental agents,* University of Pittsburgh, Center for Environmental Epidemiology, Pittsburgh, PA, pp 9-16.

Friedman, J.M. (2003) Implications of research in male-mediated developmental toxicity to clinical counsellors, regulators, and occupational safety officers. *Adv. Exp. Med. Biol.* **518**, 219-226.

Garcıa-Enguıdanosa, A., Calle, M.E., Valero, J., Luna, S. and Dominguez-Rojas, V. (2002) Risk factors in miscarriage: a review, *Eur. J. Obstet. Gynecol. Reprod. Biol.* **102**, 111–119.

Greenland, S. and Morgenstern, H. (1988) Classification schemes for epidemiologic research designs, *J. Clin. Epidemiol.* **41**, 715-716.

Greenland, S. (1992) Divergent biases in ecologic and individual-level studies, *Stat. Med.* **11**, 1209-1223.

Greenland, S. (2001) Ecologic versus individual-level sources of bias in ecologic estimates of contextual health effects, *Int. J. Epidemiol.* **30**, 1343-1350.

Haley, R.W. (1998) Point: bias from the "healthy-warrior effect" and unequal follow-up in three government studies of health effects of the Gulf War, *Am. J. Epidemiol.* **148**, 315-323.

Hatch, M. and Thomas, D. (1993) Measurement issues in environmental epidemiology, *Environ. Health Perspect.* **101**, 49-57.

Hertz-Picciotto, I., Pastore, L.M. and Beaumont, J.J. (1996) Timing and patterns of exposures during pregnancy and their implications for study methods, *Am. J. Epidemiol.* **143**, 597-607.

Hindin, R., Brugge, D. and Panikkar, B. (2005) Teratogenicity of depleted uranium aerosols: a review from an epidemiological perspective, *Environ. Health Perspect*. **26**, 4-17.

Hoffmann, W. (2001) Fallout from the Chernobyl nuclear disaster and congenital malformations in Europe. *Arch. Environ. Health* **56**, 478-484.

Hook, E.B. and Regal, R.R. (1991) Conceptus viability, malformation, and suspect mutagens or teratogens in humans. The Yule-Simpson paradox and implications for inferences of causality in studies of mutagenicity or teratogenicity limited to human livebirths, *Teratology* **43**, 53–59.

International Clearinghouse for Birth Defects Monitoring Systems (ICBDMS) (2003) *Annual Report 2003,* International Centre for Birth Defects (ed), Rome, Italy, p.23.

Jezewski, P., Vieira, A.R., Nishimura, C., Ludwig, B., Johnson, M., O'Brien, S.E., Daack-Hirsch, S., Schultz, R.E., Weber, A., Nepomucena, B., Romitti, P.A., Christensen, K., Orioli, I.M., Castilla, E.E., Machida, J., Natsume, N. and Murray, J.C. (2003) Complete sequencing shows a role for MSX1 in non-syndromic cleft lip and palate, *J. Med. Genet.* **40**, 399-407.

Johnson, P.D., Dawson, B.V. and Goldberg, S.J. (1998) A review: trichloroethylene metabolites: potential cardiac teratogens, *Environ. Health Perspect.* **106**, 995-999.

Källén, B. (1988) *Epidemiology of Human Reproduction,* CRC Press, Boca Raton.

Källén, B. (2005). Methodological issues in the epidemiological study of the teratogenicity of drugs, *Congenital Anomalies* **2**, 44-51.

Källén, B (2006). Human studies – Epidemiologic techniques in developmental and reproductive toxicology. In : Developmental and reproductive toxicology : a practical approach. RD Hood Ed, CRC Press, Boca Raton.

Kelada, S.N., Eaton, D.L., Wang, S.S., Rothman, N.R. and Khoury, M.J. (2003) The role of genetic polymorphisms in environmental health, *Environ. Health Perspect.* **111**, 1055-1056.

Khoury, M.J., Stewart, W. and Beaty, T.H. (1987) The effect of genetic susceptibility on causal inference in epidemiologic studies, *Am. J. Epidemiol.* **126**, 561-567.

Khoury, M.J., Stewart, W., Weinstein, A., Panny, S., Lindsay, P. and Eisenberg, M. (1988) Residential mobility during pregnancy: implications for environmental teratogenesis, *J. Clin. Epidemiol.* **41**, 15-20.

Khoury, M.J., Flanders, W.D., James, L.M. and Erickson J.D. (1989) Human teratogens, prenatal mortality, and selection bias, *Am. J. Epidemiol.* **130**, 361–370.

Khoury, M.J., Beaty, T.H., and Cohen, B.H. (eds). (1993) *Fundamentals of genetic epidemiology,* Oxford University Press, New York, USA, p 383.

Khoury, M.J., James, L.M. and Erickson, J.D. (1994) On the use of affected controls to address recall bias in case–control studies of birth defects, *Teratology* **49**, 273-281.

Khoury, M.J. (1994) Case-parental control method in the search for disease-susceptibility genes, *Am. J. Hum. Genet.* **55**, 414–415.

Khoury, M.J. and Flanders, W.D. (1996) Non-traditional epidemiologic approaches in the analysis of gene-environment interaction: case-control studies with no controls! *Am. J. Epidemiol.* **144**, 207-213.

Khoury, M.J. (2000) Genetic susceptibility to birth defects in humans: from gene discovery to public health action, *Teratology* **61**, 17–20.

Kleinbaum, D.G., Kupper, L.L. and Morgenstern, H. (eds) (1982) *Epidemiologic research: principles and quantitative methods*, Lifetime Learning Publications, Belmont, CA, USA.

Kline, J., Stein, Z. and Susser, M. (1989) Conception to birth epidemiology of human development. New York NY, Oxford University Press, UK.

Koren, G., Pastuszak, A. and Ito, S. (1998) Drugs in pregnancy, *N. Engl. J. Med.* **338**, 1128-1137.

Lammer, E.J., Shaw, G.M., Iovannisci, D.M. and Finnell, R.H. (2005) Maternal smoking, genetic variation of glutathione s-transferases, and risk for orofacial clefts, *Epidemiology* **16**, 698-701.

Lenz, W. (1961) Kindliche Missbildungen Nach Medicament-Einnahma Wahrend Der Gravidat? *Dtsch. Med. Wochenschr* **86**, 2555.

Little, J. and Khoury, M.J. (2003) Mendelian randomisation: a new spin or real progress? *Lancet* **362**, 930-931.

Lopez-Camelo, J.S., Orioli, I.M., da Graca Dutra, M., Nazer-Herrera, J., Rivera, N., Ojeda, M.E., Canessa, A., Wettig, E., Fontannaz, A.M., Mellado, C. and Castilla, E.E. (2005) Reduction of birth prevalence rates of neural tube defects after folic acid fortification in Chile, *Am. J. Med. Genet.* **135**, 120-125.

Macdonell, J.E., Campbell, H.and Stone, D.H. (2000) Lead levels in domestic water supplies and neural tube defects in Glasgow, *Arch. Dis. Child.* **82**, 50-53.

Mastroiacovo P., Spagnolo A., Marni E., Meazza L., Bertollini R., Segni G.and Borgna-Pignatti C. (1988) Birth defects in the Seveso area after TCDD contamination *JAMA* **18**, 1668-1672.

Mastroiacovo, P. and Rosano, A. (2001) Epidemiologic approaches for the identification of environmental risk factors for the congenital anomalies, in G. Ettorre and S. Bianca (eds) *Proceedings of 6th European symposium on the prevention of congenital anomalies*, Catania, pp 28-30.

Mayberry, J.C., Scott, W.A. and Goldberg, S.J. (1990) Increased birth prevalence of cardiac defects in Yuma, Arizona, *J. Am. Coll. Cardiol.* **16**, 1696-1700.

McBride, E.G. (1961) Thalidomide and congenital abnormalities, *Lancet* **2**, 1358.

Medical Research Council (MRC) Vitamin Study Research Group (1991) Prevention of neural tube defects: results of the Medical Research Council Vitamin Study, *Lancet* **338**, 131-137.

Morgenstern, H. (1982) Uses of ecologic analysis in epidemiologic research, *Am. J. Public Health* **72**, 1336-1344.

Morgenstern, H. and Thomas, D. (1993) Principles of study design in environmental epidemiology, *Environ. Health Perspect.* **101**, 23-28.

Munger, R.G., Romitti, P.A., Daack-Hirsch, S., Burns, T.L., Murray, J.C. and Hanson, J. (1996) Maternal alcohol use and risk of orofacial cleft birth defects, *Teratology* **54**, 27-33.

Mylchreest, E., Sar, M., Catley, R.C. and Foster, P.M.D. (1999). Disruption of androgen-regulated reproductive development by di(n-butyl) phthalate during late gestation in rats is different from flutamide, *Toxicol. Appl. Pharmacol.* **156**, 81-95.

Nelson, C.J., Holson, J.F., Green, H.G. and Gaylor, D.W. (1979) Retrospective study of the relationship between agricultural use of 2,4,5-T and cleft palate occurrence in Arkansas, *Teratology* **19**, 377-383.

Nielsen, E., Thorup, I., Schnipper, A., Hass, U., Meyer, O., Ladefoged, O., Larsen, J., Østergaard, G., Sørensen, T.L. and Larsen, P.B. (2001) *Children and the unborn child. Exposure and susceptibility to chemical substances* - an evaluation. Environmental Project, 589. Danish Environmental Protection Agency, Copenhaygen, Sweden.

OECD (1989) *Report on approaches to teratogenicity assessment (draft).* Room document no 31, Paris, French.

OECD (1999) *Guideline for the testing of chemicals*. TG 414 Prenatal developmental toxicity study; TG 416 Two-generation reproductive toxicity study; Proposal for a new guideline 426. Developmental neurotoxicity study.

Ohno, Y., Aoki, K. and Aoki, N. (1979) A test of significance for geographic clusters of disease, *Int. J. Epidemiol.* **8**, 273-280.

Paulozzi, L.J., Erickson, J.D.and Jackson, R.J. (1997) Hypospadias trends in two US surveillance systems, *Pediatrics* **100**, 831-834.

Piegorsch W.W., Weinberg C.R. and Taylor J.A. (1994) Non-hierarchical logistic models and case-only designs for assessing susceptibility in population-based case-control studies, *Stat. Med.* **13**, 153-162.

Pierik, F.H., Burdorf, A., Deddens, J.A., Juttmann, R.E. and Weber, R.F. (2004) Maternal and paternal risk factors for cryptorchidism and hypospadias: a case-control study in newborn boys, *Environ. Health Perspect.* **112**, 1570-1576.

Quataert, P.K.M., Armstrong, B., Berghold, A., Bianchi, F., Kelly, A., Marchi, M., Martuzzi, M. and Rosano, A. (1999) Methodological problems and the role of statistics in cluster response studies: a framework, *Eur. J. Epidemiol.* **15**, 821-831.

Rankin, J., Glinianaia, S., Brown, R. and Renwick, M. (2000) The changing prevalence of neural tube defects: a population-based study in the north of England, 1984-1996. Northern congenital abnormality survey steering group, *Paediatr. Perinat. Epidemiol.* **14**, 104-110.

Raphael, K. (1987) Recall bias: A proposal for assessment and control, *Int. J. Epidemiol.* **16**, 167-170.

Rasmussen, S.A. and Moore, C.A. (2001) Effective coding in birth defects surveillance, *Teratology* **64**, S3–S7.

Richardson, S., Stucker, I. and Hemon, D. (1987) Comparison of relative risks obtained in ecological and individual studies: some methodological considerations, *Int. J. Epidemiol.* **16**, 111-120.

Roberson, P.K. (1990) Controlling for time-varying population distributions in disease clustering studies, *Am. J. Epidemiol.* **13**, 131-135.

Robert, E., Vollset, S.E., Botto, L., Lancaster, P.A.L., Merlob, P., Mastroiacovo, P., Cocchi, G., Ashizawa, M., Sakamoto, S. and Orioli, I. (1994) Malformation surveillance and maternal drug exposure: the MADRE project, *Int. J. Risk Safety Med.* **6**, 75-118.

Romitti, P.A., Lidral, A.C., Munger, R.G., Daack-Hirsch, S., Burns, T.L. and Murray, J.C. (1999) Candidate genes for nonsyndromic cleft lip and palate and maternal cigarette smoking and alcohol consumption: evaluation of genotype-environment interactions from a population-based case-control study of orofacial clefts, *Teratology* **59**, 39-50.

Rosano, A., Smithells, D., Cacciani, L., Botting, B., Castilla, E., Cornel, M., Erickson, D., Goujard, J., Irgens, L., Merlob, P., Robert, E., Siffel, C., Stoll, C. and Sumiyoshi, Y. (1999) Time trends in neural tube defects prevalence in relation to preventive strategies: an international study, *J. Epidemiol. Comm. Health* **53**, 630-635.

Rothman, K.J. and Greenland, S. (1998) *Modern epidemiology*. 2nd ed. Lippincott Williams & Wilkins, Philadelphia, pp. 150-161.

Rothman, K.J. (1990) A sobering start for the cluster busters' conference, *Am. J. Epidemiol.* **132**, S6–S13.

Rothman, K.J. (1993) Methodologic frontiers in environmental epidemiology, *Environ. Health Perspect.* **101**, 19-21.

Scialli, A.R. (1992,) *A Clinical guide to reproductive and developmental toxicology*, Boca Raton, FL: CRC Press.

Shaw, G.M. and Croen, L.A. (1993) human adverse reproductive outcomes and electromagnetic field exposures: review of epidemiologic studies, *Environ. Health Perspect.* **101**, 107-119.

Shaw, G.M., Wasserman, C.R., Lammer, E.J., O'Malley, C.D., Murray, J.C., Basart, A.M. and Tolarova, M.M. (1996) Orofacial clefts, parental cigarette smoking, and transforming growth factor-alpha gene variants, *Am. J. Hum. Genet.* **58**, 551-561.

Shaw, G.M., Nelson, V., Iovannisci, D.M., Finnell, R.H. and Lammer, E.J. (2003) Maternal occupational chemical exposures and biotransformation genotypes as risk factors for selected congenital anomalies, *Am. J. Epidemiol.* **157**, 475–484.

Shaw, G.M., Iovannisci, D.M., Yang, W., Finnell, R.H., Carmichael, S.L., Cheng, S. and Lammer, E.J. (2005) Endothelial nitric oxide synthase (NOS3) Genetic variants, maternal smoking, vitamin use, and risk of human orofacial clefts, *Am. J. Epidemiol.* Nov 3; [Epub ahead of print].

Shy, C.M., Kleinbaum, D.G. and Morgenstern, H. (1978) The effect of misclassification of exposure status in epidemiological studies of air pollution health effects, *Bull. NY. Acad. Med.* **54**, 1155-1165.

Swan, S.H., Shaw, G.M. and Schulman, J. (1992) Reporting and selection bias in case–control studies of congenital malformations, *Epidemiology* **3**, 356-363.

Thomas, D.C. and Conti, D.V. (2004) Commentary: The concept of 'Mendelian Randomization', *Int. J. Epidemiol.* **33**, 21–25.

Toppari, J., Larsen, J.C., Christiansen, P., Giwercman, A., Grandjean, P., Guillette, L.J. Jr, Jegou, B., Jensen, T.K., Jouannet, P., Keiding, N., Leffers, H., McLachlan, J.A., Meyer, O., Muller, J., Rajpert-De, Meyts, E., Scheike, T., Sharpe, R., Sumpter, J. and Skakkebaek, N.E. (1996) Male reproductive health and environmental xenoestrogens, *Environ. Health Perspect.* **104**, 741-803.

Torfs, C.P., Katz, E.A., Bateson, T.F., Lam, P.K. and Curry, C.J. (1996) Maternal medications and environmental exposures as risk factors for gastroschisis, *Teratology* **54**, 84-92.

United Nations (UN) (2003) *Globally harmonized system of classification and labelling of chemicals (GHS)*, UN, Geneva and New York, USA.

Van Rooij, I.A., Groenen, P.M., van Drongelen, M., Te Morsche, R.H., Peters, W.H. and Steegers-Theunissen, R.P. (2002) Orofacial clefts and spina bifida: N-actyltransferase phenotype, maternal smoking, and medication use, *Teratology* **66**, 260–266.

Wallenstein, S., Gould, M.S. and Kleinman, M. (1989) Use of the scan statistic to detect time-space clustering, *Am. J. Epidemiol.* **130**, 1057-1064.

Warkany, J. (1978) Terhanasia, *Teratology* **17**, 187-192.

Weinberg, C.R. and Sandler, D.P. (1991) Randomized recruitment in case-control studies, *Am. J. Epidemiol.* **134**, 421-432.

Weinberg, C.R. and Wacholder, S. (1990) The design and analysis of case-control studies with biased sampling, *Biometrics* **46**, 963-975.

Wiener, S.G. (1980) Nutritional and environmental considerations in the design of drug studies during pregnancy, *Teratology* **21**, A75-A75.

Williams, L.J., Honein, M.A. and Rasmussen, S.A. (2002) Methods for a public health response to birth defects clusters, *Teratology* **66**, S50-S58.

World Health Organization Working Group (2000) Evaluation and use of epidemiological evidence for environmental health risk assessment: WHO guideline document, *Environ. Health Perspect.* **108**, 997–1002.

EUROCAT: SURVEILLANCE OF ENVIRONMENTAL IMPACT

H. DOLK
EUROCAT Project Leader
Faculty of Life and Health Sciences
University of Ulster
UNITED KINGDOM
Email: h.dolk@ulster.ac.uk

Summary

EUROCAT is a European network of population-based registries for the epidemiologic surveillance of congenital anomalies. EUROCAT started in 1979 and now covers more than a quarter of all births in the European Union. Currently, 40 registries in 19 countries of Europe survey 1.35 million births per year.

The objectives of EUROCAT are:

- To provide essential epidemiologic information on congenital anomalies in Europe,
- To facilitate the early warning of teratogenic exposures,
- To evaluate the effectiveness of primary prevention,
- To assess the impact of developments in prenatal screening,
- To act as an information and resource centre for the population, health professionals and managers regarding clusters or exposures or risk factors of concern,

P. Nicolopoulou-Stamati et al. (eds.), Congenital Diseases and the Environment, 131–145.

- To provide a ready collaborative network and infrastructure for research related to the causes and prevention of congenital anomalies and the treatment and care of affected children,

- To act as a catalyst for the setting up of registries throughout Europe collecting comparable, standardised data.

EUROCAT approaches the surveillance of environmental impact in the causation of congenital anomalies in the following 3 main ways:

1) Assessment of trends in congenital anomaly prevalence,

2) Routine detection of and response to clusters,

3) Systematic evaluation of risk related to specified environmental exposures.

Fetal life is an especially sensitive period to environmental exposures and crucial exposures often occur before the pregnancy is recognised. EUROCAT, covering more than a quarter of all births in Europe, can play an important role in a European environmental health surveillance, or envirovigilance, strategy.

1. EUROCAT: What and why?

The surveillance of congenital anomalies serves two main purposes: to facilitate the identification of teratogenic (malformation-causing) exposures and to assess the impact of primary prevention and prenatal screening policy and practice at a population level (Dolk, 2005). EUROCAT, the European network of population-based registers for the epidemiologic surveillance of congenital anomalies, now covers 1.35 million births per year, more than a quarter of births in Europe.

The aims of EUROCAT are:

1) To provide essential epidemiologic information on congenital anomalies in Europe,

2) To facilitate the early warning of teratogenic exposures,

3) To evaluate the effectiveness of primary prevention,

4) To assess the impact of developments in prenatal screening,

5) To act as an information and resource centre regarding clusters or exposures or risk factors of concern,

6) To provide a ready collaborative network and infrastructure for research related to the causes and prevention of congenital anomalies and the treatment and care of affected children,

7) To act as a catalyst for the setting up of registries throughout Europe collecting comparable, standardised data.

Funding for network co-ordination currently comes from the European Commission's Directorate General for Health and Safety, under its Public Health Programme, as a component of the European information system for rare diseases. EUROCAT is also a WHO Collaborating Centre for the Epidemiologic Surveillance of Congenital Anomalies.

Table 6. Coverage of the European population by EUROCAT registries

Country	Annual Births*	No. of EUROCAT Registries	% of country covered
Belgium	115,500	2	25.5
Croatia	39600	1	13.9
Denmark	64,800	1	7.9
Finland	57,200	1	100.0
France	789,100	4	20.2
Germany	742,500	2	2.8
Hungary	90,900	1	100.0
Ireland	65,600	3	51.2
Italy	528,300	5	36.5
Malta	4,000	1	100.0
Netherlands	195,600	1	10.4
Norway	55,200	1	100.0
Poland	343,800	1	73.7
Portugal	116,600	1	16.3
Spain	478,500	4	31.1
Sweden	99,000	1	100.0
Switzerland	74,000	1	9.2
UK	721,200	8	31.6

* 2005 World Population Data sheet, Population Reference
Full information available at www.eurocat.ulster.ac.uk/memberreg

2. The prevalence of congenital anomalies

In most Western populations, two to four percent of babies are diagnosed with one or more major congenital anomalies, with serious medical, functional or cosmetic consequences. The exact prevalence found in any survey of births depends on the rather arbitrary division between "major" and "minor" congenital anomalies and on definition and ascertainment methods of the survey (Dolk, 2004). Population-based registries, accessing multiple sources of health information and clinical specialties and covering diagnoses made prenatally and in at least the first year of life, provide the most reliable prevalence estimates. Estimates of congenital anomaly prevalence have increased during the last few decades, because of the increasing frequency of ultrasound screening for non-externally visible anomalies, such as urinary system and cardiac defects, some of which might not otherwise be revealed till later life (EUROCAT Working Group, 2002).

An important principle underlying most registries is a well defined geographical population-base of resident mothers. Basing a registry on a single hospital or selected hospitals can create selection bias, where high risk mothers are referred to or from the hospital for specialist services, thus resulting in prevalence rates which are biased upwards or downwards compared to the general population. Prenatal screening has increased the potential for such selective flow between hospitals, and emphasised the need for population-based studies.

In order to compare prevalence rates between populations, in relation to possible underlying environmental causes, it is necessary to calculate a "total" or "adjusted" prevalence rate including the terminations of pregnancy which follow prenatal diagnosis. In some countries, a high proportion of pregnancies affected by certain anomalies such as neural tube defects and Down's syndrome are terminated (Busby *et al.*, 2005b; Dolk *et al.*, 2005; Garne *et al.*, 2005; EUROCAT Working Group, 2005). Up-to-date figures can be found on the EUROCAT website (www.eurocat.ulster.ac.uk/pubdata/tables.html). However, some of these pregnancies ending in termination would otherwise have resulted in unreported spontaneous abortions and account needs to be taken of this when comparing populations in space or time (Cragan and Khoury, 2000; Dolk *et al.*, 2005).

Two congenital anomalies show a marked real increase in total prevalence since 1980 – Down's syndrome due to the increasing average maternal age across Europe (Dolk *et al.*, 2005) and gastroschisis due to unknown environmental factors (EUROCAT Working Group, 2002). The situation regarding hypospadias, of interest as a potential outcome of endocrine disrupting exposures, is unclear, although EUROCAT data do not indicate an increase in prevalence since 1980

(Dolk *et al.,* 2004). Interpretation of surveillance data on hypospadias is particularly difficult, since the anomaly ranges from minor to major forms, the more numerous minor forms being subject to variable diagnosis, treatment and reporting over time and between geographic areas. Geographic variation within Europe is evident for a number of anomalies, including oral clefts (EUROCAT Working Group 2002a, b; Calzolari *et al.,* 2004) and omphalocele (Calzolari *et al.,* 1995). The United Kingdom and Ireland used to be areas of high prevalence for neural tube defects, but prevalence rates have decreased in the 1970s and 1980s to nearly continental European levels in the 1990s (Busby *et al.,* 2005a). EUROCAT has also revealed that the potential to prevent neural tube defects (and perhaps other congenital anomalies) by folic acid supplementation is far from being fulfilled in Europe (Busby *et al.,* 2005a, Busby *et al.,* 2005b, EUROCAT Folic Acid Working Group, 2005), since there has been little or no decline in neural tube defect prevalence since 1991 when randomised trials confirmed the preventive potential of periconceptional supplementation. Many countries outside Europe have started fortifying staple foods with folic acid, likely to be the only method by which a significant lowering of neural tube defect prevalence can be achieved (De Wals *et al.,* 2003).

Prevalence data on more than 80 types of congenital anomaly reported among livebirths, stillbirths and terminations of pregnancy following prenatal diagnosis are updated each year on the EUROCAT website http://www.eurocat.ulster.ac.uk/pubdata/tables.html.

3. Environmental causes of congenital anomalies

The "environment" is a term that can be used to mean anything non-genetic, including lifestyle exposures such as smoking and maternal factors such as age, or at the other extreme, it can be used to mean very specifically exposure to the by products of industrial and agricultural development, or environmental pollution. This rather loose definition of "environment" creates confusion when the contribution of the "environment" to the causation of congenital anomalies is discussed. An extensive review of the scientific literature on environmental risk factors for congenital anomalies is available on the EUROCAT website, adopting the wider definition of "environment" but nevertheless providing a more exhaustive review of studies on environmental pollution (EUROCAT, 2004). The latter include studies of drinking water contaminants (heavy metals and nitrates, chlorinated and aromatic solvents, and chlorination by products), residence near waste disposal sites and contaminated land, pesticide exposure in agricultural areas, air pollution and industrial pollution sources, food contamination, and disasters involving accidental, negligent or deliberate chemical releases of great magnitude. There are relatively few environmental pollution exposures for which we can draw strong conclusions about the potential to cause congenital anomalies and, if so, the chemical constituents implicated. Thus, the challenge is to deal with scientific uncertainty with appropriate recourse to the "precautionary principle".

Congenital anomalies result from environmental exposures during the first trimester of pregnancy, when most organogenesis occurs. As much of organogenesis occurs before the pregnancy is even recognized, protection of the embryo cannot rely on actions taken once the woman knows that she is pregnant. Relevant preconceptional exposures may also have postconceptional effects, for example if these are indirect (e.g. effects on endocrine function) or if the chemical has a long biological half-life in the body (e.g. PCBs). The development of the brain remains subject to adverse influences well past the first trimester.

A "teratogen" is a malformation causing chemical or agent. However, whether an agent acts as a teratogen is crucially dependent on dose. Animal experiments show that a vast array of chemicals, if given in high enough doses early in pregnancy, will cause fetal malformation. The agent will be considered teratogenic if it leads to fetal malformation without significant maternal toxicity (recognizing the unique susceptibility of the fetus). For practical public health purposes, each agent can be considered to have a "threshold dose" above which it may cause fetal malformation (if exposure occurs at the sensitive time during development), and below which the developing fetus is unaffected or is able to self-regulate or repair damage. For the protection of humans, it is important to determine whether the highest exposures experienced in the population are anywhere near the estimated "threshold", taking into account the likely variation of individual thresholds, depending on other genetic and environmental factors. It follows that environmental exposure information should be concerned with the range and distribution of exposure in the population, not just the average exposure. Additional considerations are whether chemical mixtures are more teratogenic than the component chemicals, and what the additive effects are of different routes of exposure to the same chemical type.

The better established teratogens (EUROCAT, 2004) include nutritional excesses and deficiencies (e.g. low folic acid status) (MRC, 1991), maternal illness or infection (e.g. diabetes, rubella), drugs taken during pregnancy (e.g. thalidomide, valproic acid) (Schardein, 2000), chemical exposures in the workplace or home (e.g. to solvents or pesticides) (Cordier *et al.,* 1997; Garcia, 1998), and radiation (e.g. medical X-rays, and atomic bomb irradiation) (Lione, 1987; Otake and Schull, 1984).

Congenital anomalies can be presumed to be caused by a combination or interaction of genetic and environmental factors. Epidemiologic research establishes whether it is genetic or environmental factors or both which distinguish individuals with and without a congenital anomaly. Epidemiology cannot identify factors within the causal mechanism that are uniform within the study population and teratogenic only in combination with other genetic and environmental factors. We categorise as "genetic" single gene or chromosomal syndromes where individuals with and without the syndrome are distinguished by the genetic mutation alone. The example of the metabolic disease phenylketonuria reminds us that even with a "genetic" condition, environmental factors may be involved in the causal mechanism and indeed may provide the basis for therapeutic intervention. We place in the "environmental" category cases with environmental exposures known to carry a high relative and absolute risk of birth defect, such as maternal rubella, even though genetic susceptibility may play a role. Many environmental exposures raise the risk of a birth defect, but only the minority of exposed individuals are affected, depending on the existence of other genetic and environmental factors (often unknown). Part of the interest of an international system such as EUROCAT is to look in a standardised way at differences in prevalence of congenital anomalies between populations, thus expanding the scope of environmental and genetic variation that can be investigated beyond individual differences.

Congenital anomalies are one group within a range of other adverse pregnancy outcomes, including spontaneous abortions, low birthweight due to prematurity or restricted fetal growth, and poor neurodevelopmental outcomes. These differ in timing (the "sensitive periods" during pregnancy), and in the level of risk associated with different environmental exposures and at different doses. Protection of fetal health must consider the entire range of potential adverse outcomes. Congenital anomalies are not necessarily the most sensitive indicators of fetal exposure.

4. EUROCAT surveillance of environmental impact

4.1. Approaches to surveillance

Epidemiologic surveillance is the continuous scrutiny of the distribution of disease in a population in order to take and evaluate control measures. As opposed to epidemiologic research, surveillance tends to be concerned with problem detection and hypothesis generation rather than hypothesis testing; using routine data collection systems, with minimal often incomplete data as opposed to time limited, specially designed extensive and complete data collection (Thacker and Berkelman,

1988). Nevertheless, epidemiologic surveillance and epidemiologic research form a spectrum of activity, and considerable EUROCAT activity is carried out at the borderline of surveillance and research, where surveillance does test hypotheses (often less specific hypotheses than can be addressed by a full research protocol), and where attempts are made to expand routine data collection systems to fit these purposes, and to ensure higher and standardised data quality. Since the quality of data on congenital anomalies coming from population-based registries is high (and used for both surveillance and research), surveillance and research differ mainly in the quality and extensiveness of environmental and risk factor data.

EUROCAT approaches the surveillance of environmental impact in the causation of congenital anomalies in the following three ways: assessment of trends in congenital anomaly prevalence; routine detection o,f and response to, clusters; systematic evaluation of environmental exposures.

4.1.1. Assessment of trends in congenital anomaly prevalence

This has been discussed above (section 2).

4.1.2. Routine detection of and response to clusters without a well defined a priori exposure hypothesis

Each year EUROCAT conducts a statistical analysis of all new trends and temporal clusters, and a common computer software is supplied to individual registries so that a more frequent and up-to-date analysis can be done locally. Such monitoring has its origins in the thalidomide epidemic, when routine examination of congenital anomaly data was promoted as an early warning system for newly marketed drugs with previously unsuspected teratogenic effects. Since then, the sensitivity of surveillance systems has been analysed to identify under what conditions a drug related increase is most likely to be detected (Kallen, 1989; Khoury and Holtzman, 1987). Experience shows that many trends and clusters are finally attributed to changes in definition or diagnostic and reporting practice (Robert, 2003). In addition, a random distribution of cases in space and time is not a regular distribution, and there will be patches in time or space of denser concentration of cases. These random events are difficult to distinguish from clusters where the cases have a common unusual or new cause. EUROCAT has also therefore developed a web-based “Cluster Advisory Service” to collect experience in investigating and responding to clusters (www.eurocat.ulster.ac.uk/clusteradservice.html).

Clusters in time and space may also be found outside the surveillance system, perhaps outside the registry area, by the community, media or health professionals.

A community may become aware of an aggregation of cases in their area, and then seek the nearest reason such as a waste site or power line. The problem has been likened to the "Texan sharpshooter" who draws his gun and fires at the barn door, and only afterwards goes and draws the target in the middle of the densest cluster of bullet holes. Since random "clusters" are expected to occur and there are usually few cases for investigation, some argue that the likelihood of finding a common causal factor is so low that it may often be better not to investigate but instead to "clean up the mess" of the suspected contaminant without demanding causal proof (Rothman, 1990). Existing guidelines for deciding which clusters are worth investigating are described in the EUROCAT Cluster Advisory Service web pages. Distinguishing random clusters from clusters with a true local environmental cause has proved a difficult problem, whether the clusters are detected by the surveillance system or outside it. Most of the well documented instances in the literature where a cluster was observed which was subsequently established as due to environmental contaminants have been related to food exposures (Dolk and Vrijheid, 2003), involving both high numbers of cases and high relative risk, including the Minamata incident in Japan where fish and shellfish were contaminated with methylmercury, incidents of PCB contamination of cooking oil in Taiwan and Japan, and pesticide over-use at a fish farm in Hungary.

This problem of "random" clusters has also led to reluctance to conduct this type of surveillance, for fear of being "overwhelmed" with clusters to investigate. This is rather an ostrich-like policy, and no longer supported by experience. Statistical methods are now available which take into account the high level of multiple and repeated statistical testing. It is of more concern that true clusters are missed, rather than random clusters detected. A transparent surveillance, or "envirovigilance", strategy could help prioritise locally reported clusters. Moreover, examination of clustering is also an important element of data quality control, and systems are more likely to be overwhelmed by clusters caused by variations in diagnostic, reporting or registration practice than by random clusters, a situation which needs to be tackled where it arises.

4.1.3. Systematic evaluation of environmental exposures

While detection of clustering without an *a priori* specified hypothesis, as described under 4.1.2 above can be one element of an envirovigilance strategy, it is by no means enough. Surveillance will have greater sensitivity if it directly assesses specified environmental exposures: with sensitivity to find a true effect increasing the more precise and accurate the exposure assessment is. In addition, this sort of evaluation can be part of the response to individual clusters where a local exposure has been suspected, by assessing whether clustering is occurring near "similar" sources of pollution (Dolk, 1999).

The first such study to be conducted within the EUROCAT network evaluated the impact of Chernobyl on the prevalence of congenital anomalies in the Western European countries belonging to the network in 1986 (Dolk *et al.,* 1999). This study revealed that overall there was no detectable increase in congenital anomalies of the central nervous system or eye, or Down's syndrome.

EUROCAT also conducted a study of risk of congenital anomaly associated with maternal occupational exposures (Cordier *et al.,* 1997). This study in particular found further evidence of a risk associated with solvent exposure. A further study developed and tested methods of estimating exposure to potential endocrine disrupting chemical exposures, based on job title (Vrijheid *et al.,* 2003), but this has not yet been tested on EUROCAT registry data.

The EUROHAZCON study investigated risk of congenital anomaly near hazardous waste landfill sites in Europe. This study found an elevated risk of both non-chromosomal and chromosomal anomalies among residents near hazardous waste landfill sites, but no discernable relationship between the level of risk and a crude assessment of the hazard potential of the site (Dolk *et al.,* 1998; Vrijheid *et al.,* 2002a, b). More research is needed to clarify this issue, but the potential for studies based almost entirely on proximity of residence as exposure assessment is probably exhausted for landfill sites, and has already fulfilled its surveillance "warning" function.

Further study concerns the effect of socioeconomic deprivation on congenital anomaly risk (Vrijheid *et al.,* 2000). This is important in its own right (to target prevention and treatment), and also to untangle the complexities of interpreting risks associated with other environmental exposures. Typically, for example, people of lower social status are more highly exposed to pollution, either because they move where housing prices are lowest, have less power or advocacy skills to prevent pollution sources being sited near them, have less access to environmental health information, or because aspects of lifestyle associated with greater deprivation (such as ability to buy bottled water) lead to higher exposure. Although information on the extent to which congenital anomaly prevalence is linked to social status is rather limited, current evidence does suggest that more socio-economically deprived groups have higher non-chromosomal congenital anomaly rates, and part of this may be explained by nutritional status. Thus we have to take this into account when interpreting an association between, for example, residence near an industrial site and a raised prevalence of congenital anomaly, as a causal effect of industrial releases. The EUROHAZCON study included a crude measure of socioeconomic status and found that the effect of residence near landfill sites was independent of any socioeconomic effect. Socioeconomic status may also modify the extent to which a pollutant presents a teratogenic risk, perhaps through interaction with nutritional or other environmental conditions. In addition, demands of environmental justice necessitate an assessment of whether deprived communities have higher congenital anomaly risks due to their environmental conditions.

Numerous other studies of environmental exposures, ranging from recreational and therapeutic drugs to dioxin contaminated land and incinerators, have been conducted by individual member EUROCAT registries or countries, and references can be found on the EUROCAT website (www.eurocat.ulster.ac.uk/pubdata/publications.html).

One of the challenges is to turn this strand of surveillance, now consisting of *ad hoc* pieces of research, into a truly systematic surveillance approach, with a clear system of identification and prioritisation of exposures to be evaluated.

4.2. Envirovigilance

Postmarketing drug surveillance, or pharmacovigilance, is a vital part of congenital anomaly surveillance. Statistical monitoring may pick up a change in frequency of a specific type(s) of congenital anomaly, the surveillance system can provide a rapid population-based response to the observations of the "alert clinician" or the surveillance system can be used to assess how well the use of known teratogenic drugs is avoided in pregnancy. To be truly effective however we need to operate on the borderline between surveillance and research, investing in registry-based case-control approaches with extensive and accurate data on women's medication history during pregnancy, and linkages between congenital anomaly registries and clinical databases of women with specific diseases (epilepsy, diabetes). Some individual registries have been active in pharmacovigilance, but ongoing collaborative activity in this area within EUROCAT has only recently begun. An additional issue growing in prominence is the use of assisted reproductive technologies (ART) (Hansen *et al.*, 2002), where again linkages between congenital anomaly registers and ART registers should provide more precise and accurate data on congenital anomaly risks.

In EUROCAT, we have suggested that an analogous form of surveillance to pharmacovigilance should be "envirovigilance", the post-licensing surveillance of environmental chemicals resulting from industrial and agricultural practices. The elements of the envirovigilance strategy for EUROCAT are essentially as described above (sections 4.1.1, 4.1.2, and 4.1.3). Whereas pharmacovigilance finally leads to a better balancing of costs and benefits for the individual, envirovigilance would lead to a better balancing of costs and benefits of industrial and agricultural practices for the community.

The EUROHAZCON study of hazardous waste landfill sites was a feasibility study in relation to building an envirovigilance system (Dolk *et al.,* 1998; Vrijheid *et al.,* 2002a, b). It showed that the standardisation of registry methods and coding worked on by EUROCAT since 1979 made pooling congenital anomaly data across Europe relatively straightforward. The main difficulties with proceeding to routine surveillance lay in variable access in different countries to geocoding (to high geographic resolution) of cases and controls, variable access to and categorisation of socio-economic status as a confounding factor, and variable and generally insufficient access to appropriate detailed environmental exposure information. Envirovigilance of the future therefore, now depends crucially on building environmental information systems in Europe which can be routinely accessed.

One would not expect epidemiologic surveillance on its own to establish beyond doubt the level of risk associated with an environmental exposure. Where there is an indication of a potential problem from surveillance, specially designed epidemiological research studies are needed with extensive exposure assessment. Complementary toxicological and teratological approaches are also needed for problem detection and investigation.

Surveillance requires the maintenance of high quality complete data on fetuses and babies born with congenital anomalies. The public may assume that its health is being protected by routine surveillance, but in fact this activity is at risk from the logistical (and associated financial) difficulties imposed by requirements in some countries to ask for parental consent for the inclusion of children in any database (Busby *et al.,* 2005c). The duty to obtain consent is often not balanced by the duty to give parents the opportunity to participate in the future protection of child health.

References

Busby, A., Abramsky, L., Dolk, H., Armstrong, B. and EUROCAT Folic Acid Working Group (2005a) Preventing neural tube defects in Europe – population-based study, *Br. Med. J.* **330**, 574-575.

Busby, A., Abramsky, L., Dolk, H., Armstrong, B., Addor, M.C., Anneren, G., Armstrong, N., Baguette, A., Barisic, I., Berghold, A., Bianca, S., Braz, P., Calzolari, E., Christiansen, M., Cocchi, G., Daltveit, A.K., De Walle, H., Edwards, G., Gatt, M., Gener, B., Gillerot,

Y., Gjergja, R., Goujard, J., Haeusler, M., Latos-Bielenska, A., McDonnell, R., Neville, A., Olars, B., Portillo, I., Ritvanen, A., Robert-Gnasia, E., Rosch, C., Scarano, G. and Steinbicker, V. (2005b) Preventing neural tube defects in Europe: a missed opportunity, *Reprod. Toxicol.* **20**, 393-402.

Busby, A., Ritvanen, A., Dolk, H., Armstrong, N., De Walle, H., Riano-Galan, I., Gatt, M., McDonnell, R., Nelen, V. and Stone, D. (2005c) Survey of informed consent for registration of congenital anomalies in Europe, *Br. Med. J.* **331**, 140-141.

Calzolari, E., Bianchi, F., Dolk, H., Milan, M. and EUROCAT Working Group (1995) Omphalocele and Gastroschisis in Europe: a survey of 3 million births 1980-90, *Am. J. Med. Gen.* **58**, 187-194.

Calzolari, E., Bianchi, F., Rubini, M., Ritvanen, A., Neville, A. and EUROCAT Working Group. (2004) Epidemiology of Cleft Palate in Europe: Implications for Genetic Research Strategy", *The Cleft Palate-Craniofacial J.* **41**, 244-249.

Cordier, S., Bergeret, A., Goujard, J., Ha, M-C., Ayme, S., Bianchi, F., Calzolari, E., De Walle, H., Knill-Jones, R., Candela, S., Dale, I., Danaché, B., De Vigan, C., Fevotte, J., Kiel, G. and Mandereau, L., (for the Occupational Exposure and Congenital Malformations Working Group). (1997) Congenital malformations and maternal occupational exposure to glycol ethers, *Epidemiology* **8**, 355-363.

Cragan, J.D. and Khoury, M.J. (2000) Effect of prenatal diagnosis on epidemiologic studies of birth defects, *Epidemiology* **11**, 695-699.

De Wals, P., Rusen, I.D., Lee, N.S., Morin, P. and Niyonsenga, T. (2003) Trend in prevalence of neural tube defects in Quebec. *Birth Defects. Res. Part A. Clin. Mol. Teratol.* **67**, 919-923.

Dolk, H., Nichols, R and a EUROCAT Working Group (1999) Evaluation of the impact of Chernobyl on the prevalence of congenital anomalies in 16 regions of Europe, *Int. J. Epdiemiol.* **28,** 941-948.

Dolk, H. (1999) The role of the assessment of spatial variation and clustering in the environmental surveillance of birth defects, *Eur. J. Epid.* **15**, 839-845.

Dolk, H. (2004) Epidemiologic approaches to identifying environmental causes of birth defects. Seminars in Medical Genetics: Public Health Issue 2004. Eds Rasmussen SA and Moore CA, *Am. J. Med. Gen. Part C: Seminars in Med. Gen.* **125C**, 4-11.

Dolk, H. (2005) EUROCAT: 25 years of European surveillance of congenital anomalies, *Arch. Dis. Child. Fetal. Neonatal. Ed.* **90**, F355-F358.

Dolk, H. and Vrijheid, M. (2003) The impact of environmental pollution on congenital anomalies, In "The impact of environmental pollution on health", *Br. Med. Bull.* **68**, 25-45.

Dolk, H., Loane, M., Garne, E., De Walle, H., Queisser-Luft, A., De Vigan, C., Addor, M.C., Gener, B., Haeusler, M., Jordan, H., Tucker, D., Stoll, C., Feijoo, M., Lillis, D. and Bianchi, F. (2005) Trends and geographic inequalities in the livebirth prevalence of Down Syndrome in Europe 1980-1999, *Revues Epidem. Sante. Publique*, **53**, 1-9. EUROCAT Working Group 2002.

Dolk, H., Vrijheid, M., Armstrong, B., Abramksy, L., Bianch,i F., Garne, E., Nelen, V., Robert, E., Scott, J.E., Stone, D. and Tenconi, R. (1998) Risk of Congenital Anomalies near Hazardous Waste Landfill Sites in Europe: the EUROHAZCON study, *Lancet* **352**, 423-427.

Dolk, H., Vrijheid, Scott, J.E.S., Addor, M.C., Botting, B., de Vigan, C., de Walle, H., Garne, E., Loane, M., Pierini, A., Garcia-Minaur, S., Physick, N., Tenconi, R., Wiesel, A., Calzolari, E. and Stone, D. (2004) Towards the effective surveillance of hypospadias, *Environ. Health Perspect.* **112**, 398-402.

EUROCAT Folic Acid Working Group (2005) *Prevention of neural tube defects by periconceptional folic acid supplementation in Europe*, University of Ulster. Online : [http://www.eurocat.ulster.ac.uk/pubdata/publications.html.][last accessed 03/04/06].

EUROCAT special report (2004) *A review of environmental risk factors for congenital anomalies*, EUROCAT Central Registry, University of Ulster. ISBN 1-85923-187-X. Online: [http://www.eurocat.ulster.ac.uk/pubdata/publications.html.][last accessed 03/04/06].

EUROCAT Working Group (2002a) *EUROCAT Report 8: Surveillance of Congenital Anomalies in Europe 1980-1999*, University of Ulster.

EUROCAT Working Group (2002b) *EUROCAT Special Report: EUROCAT and Orofacial Clefts: The Epidemiology of Orofacial Clefts in 30 European regions*", EUROCAT Central Registry, University of Ulster; University of Ferrara, Italy and the CNR Institute of Clinical Physiology, Pisa, Italy. Online : [http://www.eurocat.ulster.ac.uk/pubdata/publications.html.][last accessed 03/04/06].

EUROCAT Working Group (2005) *EUROCAT special rReport: prenatal screening policies in Europ,.* EUROCAT Central Registry, University of Ulster. Online : [http://www.eurocat.ulster.ac.uk/pubdata/publications.html.][last accessed 03/04/06].

Garcia, A. M. (1998) Occupational exposure to pesticides and congenital malformations: A review of mechanisms, methods, and results, *Am. J. Ind. Med.* **33**, 232-240.

Garne, E., Loane, M., Dolk, H., de Vigan, C., Scarano, G., Tucker, D., Stoll, C., Gene,r B., Pierini, A., Nelen, V., Rosch, C., Gillerot, Y., Feijoo, M., Tincheva, R., Queisser-Luft, A., Addor, M.C., Mosquera, C., Gatt, M. and Barisic I. (2005) Prenatal diagnosis of congenital malformations in Europe, *Ultras. Obstet. Gynecol.* **25**, 6-11.

Hansen, M., Kurinczuk, J.J., Bower, C. and Webb, S. (2002) The risk of major birth defects after intracytoplasmic sperm injection and *in vitro* fertilisation, *N. Engl. J. Med.* **346**, 725-730.

Kallen, B. (1987) Population surveillance of congenital malformations. Possibilities and limitations, *Acta Paediat. Scand.* **78**, 657-663.

Khoury, M.J. and Holtzman, N.A. (1987) On the ability of birth defects monitoring to detect new teratogens, *Am. J. Epidemiol.* **126**, 136-143.

Lione A. (1987) Ionizing radiation and human reproduction, *Reprod. Toxicol.* **1**, 3-16.

MRC Vitamin Research Group (1991) Prevention of neural tube defects: results of the Medical Research Council Vitamin Study, *Lancet* **338**, 131-137.

Otake, M. and Schull, W.J. (1984) *In utero* exposure to A bomb radiation and mental retardation; a reassessment, *Br. J. Radiol.* **57**, 409-414.

Robert-Gnasia, E. (2003) *Acting on clusters arising from routine surveillance*. Online : [http://www.eurocat.ulster.ac.uk/pubdata/clusteradservice.html.][last accessed 03/04/06].

Rothman, K.J. (1990) A sobering start for the cluster busters' conference, *Am. J. Epidemol.* **132**, suppl 1, 6-13.

Schardein, J.L. (2000) *Chemically induced birth defects* (3rd Ed), Marcel Dekker, New York.

Thacker, S.B. and Berkelman, R.L. P (1988) Public heath surveillance in the United States, *Epidemiol. Rev.* **10**, 164-190.

Vrijheid, M., Dolk, H., Stone, D., Abramsky, L., Alberman, E. and Scott, J.E.S. (2000) Socio-economic inequalities in risk of congenital anomaly, *Arch. Dis. Childh.* **82**, 349-352.

Vrijheid, M., Dolk, H., Armstrong, B, Abramsky, L., Bianchi, F., Fazarinc, I., Garne, E., Ide, R., Nelen, V., Robert, E., Scott, J.E., Stone, D. and Tenconi, R. (2002a) Risk of chromosomal congenital anomalies in relation to residence near hazardous waste landfill sites in Europe, *Lancet* **359**, 320-322.

Vrijheid, M., Dolk, H., Armstrong, B., Boschi, G., Busby, A., Jorgensen, T., Pointer, P. and EUROHAZCON Collaborative Group (2002b) Hazard potential ranking of hazardous waste landfill sites and risk of congenital anomalies, *Occup. Env. Med.* **59**, 768-776.

Vrijheid, M., Armstrong, B., Dolk, H., van Tongeren, M. and Botting, B. (2003) Risk of hypospadias in relation to maternal occupational exposure to endocrine disrupting chemicals, *Occup. Environ. Med.* **60**, 543-550.

CLINICAL TERATOLOGY

M. CLEMENTI* AND E. DI GIANANTONIO
CEPIG, Genetica Clinica ed Epidemiologica
Dipartimento di Pediatria
Università di Padova
Via Giustiniani 3
Padova
ITALY
(author for correspondence, Email: maurizio.clementi@unipd.it)*

Summary

Teratology is the branch of medical science which studies the contribution of the environment to abnormal prenatal growth as well as morphological or functional developmental defects. Despite some pioneering research, the clinical and scientific interest in clinical teratology only developed because of the rubella pandemics in 1941 and the thalidomide tragedy in the late 1950s. The following decades have seen a definition of criteria for proof of human teratogenicity, the classification of drugs used in pregnancy, and the development of the "Teratogen Information Services" in developed countries.

Teratogens are chemical, physical or infectious agents, or a maternal status/disease whose prenatal exposure during the pregnancy can provoke developmental defects. Susceptibility to teratogenic agents depends on the combination of several factors, including the genotype of the mother and/or of the fetus, dosage, the gestational period at exposure, pharmacokinetics and the pharmacodynamic of the substance.

The methodology for the identification of teratogens includes animal studies, prospective epidemiologic studies and retrospective epidemiologic studies. The criteria for "proof" of human teratogenicity have not been defined although some criteria have been proposed and applied in the clinical/animal/epidemiologic studies. At present, some international classifications of drugs used in pregnancy are

P. Nicolopoulou-Stamati et al. (eds.), Congenital Diseases and the Environment, 147–160.

available, but their usefulness and reliability have been criticized. For this reason, in the 1980's Teratogen Information Services were set up in order to provide public and health professionals with appropriate information on drug risk during pregnancy.

1. Introduction

Teratology is the branch of medical science which studies the contribution of the environment to abnormal prenatal growth as well as morphological or functional developmental defects (Smithells, 1980). The term teratology derives from the Greek word *τερασ*, which means *terrible vision, monster*.

Although a few pioneering studies were conducted in the early 20th century (Hale, 1933; Warkany and Nelson, 1940), the interest of the medical and public media only developed following the evidence that rubella was responsible for a well-defined phenotype of embryopathy (Greg, 1941); and the thalidomide tragedy in the late 1950s and early 1960s (Smithells, 1962). The latter definitely demonstrated that the environment plays a significant role in determining congenital malformations.

The consequences were an increase of basic, clinical and epidemiological research in the fields of environmental teratogenesis, and the worldwide launch of restrictions and controls (experimental, epidemiological, and clinical) with the intention of avoiding the introduction of teratogenic agents into the human environment. Furthermore, teratogen information services (TIS) were introduced in the 1980's; consequently, information resources such as Reprotox and TERIS (Friedman and Polifka, 1994; Scialli *et al.,* 1995) were available.

Congenital malformations are a heterogeneous group of individually rare conditions. The birth prevalence of congenital malformations is about 3-5%. Both genetic and environmental factors can cause birth defects. However, the causes of about 60 to 70 percent of birth defects currently are unknown. Only 2-3% of birth defects are classified as teratogen-induced malformations, and they appear to be caused by a combination of one or more genes and environmental factors such as drug or alcohol abuse, infections, or exposure to certain medications or other chemicals.

The aim of this chapter is to discuss the principles of teratology and to present the characteristics of teratogens and the methodology used to identify new teratogens. Almost all of the data presented here is focused on drug exposure (De Santis *et al.,* 2004) and infectious diseases, which are the main interests of clinical teratology. In fact, limited data is available on environmental factors such as air pollution, water, soil, and food. The studies on these latter issues regard mainly the dioxin exposure (in particular the Seveso disaster) and waste sites, which are the subjects of other chapters of this book. It is very difficult to define, according to the methodology used in clinical teratology, the exact exposure in pregnant women as well as the control samples. It is difficult to compare experimental animal data to human exposure.

2. Teratogens

Teratogens are environmental agents, which can affect normal prenatal development. Teratogens can be physical agents (for example X-rays or temperature), chemicals (for example drugs, pesticides or vitamins), infectious agents (for example the rubella virus) or diseases which the mother is suffering from (for example diabetes) (table 7).

Table 7. Characteristics of teratogens

Agent
Physical, chemical, infectious agent, maternal status/disease
Sensitivity period
1st trimester (cell growth and cell differentiation, embryogenesis),
2nd & 3rd trimester (histogenesis, function and growth)
Factors determining teratogenicity
Dose, genetic predisposition, interaction with other agents
Teratogenic general effects
• Structural abnormalities
- Malformations
• Fetal and infant mortality
- Miscarriage, embryolethality, stillbirth
• Impairment of physiologic function
- Endocrinopathy, deafness, blindness, neurodevelopmental effects, impairment of reproduction function
• Altered growth
- Growth retardation or enhancement, delayed or early maturation
Teratogenic specific effects
Specific anomalies, syndromes
Risk
Absolute risk, relative risk

Teratogenic exposure can affect prenatal development by altering gene expression, cell proliferation or migration, programmed cell death, function and production of proteins and metabolism. Some agents can act directly on the embryo/fetus, whereas others operate by altering the maternal metabolism. The outcome can be a morphological/structural defect or a functional anomaly; it may already be evident at birth or appear later in childhood or adulthood.

In the years after the thalidomide tragedy, several aspects of teratology were developed:

- Basic principles of teratology,
- Methodology for preclinical studies, in particular animal studies,
- Pharmacokinetics in pregnancy,
- Classification of drugs used in pregnancy,
- Institution of Teratogen Information Services (TIS).

2.1. Basic principles of clinical teratology

After Wilson's definition (1977), the basic principles which determine the potential of an environmental agent to induce morphological (i.e. structural malformation) or functional anomalies in an exposed embryo/fetus have been developed (Finnel, 1999).

The principles are:

a. Susceptibility to teratogenic agents depends on the genotype of the mother and/or the fetus. Such gene-teratogen interactions may explain why the range of effects seen after identical exposures can be so broad. The best-known example is the thalidomide tragedy. It has been estimated that only about 50% of the exposed fetuses developed the thalidomide embryopathy.

b. Susceptibility to teratogens depends on the stage of pregnancy during which teratogenic exposure occurs. This means that the same agent can produce specific abnormalities at specific times during gestation. Furthermore, there is a particularly sensitive time period for every teratogenic agent (for example thalidomide, coumarin or the rubella virus) and consequently a teratogen does not provoke malformations if there is not an exposure during the sensitive period.

c. Each teratogenic agent acts in a specific way.

d. There is a dose-effect relationship ranging from no effects up to a lethal outcome for most teratogenic agents. However, some exceptions are known, and among these thalidomide is the best example.

e. Teratogenic effects are death, malformations, intrauterine growth retardation and functional anomalies.

The first principle has largely been developed within the last decades. Many studies have demonstrated genetic susceptibility as a major factor in determining the teratogenic effect of environmental factors (Streissguth and Dehane, 1993; Hall *et al.,* 1997; Finnel *et al.,* 2002). In the last few years, the genomic sequences of different organisms have become available and subsequently it is now possible to search for changes in gene-expression in early stages of development related to specific exposures (Lee *et al.,* 2004; Nemeth *et al.,* 2005).

2.2. Animal studies

Animal studies are important because, in some instances, they have shed light on the mechanisms of teratogenicity because when an agent causes similar patterns of anomalies in several species, human teratogenesis should also be suspected.

The various endpoints considered in these studies are fertility, embryo lethality, organogenesis/teratogenesis, growth retardation, and postnatal physiological, biochemical, developmental, and behavioural effects. Different species (i.e. rats, mice, rabbits and dogs) should be investigated because of a well-known species-specific sensitivity.

Mechanism of action studies should always be initiated to determine the active metabolites that result in deleterious effects, if any, and in order to find out whether animals and humans do respond similarly to the toxin and its metabolites.

In animal studies, toxic exposure should occur in the same way as it occurs in humans. Animals should be exposed to different dosages, including those affecting humans. Pharmacokinetics, metabolism, excretion and tissue concentration of the substance should be determined.

However, the results of animal studies can not be used to determine the risk in humans (Brent, 2004), and it must be taken into consideration that:

a. the side-effects could be different in animals and humans,

b. an agent proven safe in animals could be deleterious in humans,

c. an agent deleterious in animals could be safe in humans.

2.3. Pharmacokinetics in pregnancy

Maternal physiologic changes in pregnancy may alter the pharmacokinetics of drugs and result in a decrease of the blood concentration of most pharmaceutical products. In fact the total body water increase causes a shift of the concentration of protein-binding drugs and the drugs themselves. In addition, many physiological modifications in the maternal body modify the clearance rate, mainly in late pregnancy, of many substances (in particular drugs) (Loebstein *et al.,* 1997). These modifications include the following:

- changes in total body weight and body fat composition,
- delayed gastric emptying and prolonged gastrointestinal transit time,
- increase in extra cellular fluid and total body water,
- increased cardiac output, increased stroke volume, and elevated maternal heart rate,
- decreased albumin concentration with reduced protein binding,
- increased blood flow to the various organs (e.g. kidneys, uterus),
- increased glomerular filtration rate,
- changes in hepatic enzyme activity, including phase I CYP450 metabolic pathways (e.g. increased CYP2D6 activity), xanthine oxidase, and phase II metabolic pathways (e.g. N-acetyltransferase).

In addition, the placenta metabolism, including the distribution within the fetus, has to be taken into consideration (Shiverick *et al.,* 2004).

All of these subjects have already been studied in animals.

A significant amount of pharmacological research has been conducted to improve the quality and quantity of the data available for other altered physiologic states (e.g. in patients with renal and hepatic disease) and for other patient subpopulations (e.g. pediatric patients). The need for pharmacokinetic, toxicokinetic, pharmacodynamic studies in pregnancy is no less than for these populations, nor is the need for the development of therapeutic treatments for pregnant women.

2.4. *Methods to identify human teratogens*

There are different theoretical approaches to the identification of an environmental agent as a human teratogen: laboratory screening methods, animal studies, case-reports, clinical observations and epidemiological studies.

Seldom, if ever, have teratogens been identified following designed epidemiological studies. Usually an increased prevalence of particular birth defects leads to the discovery of a teratogenic agent.

The best-known example is the thalidomide tragedy; independently identified by clinicians in Australia, Germany and England in the early 1960s (McBride, 1961; Lenz, 1962; Smithells, 1962) and confirmed by epidemiological studies 2 years later. Recently the methimazole teratogenicity was first recognized by clinicians (Clementi *et al.,* 1999) and only later confirmed by a multicentrical study (Di Gianantonio *et al.,* 2001). The list of human teratogens identified by clinicians includes: rubella virus, aminopterin, trimethadione, PCBs (polychlorinated biphenyl), coumarin, alcohol, lithium, diethylstillbesterol, misoprostol, and many more.

Case-reports are often considered anecdotal, but they stimulate further "*ad hoc*" research. They carefully describe defects in children with proven exposure to particular agents at the sensitive stage of pregnancy (i.e. when the organ is developing). Epidemiological studies, on the other hand, often have the statistical power to confirm or exclude hypotheses. Including large numbers of exposed individuals, they analyse confounding factors, and use appropriate control groups.

Epidemiological studies may either be retrospective in nature or prospective cohorts' studies. Prospective studies minimise recall bias, but are time-consuming. It takes a lot of time to collect enough data to give statistically significant results and they are not useful for evaluating the relative risks of rare malformations. Even large multicentrical prospective studies often fail to provide conclusions for low-risk teratogens, i.e. an agent producing congenital defects in less than 10 infants among 1,000 prenatal exposures (Shepard, 2002).

On the other hand, prospective data permits an evaluation of almost all environmental effects; covering the entire spectrum of developmental outcomes and including the frequency of spontaneous abortion.

Retrospective case-control studies (data usually obtained from congenital malformation registries or hospital records) are less costly and easier to conduct, but they have other weaknesses, such as the inaccuracy of data collected from medical records and recall bias. However, for rare malformation/rare exposure, the case report method is commonly used. The two methodologies can complement one another (Martinez-Frias and Rodriguez-Pinilla, 1999; Ornoy and Mastroiacovo, 2000).

2.5. Classification of drugs used in pregnancy

About 80% of pregnant women all over the world use at least one drug on prescription or drugs bought over the counter. In many cases, the exposure is very early, especially for unplanned pregnancies, where the mother is unaware of the pregnancy.

Proper prescribing of drugs in pregnancy is a challenge and should provide maximum safety to the fetus as well as therapeutic benefit for the mother. However, the prescribing physicians often find themselves in difficulty. Safety warnings on package leaflets as well as databases, such as the "Physicians' Desk Reference" are out-dated and in some cases even misleading. The label "contraindicated in pregnancy" can signify two things: toxic for the embryo/fetus or no scientific data on humans available so far.

Since the early 1980s, classification systems in Sweden (1978), the USA (1979), and Australia (1989) have set up risk assignment systems with the intention of guiding physicians to a better assessment of fetal risk. These classifications have been based on data obtained from human and animal studies. The three systems comprise of different categories: the "safest drugs", drugs only used by a limited number of pregnant women but without known increase of reproduction toxicity in animals; drugs that may have certain risks for the fetus; and drugs demonstrated to be teratogenic.

These systems have been useful in earlier years for easy and fast consultations. However, the given information was found to be insufficient for individual assessments of embryo-fetal risks. Scarce and ambiguous information still renders it difficult to estimate teratogenic risks and developmental toxicity. Recently the three different systems have been criticized for differences in the sub classification of drugs; and thus are limited regarding usefulness and reliability (Addis *et al.,* 2000; Merlob and Stalh, 2002).

2.6. *Criteria for proof of human teratogenicity*

The criteria for "proof of human teratogenicity" has not been defined, although an amalgamation of criteria has been proposed and published in each edition of the *Catalog of Teratogenic Agents* since 1986 (Shepard, 2002) (table 8).

Table 8. Major criteria for proof of human teratogenicity (modified from Shepard, 2002)

1. Proven exposure to agent at critical gestational period
2. Consistent findings by two or more epidemiologic (possibly prospective) studies including exclusion of bias, control of confounding factors, large samples, careful definition of major/minor defects
3. Definition of clinical findings for rare defects associated with rare exposures
4. Teratogenic effect should make biologic sense

There must be proven exposure to a specific agent at critical times in prenatal development and at least two independent epidemiological studies should have evidence of a statistical correlation between exposure and the increase of specific congenital malformations (isolated or associated in a recognizable syndrome). The studies should comprise of large numbers of exposed individuals, random control groups and should try to eliminate bias and confounding factors.

For rare exposures associated with rare malformations, the presence of well-delineated single case reports is usually considered sufficient (for example the exposure to methimazole associated with choanal and/or oesophageal atresia) (Clementi *et al.,* 1999; Di Gianantonio *et al.,* 2001).

2.7. *Teratogen Information Services (TIS) in Europe*

TIS were set up in the 1980s in order to provide public and health professionals with appropriate information on drug risks during pregnancy. Until then, data on prenatal exposure to possible teratogenic agents, and drugs in particular, had not been available easily and were difficult to interpret (Clementi *et al.,* 2002; Felix *et al.,* 2004; Schaefer *et al.,* 2005).

Pregnant women and health professionals of various disciplines (in particular general practitioners, obstetricians, and midwives) do seek advice from TIS. Counselling can be requested under various circumstances, requiring different approaches to risk assessment and communication:

1. if pharmacotherapy is intended in pregnancy, usually because of a maternal chronic disease,
2. when exposure has taken place in pregnancy,
3. when a newborn, exposed prenatally to a particular agent, is affected with a malformation and a specific diagnosis is required.

TIS units vary in staff size, methodology and the population size they account for. Some provide information mostly to physicians, others to the public. Different ways of counselling have been established within the various TIS, with different emphasis on written or oral counselling. In some units, follow-up is limited to special exposures, whereas others try to follow every eventual outcome.

TIS carry out studies with data prospectively ascertained from earlier pregnancies and can give results on morphological defects (major malformations), pregnancy outcomes (spontaneous abortions, late fetal death, stillbirth, and live births), intrauterine growth, and neonatal effects (such as sedation, hypoglycaemia, heart rhythm anomalies, and withdrawal or toxicity symptoms).

In 1990, members of 13 European programmes engaged in pregnancy counselling and risk evaluation met in Milan and founded the European Network of Teratology Information Services (ENTIS). This institution is registered in the Netherlands, at the National Institute of Public Health and Environmental Protection (RIVM) in Bilthoven. At present, it is made up of 33 active members.

The main objectives and functions of ENTIS were defined as follows:

1. to recognise, to detect and to prevent causes of birth defects,
2. to stimulate and facilitate the exchange of experience between the different TIS in Europe,
3. to improve skills in counselling and risk assessment in pregnancy,

4. to provide training by giving information on teratology,

5. to create joint solutions for counselling procedures, databases and terminology,

6. to hold annual meetings, in order to enhance scientific collaboration and improve TIS specific organisational procedures,

7. to create and maintain an ENTIS website (www.entis-org.com) for the public and health professionals and for communication among ENTIS members in a password protected forum.

Since its foundation, ENTIS has successfully conducted several studies on drug risks in pregnancy. Nevertheless, for many groups of drugs, a substantial teratogenic risk has not been able to be established so far. Among these are: calcium channel blockers during first trimester, tricyclic antidepressants, selective serotonin reuptake inhibitors, proton pump inhibitors, haloperidol/benfluridol, H2-blocker, mesalazine, Yellow fever vaccination during first trimester, quinolones, and vitamin A supplementation.

One of the major results obtained by the TISes is to alleviate unnecessary and unrealistic fears of pregnant women and to reduce the rate of induced abortions because of involuntary prenatal exposure to drugs (Garbis, 1990).

3. Conclusions

It goes without saying that some pregnant women need to be on medication and the avoidance of any treatment after conception may often be unwise for maternal well-being. Therefore the prescription of drugs in pregnancy should provide maximal safety to the fetus as well as therapeutic benefit to the mother.

The embryo or fetus represents an additional "patient" with specific and defined pharmacometabolical needs. In the case of maternal treatment, the fetus, although not requiring therapy itself, is exposed to the therapy and may suffer pharmacological effects.

Therefore, the first rule in treating pregnant women/women of reproductive age is to only prescribe drugs sufficiently tested for their eventual toxicity and teratogenicity.

In recent decades, clinical teratology has been a field of interest for geneticists, embryologists, obstetricians, pharmacologists and many other specialists.

The list of known teratogens includes low-risk teratogenic agents and therefore accurate counselling, risk assessment and appropriate backup therapy can be offered to pregnant women. Many web sites provide information to public and physicians (table 9).

Table 9. Websites useful on clinical teratology

Address	Characteristics
www.entis-og.com	Web site of ENTIS
www.otispregnancy.org	Website of OTIS
www.teratology.org	Teratology society
www.etsoc.com	European Teratology Society
www.motherisk.org	Motherisk program, Canada
www.ifts-atlas.org/ifts/index.html	Atlas of congenital malformations

However, new substances and drugs are being introduced into the market every day. Women suffering from chronic diseases or malformations are able to become pregnant, thanks to substantial medical progress (for example transplant recipients, patients affected with phospholipid syndrome). Unfortunately, in these cases, laboratory tests and animal studies are of no help. Epidemiological studies and careful clinical evaluations will eventually provide information.

New molecular technologies will initiate a different development of clinical teratology (Polifka and Freidman, 2002). In recent years, genetic polymorphisms and variants of drug-metabolising enzymes that may produce significant alteration in clinical responses to drug treatment have been identified in humans. Genetic polymorphisms that result in changes of the elimination of drugs or other agents could be expected to increase teratogenicity.

In the future, pharmacogenomics will unearth new evidence for the understanding of how certain genetic polymorphisms influence teratogenic processes in humans.

References

Addis, A., Sharabi, S. and Bonati, M. (2000) Risk classification systems for drug use during pregnancy. Are they a reliable source of information? *Drug Safety* **23**, 245–253.

Brent, R.L. (2004) Utilization of animal studies to determine the effects and human risks of environmental toxicants (drug, chemicals, and physical agents), *Pediatrics* **113**, 984-995.

Clementi, M., Di Gianantonio, E. and Ornoy, A. (2002) Teratology information services in Europe and their contribution to the prevention of congenital anomalies, *Comm. Genet.* **5**, 8-12.

Clementi, M., Di Gianantonio, E., Pelo, E., Mammi, I. and Tentoni, R. (1999) Methimazole embryopathy. Delineation of the phenotype, *Am. J. Med. Genet.* **83**, 43-46.

De Santis, M., Straface, G., Carducci, B., Cavaliere, A.F., De Santis, L., Lucchese, A., Merola, A.M. and Caruso, A. (2004) Risk of drug-induced congenital defects, *Eur. J. Obst. Gynec Repr. Biol.* **117**, 10-19.

Di Gianantonio, E., Schaefer, C., Mastroiacovo, P.P., Cournot, M.P., Benedicenti, F., Reuvers, M., Occupati, B., Robert, E., Bellemin, B., Addis, A., Arnon, J. and Clementi, M. (2001) Adverse effects of prenatal methimazole exposure, *Teratology* **64**, 262–266.

Felix, R.J., Jones, K.L., Johnson, K.A., McCloskey, C.A., Chambers, C.D. and For the Organization of Teratology Information Services (OTIS) Collaborative Research Group, (2004) Postmarketing surveillance for drug safety in pregnancy: the organization of teratology information services poject, *Birth Defects Res.* **70**, 944–947.

Finnell, R.H. (1999). Teratology: general considerations and principles, *J. Allergy Clin. Immunol.* **103** Suppl 2, 337-342.

Finnell, R.H., Waes, J.G.V., Eudy, J.D. and Rosenquist, T.H. (2002). Molecular basis of environmentally induced birth defects, *Ann. Rev. Pharmacol. Toxicol.* **42**, 181-208.

Friedman, J.M. and Polifka, J. (1994) *Teratogenic effects of drugs: A resource for clinicians (TERIS)*. Baltimore, MD: Johns Hopkins University Press, USA.

Garbis, J.M., Robert, E. and Peters, P.W.J. (1990). Experience of two teratology information services in Europe, *Teratology* **42**, 629–634.

Gregg, N.M. (1941) Congenital cataract following German measles in the mother, *Trans. Ophthalmol. Soc. Aust.* **3**, 35-46.

Hale, F. (1933) Pigs born without eyeballs, *J. Hered.* **24**, 105-106.

Hall, J.L., Harris, M.J. and Juriloff, D.M. (1997) Effect of multifactorial genetic liability to exencephaly on the teratogenic effect of valproic acid in mice, *Teratology* **55**, 306-313.

Lee, R.D., Rhee, G.S., An, S.M., Kim, S.S., Kwack, S.J., Seok, J.H., Chae, S.Y., Park, C.H., Yoon, H.J., Cho, D.H., Kim, H.S. and Park, K.L. (2004) Differential gene profiles in developing embryo and fetus after *in utero* exposure to ethanol, *J. Toxicol. Environ. Health Part-A* **67**, 2073-2084.

Lenz, W. (1962) Thalidomide and congenital malformations, *Lancet* **1**, 271-272.

Loebstein, R., Lalkin, A. and Koren, G. (1997) Pharmacokinetic changes during pregnancy and their clinical relevance, *Clin. Pharmacokinet.* **33**, 19-24.

Martinez-Frias, M.L. and Rodriguez-Pinilla, E. (1999), The problems of using data from teratology information services (TIS) to identify putative teratogens, *Teratology* **60**, 54–55.

McBride, W.G. (1961) Thalidomide and congenital malformations, *Lancet* **16**, 79-82.

Merlob, P. and Stahl, B. (2002) Classification of drugs for teratogenic risk: An anachronistic way of counselling, *Teratology* **66**, 61-62.

Nemeth, K.A., Singh, A.V. and Knudsen, T.B. (2005) Searching for biomarkers of developmental toxicity with microarrays: Normal eye morphogenesis in rodent embryos, *Toxicol. Appl. Pharmacol.* **206**, 219-228.

Ornoy, A. and Mastroiacovo, P. (2000) More on data from teratogen information systems (TIS), *Teratology* **61**, 327–328.

Polifka, J.E. and Friedman, J.M. (2002) Medical genetics: 1. Clinical teratology in the age of genomics, *Can Med. Assoc. J.* **167,** 265-273.

Schaefer, C., Hannemann, C. and Meister, R. (2005) Post-marketing surveillance system for drugs in pregnancy - 15 years experience of ENTIS, *Repr. Toxicol.* **20**, 331-343.

Scialli, A.R., Lione, A. and Boyle Padgett, G.K. (1995) *Reproductive effects of chemical, physical and biologic agents. Reprotox*, Baltimore, MD: Johns Hopkins University Press.

Shepard, T.H. (2002) Annual commentary on human teratogens, *Teratology* **66**, 275-277.

Shiverick, K. T., Slikker, Jr W., Rogerson, S. J. and Miller, R. K. (2004) Drugs and the Placenta—A Workshop Report, *Placenta* **24**, S55-S59.

Smithells, R.W. (1962) Thalidomide and congenital malformations in Liverpool, *Lancet* **1,** 1270-1273.

Smithells, R.W. (1980) The challenges of teratology, *Teratology* **22**, 77-85.

Streissguth, A.P. and Dehaene, P. (1993) Fetal alcohol syndrome in twins of alcoholic mothers: Concordance of diagnosis and IQ, *Am. J. Med. Genet.* **47**, 857-861.

Warkany, J. and Nelson, R.C. (1940) Appearance of skeletal abnormalities in the offspring of rats retarded on a deficient diet, *Science* **92**, 383-384.

Wilson, J.D. (1977) *Embryotoxicity of drugs in man*, In J.D. Wilson, and F.C. Frazer (eds) Handbook of teratology, New York, USA, Plenum press, 309-355.

SECTION 2:

TERATOGENS

DIOXINS AND CONGENITAL MALFORMATIONS

J.G. KOPPE[1*], M. LEIJS[2], G. TEN TUSSCHER[2,3] AND P.D. BOER[1]
[1] *Emeritus Professor of Neonatology*
Ecobaby Foundation
Hollandstraat 6
Loenersloot
The NETHERLANDS
[2] *Department of Paediatrics and Neonatology*
Emma Children's Hospital Academic Medical Centre
University of Amsterdam
Amsterdam
The NETHERLANDS
[3] *Department of Paediatrics and Neonatology*
Westfries Gasthuis
Maelsonstraat 3
Hoorn
The NETHERLANDS
(author for correspondence, Email: janna.koppe@inter.nl.net)*

Summary

Hot spots of pollution have shown that dioxins can cause congenital malformations, prematurity and intra-uterine growth retardation. In regions with background concentrations, like in Europe, there has generally been no indication of an increase in congenital malformations in relation to dioxins.

Specific types of malformations, such as cleft lip and palate, were found to be associated with exposure to a mixture of chemicals in clouds formed by open burning of chemical waste in Amsterdam, The Netherlands. While such defects have been seen in rodents, occurrence in humans had previously not been documented. In a similar fashion, renal abnormalities like agenesis and dysgenesis were found in Chapaevsk, also previously only noted in animal experiments (See sections on Amsterdam Diemerzeedijk and Chapaevsk).

P. Nicolopoulou-Stamati et al. (eds.), Congenital Diseases and the Environment, 163–181.

The finding in Chapaevsk of a high incidence of *congenital hydrocephaly* without spina bifida 4/1,000 births is remarkable. Congenital hydrocephaly is a rare condition, in general 2.0-3.9/10,000 births (Abbott, 1995). However, this specific malformation is also found in the offspring of mothers with type 1 diabetes. Abnormal glycosylation is assumed to play a role and cause congenital malformations as expressed in an increased level of Hemoglobin A 1c (glycosylated hemoglobin) in mothers with type I diabetes.

We suggest that in both situations (Chapaevsk and pregnant women with type 1 diabetes) the same metabolic mechanism might be the cause of malformations.

Animal experiments have shown that dioxins can deplete vitamin A stores in the liver. In other animal experiments, vitamin A can counteract negative effects of PAHs, including dioxins. This vitamin is necessary for glycosylation. And the vitamin might be important as a *prevention*, the more so because recent research in the Netherlands revealed a 48% deficiency of this vitamin in the diet of women trying to fall pregnant. In general, more attention to keeping a balanced vitamin A-status in general, and in women with type 1 diabetes especially, is recommended.

1. Introduction

Dioxins and dioxin-like polychlorinated biphenyls (PCBs) are halogenated hydrocarbons, members of a group of structurally similar environmental pollutants, the Persistent Bioaccumulating Toxicants (PBTs). Other PBTs include polybrominated biphenyls (PBBs) and perfluoro-octane-sulfonate (PFOS). PCBs were commercially mainly used as additives to hydraulic oils and plasticisers. The industrial manufacture of PCBs was essentially banned in 1977 but they are also by-products of many chemical and thermal reactions containing organic substances and chlorine, and are produced by smelt furnaces and waste incinerators. They are highly persistent in the environment and found in many levels of the food chain, notably in fish and animal fat (Koppe and Keys, 2001). Some PCB congeners have effects like dioxins and are thus referred to as dioxin-like PCBs.

The term "dioxins" refers to the group of polychlorinated dibenzo-p-dioxins (PCDDs) and dibenzofurans (PCDFs). The most toxic form of dioxin is 2,3,7,8-tetrachlorodibenzo-p-dioxin (2,3,7,8 TCDD). Dioxins are not produced commercially, with the exception of small quantities for research, but are formed as unwanted by-products during combustion processes involving chlorine, such as waste incinerators, in metal processing or in industrial processes like the production of insecticides. As a result of reduction measures, especially for incinerators at the end of the 1980s, dioxin emissions in Western Europe have decreased (Zeilmaker *et al.,* 2002).

The toxicity of dioxins and the dioxin-like PCBs are generally expressed in Toxic Equivalent (TEQ). Every congener has a specific Toxic Equivalency Factor (TEF) relative to 2,3,7,8 TCDD, which has a TEF of 1. This is a system for ranking the relative potencies of the dioxin-like activity of the different congeners so that the toxic effect of mixtures of dioxin-like chemicals can be summated. I-TEQs refer to international TEQs; however, WHO TEQs are currently used.

It must be borne in mind, however, that in the environment and in animals and humans, most PCBs are non-dioxin-like, and hence do not contribute to the dioxin TEQ. These latter chemicals do not use the aryl hydrocarbon receptor pathway, which is used by the dioxin-like PCBs.

In Europe, most studies address the exposure to *background levels of dioxins* as found in the European countries. Various abnormalities have been found, some transient - like the increase in liver transaminases in relation to postnatal exposure and reduced thyroid hormone, as expressed by an increase in thyroid stimulating hormone (TSH) in relation to prenatal exposure. Others effects seem to be persistent, like an abnormal lung function (restriction and obstruction) and neurophysiologic abnormalities, like an increase in the P300 latency after visual evoked potentials. Furthermore, a persistent (until the age of 8 years) decrease in the number of thrombocytes, and immunologic abnormalities, in different mother-baby cohorts have been documented (ten Tusscher and Koppe, 2004; ten Tusscher *et al.,* 2003; ten Tusscher *et al.,* 2001; ten Tusscher, 2002; Patandin *et al.,* 1999; Patandin *et al.,* 1998; Koopman-Esseboom *et al.,* 1994; Vreugdenhil *et al.,* 2002). However, no overt congenital malformations have been described in relation to these background levels.

In contrast, higher doses of dioxins, as found in hot spots, have been related to congenital malformations in human beings. In striking similarity to children of diabetic mothers, high dioxin exposure has been associated with congenital hydrocephaly without spina bifida. This might indicate a similar aetiology in mothers of type I diabetes and mothers exposed to high levels of dioxins (discussed later in this paper).

Acute dioxin poisoning will first be addressed, followed by a review of hot spots of dioxin pollution in relation to congenital malformations.

2. Acute dioxin poisoning

2.1. Animal studies

The most commonly seen acute effects of exposure to TCDD in animals are lethality, wasting syndrome, enlargement of the liver, thymic atrophy, skin abnormalities and endocrine effects. Animals die 1-6 weeks after exposure to TCDD (McConnell, 1984).

The wasting syndrome is characterized by weight loss or reduced weight gain and a depletion of adipose tissue. Rats exhibit a tendency to hypothermia.

There is an enlargement of the liver based on hyperplasia and hypertrophy of parenchymal cells and there is an increase in microsomal mono-oxygenase activity.

Most TCDD is excreted in the feces. Abnormal hyperplasia of the gastric epithelium and the urothelium is described.

In all animals studied, TCDD affects immune function, and thymus atrophy is well known (Vos, 1984).

One of the most pronounced endocrine effect of TCDD in rats is the strong decrease in thyroxine levels. There is depletion in the liver of vitamin A stores and effects on the skin; for example hyperkeratosis and immune suppression resemble vitamin A deficiency. There is a good relation between acute toxicity of TCDD and its effect on vitamin A status (depletion) in rats, guinea pigs and hamster (Thunberg, 1984). Congenital malformations described in animals are hydronephrosis in mice and rats and cleft palate in mice (Pratt, 1984).

2.2. Acute dioxin toxicity in man

2.2.1. Yushchenko

Chloracne is seen as the hallmark of acute dioxin poisoning, recently illustrated in the well known case of the poisoning of Yushchenko.

In 2004 Yushchenko was intoxicated with 1-2 mg of dioxin. He suffered gastro-intestinal complaints, developed severe chloracne and disturbances of many vital systems in his body. It takes some time to develop chloracne, as is seen in victims of Seveso, and this makes the date of the poisoning uncertain. It was told by the

Austrian doctors that many systems in his body were dysregulated. However, precise information is not available.

2.2.2. *Two secretaries*

Other examples of severe poisoning are the cases of two secretaries in Austria. In the autumn of 1997, two secretaries commenced working in a new building. Shortly thereafter both developed signs of dioxin intoxication (Geusau, 2001). A 30 year old woman developed a centro-facial chloracne after intoxication with about 1.5mg dioxin, while the other woman, with a lower intake of 0.4mg, had gastro-intestinal symptoms during several months following the autumn of 1997. Chloracne spreads over the whole body and is very itchy. A year after intoxication, Patient 1 had only a few pustules on her back, but a year later her whole back was covered with many severely inflamed cysts. The level of dioxin in the first secretary was 144.00 pg/g fat in blood, and of the second 26.000 pg/g fat in blood.

In patient 1, a moderate elevation of blood lipids (cholesterol and triglycerides), leukocytosis, anaemia, secondary amenorrhoea and a decrease in the number of thrombocytes (74 x 109 /l) was found.

In patient 2, these parameters were almost normal, although she also demonstrated a decline in the number of thrombocytes: lowest measured 154 x 10^9 /l.

Chloracne is a severe skin condition with both hyperkeratotic and hyperproliferative responses, that is, both altered differentiation and proliferation of the epidermis. In addition, an inflammatory response is often seen in the dermis. A proliferative response of multiple cell types in the dermis is involved.

2.3. ***Hormones***

The endocrine system seems to be vulnerable to chemical effects. Free thyroxine (FT4) was decreased in both secretaries, while in the less intoxicated secretary the thyroid stimulating hormone (TSH) was increased. DHEAS (dehydro-epi-androsterone), FSH (follicle stimulating hormone), estradiol, progesterone and testosterone were all decreased to subnormal levels in patient 1; patient 2 was normal in this respect except for progesterone. Prolactin was increased in both patients.

The scales of the skin were highly polluted with dioxins and both secretaries probably lost much dioxin via this route (Geusau *et al.,* 2002). The half-life of their dioxin concentration was much shorter than usual, probably because of this effect and possibly through increased (stimulated) liver enzyme activity (1.5 and 2.9 years respectively in secretary 1 and 2 versus 7 years at low background concentrations) (Geusau *et al.,* 2002).

2.4. Thrombocytopenia

Thrombocytopenia is an early characteristic parameter of dioxin intoxication. Thrombocytopenia was seen in BASF workers in 1953 (Ott *et al.,* 1994), in Japanese workers (Watanabe *et al.,* 2001), in the above mentioned secretaries and in The Netherlands in relation to background exposure to dioxins (ten Tusscher *et al.,* 2003). This might be explained by a shift in the bone-marrow progenitor cells from megakaryocytes to erythroblasts caused by an accumulation of delta-aminolaevulinic acid (ALA, figure 2). Figure 2 is based on a compilation of information in the articles of Piomelli and of Sassa (Piomelli, 1993; Sassa *et al.,* 1984).

The figure of the heme synthesis pathway demonstrates the route. PCBs and dioxins induce ALA-synthetase, resulting in an increase in ALA, which is further enhanced by the inhibiting effect of dioxin on the URO-decarboxylase

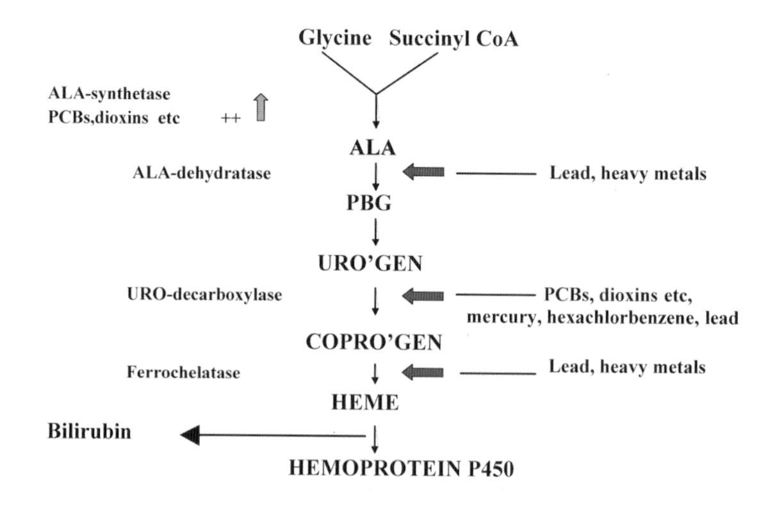

Figure 2. Heme synthesis

The prognosis of acute poisoning is not good. Chloracne disappears, after an initial worsening, slowly during 25 years. Life expectancy is shortened by about 8 years. Dioxins increase the aging process (Karamova *et al.*, 2001).

3. Hotspots of dioxin pollution

3.1. Amsterdam Diemerzeedijk 1961-1973

In 1961, open chemical combustions commenced at the "Diemerzeedijk". In the period until 1970, between 4,000 and 14,000 tons per year were burned. The chemical waste came from all over Europe, including Germany, Czechoslovakia and Finland. Every three months the fires were started and many complaints from the inhabitants in the surrounding areas were voiced, regarding the smell and the sedimentation powders that fell over their areas. The site turned out to be highly polluted with dioxins. Serendipitously, an increase in the number of children with non-syndromal orofacial clefts was found in this area, close to the place of incineration. Seven per thousand orofacial clefts were seen in the study area versus one to two per thousand in the control area. Levels of dioxins were not measured at the time. The sharp rise in the incidence of cleft lip and palate came as an unpleasant

surprise. In animal studies, dioxins have been shown to cause cleft lip and palate (ten Tusscher *et al.*, 2000; Abbott, 1995).

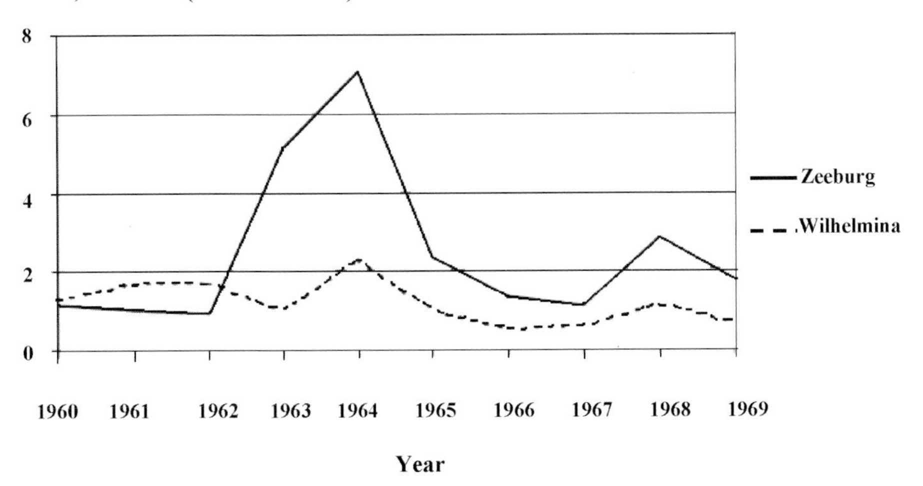

Figure 3. Data from Zeeburg maternity clinic in proximity to open burning of chemicals, Wilhelmina Hospital far west of the incineration site

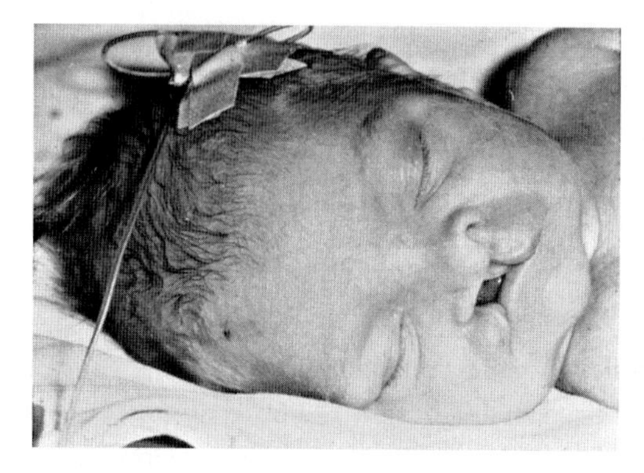

Figure 4. Baby with non-syndromal orofacial cleft lip and palate (provided by Janna G. Koppe)

3.2. Yusho-disaster 1968

In 1968 an epidemic of a strange disease occurred in Western Japan. The epidemic concerning 18,000 people was caused by a brand of commercial rice-oil contaminated with PCBs and related compounds. Clinical signs in adults were pigmentations of skin, nails and mucous membranes, together with acne-like eruptions on the skin and an increased eye discharge. Some victims also experienced vomiting, diarrhoea and jaundice. The disease was named "Yusho", meaning "oil disease". 13 pregnant women, intoxicated during pregnancy, transferred the toxic compounds to their fetuses with serious effects in the offspring: an increase of congenital malformations and perinatal deaths. The pronounced skin pigmentation of some affected newborns resulted in their name: "Coca-Cola" babies. This often occurred in combination with pigmented nails and gingiva. The babies were growth retarded. Two of the thirteen babies were stillborn. At follow-up, these children were apathetic and dull with I.Q.s in the seventies (normal is 100).

3.3. Yucheng-disaster 1978

Ten years later, in 1978, a similar disaster took place on Taiwan with more pregnant women involved. Once again around 2,000 people were intoxicated. An epidemiologic study was done in 117 children born to women with Yucheng (Rogan *et al.,* 1988). At birth the children were generally growth retarded, had dysplastic nails (similar to that seen in phenytoin syndrome), antenatal teeth, spotty calcifications of the skull, widely open fontanels and conjunctivitis. Many developed severe respiratory problems in the first year of life and 20 percent died before the age of four due to this. The children have poorer cognitive development and the

P300 latencies were significantly longer, suggesting that a higher cortical function was affected than the sensory pathway. (Chen, 1990; Chen *et al.*, 1992). There was a catch-up growth in these children. However, hearing impairment and abnormal sexual development (smaller penis) have been described (Guo *et al.*, 1993; Guo *et al.*, 2004). Nail changes were persistent until the age of 11 years. Besides PCBs, the rice-oil was also contaminated with dioxins, formed after heating of the oil, similar to in the YUSHO disaster. Abnormalities are probably not only related to PCBs, but to dioxins formed by heating of the PCB-oil.

3.4. Seveso-disaster 1976

Near the small town Seveso, in Northern Italy, the ICMESA chemical plant was located. It produced intermediate compounds for the cosmetics and pharmaceutical industry, including 2,4,5 trichlorophenol (TCP), a toxic, nonflammable compound used for the chemical synthesis of herbicides. On Saturday, 10 July 1976, a major chemical accident occurred: a hazardous emission during a period of twenty minutes. A large area including the communities of Seveso, Meda, Desio, Cesano and Maderno (all part of the province of Milan) were contaminated. The air emission contained a total of about 300 kg, including between 300 g and 130 kg of dioxin, pure 2,3,7,8 tetrachlorodibenzo-dioxin. Three days later, it was learned that many small animals such as cats and poultry had died in the area, while dogs were resistant. A week after the incident, Prof. Aldo Cavallaro made the first diagnosis of a dioxin intoxication. The most striking clinical sign was chloracne as demonstrated in the child shown in figure 5.

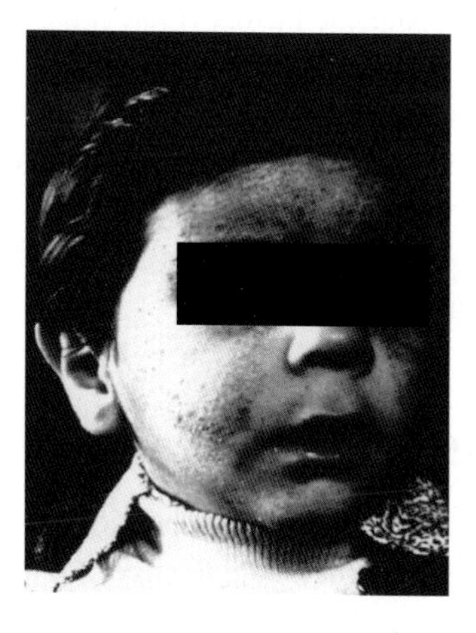

Figure 5. Child living in Seveso showing the characteristic chloracne, developing in the weeks after an acute intoxication (modified from Bertazzi, 1998)

Many pregnant women chose to abort their fetus out of fear of congenital malformations. In a later study, no increase in congenital malformations was noted. However, in the years 1976-1978, the incidence of aberrant cytogenetic findings in miscarriages was increased (Bertazzi, 1998).

A new finding was the change in sex ratio, in the offspring of men intoxicated with dioxins. Sex ratios of the children from exposed fathers and/or mothers were compared. Serum samples were collected from 239 men and 296 women. 346 girls and 328 boys were born to potentially exposed parents between 1977 and 1996, showing an increased probability of female births (lower sex ratio) with increasing TCDD concentrations in the serum samples from the fathers ($p = 0.008$) The effect starts at 15-20 ng per kg bodyweight in fathers and they had to be exposed before the age of nineteen. Then the sex ratio was 0.38 instead of the expected one of 0.514.

On average, the sex-ratio of the children decreases with increasing paternal TCDD concentration (Mocarelli *et al.,* 2000): a dose-response effect.

At follow-up of the Seveso population, an increase in various forms of cancer was seen, although recently this has been denied. Strikingly, in the publication of the book "Twenty years after Seveso", an increase in rectal cancer (Bertazzi, 1998) was described, similar to a Lancet publication of an 11-year old boy exposed to dioxins, in Times Beach Missouri, where he resided from 1979-1983. Anorectal cancer at this age is very rare (1 in 10 million). In 1982 a widespread contamination of the soil took place in Times Beach, due to spraying of unpaved roads with oil containing waste products (TCDD) of a hexachlorophene factory (Pratt *et al.,* 1987).

In a recent publication, an increase in breast cancer is published in the Seveso women, contrary to earlier publications. There is a dose related increase in incidence and especially in premenopausal women (Koopman-Esseboom *et at., 1994*). In addition and epidemiological study on the IARC chlorphenol cohort indicates an increase in breast cancer (Vreugdenhil *et al., 2002*).

3.5. ***Vietnam: Agent Orange spraying 1962-1971***

During the Vietnam war in the period 1962-1971 the herbicide Agent Orange was used by the American and Allied forces to defoliate the trees, in order to better spot the enemy. Spraying took place especially around the outer circumference of an installation/base. Agent Orange, a mixture of the herbicide 2,4,5-trichloro-phenoxyacetic acid (2,4,5-T) and 2,4 dichlorophenoxy-acetic acid (2,4-D), was contaminated with dioxins. TCDD content was not measured, but probably varied between 0.02 and 47 ppm. There is a significant amount of uncertainty about the quantity of dioxins spread. The total quantity of herbicides disseminated in South Vietnam is unknown. In total about 67 million litres were used, of which 45 million were of Agent Orange. In a group of American Vietnam War veterans, the so-called Ranch Hand cohort, a telephone interview was held in 2000, amongst 1499 Vietnam War veterans and 1428 non-Vietnam veterans. Hepatitis, poor current health, limited type and amount of work, and all cancers were significantly increased in the group of Vietnam War veterans. A significant increase in diabetes was seen in relation to current dioxin levels in the blood of 385 participants (332 Vietnam veterans and 53 Non-Vietnam veterans, 7.49 ppt in the high group versus 1.79 ppt in the low group in serum). In contrast, other health conditions mentioned by the interviewed group were not related to current dioxin levels (Michalek *et al.,* 2001a, b).

In an Australian publication of 8,517 case-controls studied, no increase in congenital malformations was found in the group of veterans of the Vietnam Army Service (Donovan *et al.,* 1984).

In the weekly published journal Time (2002), a publication of congenitally malformed children was published, relating this to the spraying of herbicides under the civilians in Vietnam. Short stature and eye abnormalities leading to blindness were seen. The published outcomes have remained controversial, under the guise that the publication was not scientific, coming from a communist country. For more information on Vietnam, see the article in this book addressing Agent Orange, Health Outcomes and Birth Defects.

3.6. ***Bashkortostan: 1965-1985***

Another well-known hot spot is the city of Bashkortostan close to the Russian Urals. From 1965 to 1985 large scale production of the herbicide 2,4,5-T took place in the CHIMPROM factory. Workers came into contact with high doses of dioxins, through the herbicides containing between 30 and 100 mg pure TCDD. In the adults, the typical clinical signs seen were chloracne, increase in cholesterol and triglycerides, hypertension and a well documented state of hypercoagulation and an increase in the hematocrit. Exposed workers died 8 years younger from cancer compared to a control city, UFA, in the neighbourhood, and mortality in general was 8 years younger, indicating that dioxin stimulates the aging process.

Reproduction was disturbed. There were more spontaneous miscarriages: 6.9% versus 5.7% in Ufa. There was a preponderance of girls among the first infants, 120 versus 100: that is, a changed sex ratio similar to that seen in Seveso. A tendency to chromosome aberrations was found in the peripheral blood lymphocytes (Karamova *et al.,* 2001).

3.7. ***Chapaevsk 1967-1987***

This Russian city in the Samara Region has one of the largest documented environmental polluters: the Middle Volga chemical plant. From 1967 to 1987 the factory produced hexachlorocyclohexane (lindane) and its derivatives. Currently, dioxins are found everywhere. The town's drinking water contains between 28,4-74,1 pg/litre. Cow's milk contains levels of 17.32pg TEQ/g fat of pure TCDD. The mean concentration in human milk was 42 pg TEQ/g fat, but in female factory workers, levels in blood were 412 pg TEQ/g fat. Both lung cancer and breast cancer are increased.

Effects on reproduction entailed:

3.7.1. *Prematurity*

Significant disruptions in reproductive function were detected. The spontaneous abortion (miscarriage) rate was 24 %. Premature labour was 45.7% in Chapaevsk, with half of all women delivering prematurely.

In this respect, the finding of devlieger in Belgium is very interesting. He describes an unexplained 40% rise in the incidence of prematurity in the period 1991-2001. This was also seen in France, Norway and the USA. Environmental factors (as is seen in Chapaevsk) are hypothesized to play a role in this unexplained rise and research in this area is urgently needed (Devlieger *et al.,* 2005).

3.7.2. *Intra-uterine growth retardation*

In Chapaevsk, the number of low birth weight babies was 7.4%, higher but not statistically significantly higher than the 5.1 – 6.2% of surrounding cities in the samara region.

3.7.3. *Congenital morphogenetic conditions*

The most frequent congenital morphogenetic conditions seen were epicanthus, shawl scrotum for boys, clinodactyly and broad first fingers of hands. The average rate of dysmorphic signs was 4.4 per child on a scale of 1-10.

3.7.4. Congenital malformations

The average frequency of congenital malformations is 11.8 per 1,000 births and this is lower than published by Eurocat in Europe. But there is a strikingly significant increase in *congenital hydrocephalus*, without spina bifida, and an increase in *agenesis and dysgenesis of the kidney*. The latter being a well known congenital anomaly in relation to dioxin exposure in animals (Revich *et al.,* 2001).

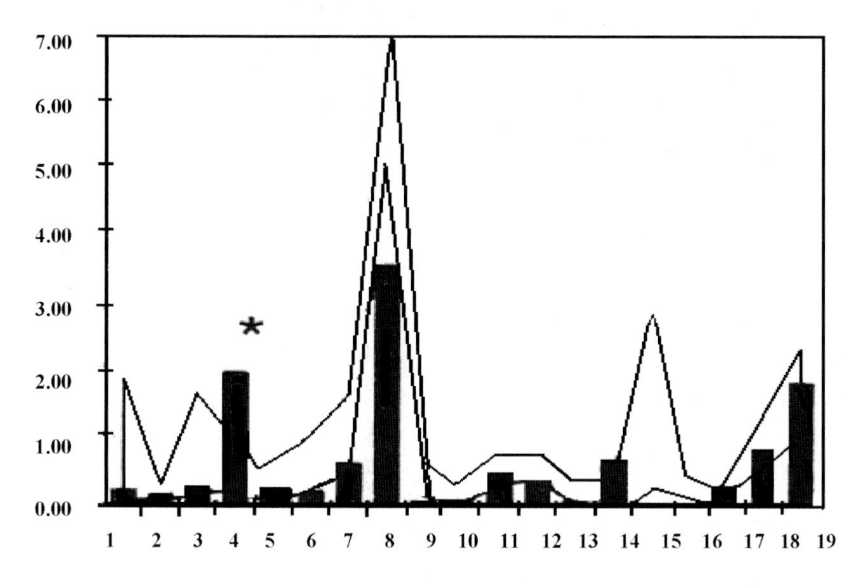

Figure 6. Congenital malformations in Chapaevsk per 1,000 births. Range (Band) Eurocat 1997. Columns Chapaevsk (1982-1997) No 4 = congenital hydrocephaly, "*" = significant different from the Eurocat data, no 14 kidneys: a/dysgenesis (Modifed from Revich *et al.,* 2001)

3.7.5. Congenital hydrocephaly without spina bifida

The specific finding of congenital hydrocephaly (4/1,000 births) in Chapaevsk is interesting, since the abnormality is rare and in general 2.0-3.9/10,000 births. Congenital hydrocephaly without spina bifida is, however, a well-known specific anomaly seen in the offspring of mothers with type I diabetes (figure 7).

In 1885, Lecorché documented two cases of congenital hydrocephalus in the offspring of two mothers with type I diabetes (Lecorché, 1885) long before insulin was used. In Amsterdam, in a group of 173 women with type I diabetes giving birth, a 10% congenital malformation rate was seen during the years 1963-1979 (Koppe *et al.,* 1983) and between 16 malformed children, two had a congenital hydrocephalus.

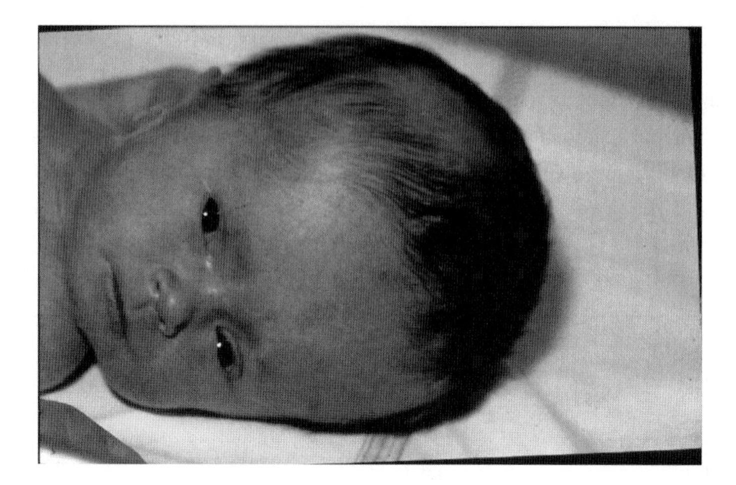

Figure 7. A child of a diabetic mother, exhibiting congenital hydrocephaly (Provided by Janna. Koppe)

According to the Priscilla White classification (the higher in the alphabet the more severe the maternal type I diabetes), the congenital malformations in this group of mothers were the following:

- Group A: one case: Down's syndrome,
- Group B: three cases: inguinal hernia, Down's syndrome, heart defect + costovertebral anomaly,
- Group C: five cases: heart defect + situs inversus, large haemangioma, Pierre-Robin sequence, sacral agenesis, ***congenital hydrocephaly,***
- Group D: seven cases: heart defect + sacral agenesis + anus atresia + rectal atresia, ***misformed ribs and congenital hydrocephaly,*** misformed ears, urethral stenosis, double scrotum with repeated urinary tract infections, accessory ears, sacral agenesis (Koppe *et al.,* 1983).

Congenital malformations are still a problem in the offspring of mothers with diabetes. Meticulous control of blood sugar levels three months before conception helps to lower the rate, but not all can be prevented.

3.7.6. Increasing type 1 diabetes

The incidence of type I diabetes has increased over the last decade, probably as a result of environmental factors (Gillespie *et al.,* 2004). This will probably lead to an increase in congenital malformations in the near future.

4. The role of vitamin A

Vitamin A is necessary for the mono-glycosylation of proteins. Animals deficient in vitamin A are more susceptible to negative effects of poly-aromatic-hydrocarbons (PAHs), like dioxins. In animal experiments the effect of PAHs are reversed with the help of extra vitamin A (Aust, 1984).

An aberrant glycosylation is also hypothesised to be the cause of congenital malformations in the offspring of mothers with type I diabetes. The higher her concentration of HbA1c (glycosylated hemoglobin) the more chance there is that the child will be malformed.

An interesting article in science focuses, on retinoic acid (Keegan *et al.,* 2005).The authors emphasise the importance of vitamin A already during the phase of gastrulation, for a balance between cardiac and non-cardiac identities. This is important for the prevention of cardiac malformations and abnormalities in the skeleton, vertebrae and ribs (Box 1). Combinations of these anomalies are often found. See also the congenital malformations summated above in the Amsterdam group of babies of mothers with type I diabetes.

Box 1. Relation between retinoic acid, and cardiac differentation in Zebra-fish. Reprinted abstract with permission from Keegan *et al.* Retinoic Acid Signaling Restricts the Cardiac Progenitor Pool. Science 2005; 307:247-9. Copyright 2005 AAAS.

> "*Organogenesis begins with specification of a progenitor cell population, the size of which provides a foundation for the organ's final dimensions. Here, we present a new mechanism for regulating the number of progenitor cells by limiting their density within a competent region. We demonstrate that retinoic acid signaling restricts cardiac specification in the zebra-fish embryo. Reduction of retinoic acid signaling causes formation of an excess of cardiomyocytes, via fate transformations that increase cardiac progenitor density within a multipotential zone. Thus, retinoic acid signaling creates a balance between cardiac and non-cardiac identities, thereby refining the dimensions of the cardiac progenitor pool.*"

It is well known that dioxins and other PAHs lower the levels of, and interfere with, retinoic acid metabolism and depletion in the liver is found after dioxin intoxication (ten Tusscher *et al.,* 2003).

Vitamin A is an important vitamin, with a surplus and a deficit both being detrimental to the developing embryo.

In the light of the recently published knowledge in Science, it is of concern that a Dutch study of preconceptional nutritional intake, found that 48 % of women, trying to fall pregnant, have a *diet deficient in vitamin A* (de Weerd, 2003).

5. Conclusion

Dioxins cause congenital malformations, prematurity and intra-uterine growth retardation, as seen in various pollution hot spots. In regions with lower background concentrations, like Europe, there is no indication of an increase in congenital malformations in relation to dioxins.

Analyses of individual congenital malformations, cleft lip and palate and renal abnormalities, like agenesis and dysgenesis, are described in humans, and in animal experiments (see Amsterdam Diemerzeedijk and Chapaevsk).

The novel finding in Chapaevsk of a high incidence of *congenital hydrocephaly* without spina bifida is alarming. This rare condition is also seen in the offspring of mothers with type I diabetes. Abnormal glycosylation is assumed to play a role in causing the malformations in mothers with type I diabetes as expressed a.o. in an increased level of Hemoglobin A1c. In animal experiments, vitamin A may counteract the negative effects of PAHs, including dioxins.

A meticulous care to establish a balanced vitamin A status periconceptionally is recommended.

References

Abbott, B.D. (1995) Review of the interaction between TCDD and glucocorticoids in embryonic palate, *Toxicology* **105**, 365-373.

Aust, S.D. (1984) On the mechanism of anorexia and toxicity of TCDD and related compounds, in A. Poland, and R.D. Kimbrough (Eds) *Banbury Report 18*, Cold Spring Harbor Laboratory, pp. 309-319.

Bakker M.K., van den berg M.D., Boonstra A., van Diem M.T., Meijer W.M., ter Beek L., Siemensma- Múhlenberg N., Priester-Engel E., van der Werf m.c.r., Verheij J.B.G.M. and de Walle H.E.K. (2005) *Eurocat Northern Netherlands Tables 1981-2002*. Groningen: Department of Medical Genetics, University Medical Centre Groningen, The Netherlands.

Bertazzi, P.A. (1998) Study on mortality and cancer incidence, in M. Ramondetta, and A. Repossi (Eds) *Seveso 20 years after*. Fondazione Lombardia per l'Ambiente ISBN 88-8134-038-0, 73-78.

Birnbaum, L.S., Harris, M.W., Stocking, L.M., Clark, A.M. and Morrissey, R.E. (1989) Retinoic acid and 2,3,7,8 - tetrachlorodibenzo-p-dioxin selectively enhance teratogenesis in C57BL/6N mice, *Toxicol. Appl. Pharmacol.* **98**, 487-500.

Chen, H.S. (1990) A 6 year follow-up of behaviour and activity disorder in the Taiwan Yu-cheng children, *Am. J. Public Health* **84**, 415-421.

Chen, Y.C., Guo, Y.L., Hsu, C.C. and Rogan, W.J. (1992) Cognitive development of Yu-Cheng ("oil disease") children prenatally exposed to heat-degraded PCBs, *JAMA* **268**, 3213-3218.

de Weerd, S. (2003) *Preconception counselling; chapter 6: preconception nutritional intake and life-style factors: first results of an explorative study*, University of Nijmegen ISBN: 90-6734-107-X, Nymegen, The Netherlands.

Devlieger, H., Hanssens, M. and Martens, G. (2005) Why is prematurity on the rise in the northern part of Belgium (The Flanders)? *Int. Perinatal Collegium*, abstract 21.

Donovan, J.W., McLennan, R. and Adena, M. (1984) Vietnam service and the risk of congenital malformations. A case-control study, *Med. J. Aust.* **140**, 394-397.

Geusau, A., Abraham, K., Geissler, K., Sator, M.O., Stingl, G. and Tschachler, E. (2001) Severe 2,3,7,8-tetrachlorodibenzo-p-dioxin (TCDD) intoxication: clinical and laboratory effects, *Environ. Health Perspect.* **109**, 865-869.

Geusau, A., Schmaldienst, S., Derfler, K., Pápke, O. and Abraham, K. (2002) Severe 2,3,7,8-tetrachlorodibenzo-pdioxin (TCDD) intoxication: kinetics and trials to enhance elimination in two patients, *Arch. Toxicol.* **76**, 316-325.

Gillespie, K.M., Bain, S.C., Barnett, A.H., Bingley, P., Christie, M.R., Gill, G.V. and Gale, E.A.M. (2004) The rising incidence of childhood type 1 diabetes and reduced contribution of high-risk HLA haplotypes, *Lancet* **364**, 1699-1700.

Guo, Y.L., Lai, T.J., Ju, S.H., Chen, Y.C. and Hsu, C.C. (1993) Sexual developments and biological findings in Yucheng children, *Organohal. Com.* **14**, 235-237.

Guo, Y.L., Lambert, G.H., Hsu, C.C. and Hsu, M.M.L. (2004) Yucheng: health effects of prenatal exposure to polychlorinated biphenyls and dibenzofurans, *Int. Arch. Occup. Environ. Health* **77**, 153-158.

Harada, M. (1976) Clinical and epidemiological studies and significance of the problem, *Bull. Inst. Constitutional Med., Kumamoto Univ.* **25**, 1-60.

Karamova, L., Basharova, G. and Pyanova, F. (2001) Medico-biological results of dioxin exposure, *Organohal. Com.* **52**, 222-225.

Keegan, B.R., Feldman, J.L., Begemann, G., Ingham, P.W. and Yelon, D. (2005) Retinoic Acid Signaling Restricts the Cardiac Progenitor Pool, *Science* **307**, 247-249.

Kogevinas, M., Becher H., Benn T., Bertazzi P.A., Boffetta P., Bueno-de-Mesquita H.B., Coggon D., Colin D., Flesch-Janys D., Fingerhut M., Green L., Kauppinen T., Littorin M., Lynge E., Mathews J.D., Neuberger M. and Pearce N. Saracci R. (1997) Cancer mortality in workers exposed to phenoxy herbicides, chlorophenols and dioxins. An expanded and updated international cohort study. *Am. J. Epidemiol.* **145**, 1061-1075.

Koopman-Esseboom, C., Morse, D.C., Weisglas-Kuperus, N., Lutkeschipholt, I.J., van der Paauw, C.G., Tuinstra, L.G., Brouwer, A. and Sauer, P.J. (1994) Effects of dioxins and polychlorinated biphenyls on thyroid hormone status of pregnant women and their infants, *Pediatr. Res.* **36**, 468-473.

Koppe, J.G. and Keys, J. (2001) PCBs and the precautionary principle, in Harremoes, P., Gee, D., MacGarvin, M., Stirling, A., Keys, J., Wynne, B. and Vaz, S.G. (Eds) *Late lessons from early warnings: The precautionary principle 1896-2000*, (2001) Environmental issue report No. 22, European Environmental Agency, Office of the official publications of the European communities. pp.64-75.

Koppe, J.G., Smorenberg-Schoorl, M.E., van den Berg-Loonen, E.M. and Mills, J.L. (1983) Diabetes, congenital malformations, and HLA-type, in L. Stern, H. Bard. and B. Friis-Hansen (Eds) *Intensive Care in the Newborn, IV*, New York: Masson Publishing, pp.15-18.

Lecorché, Dr. (1885) Du diabéte. Dans ses rapports avec la vie utérine, la menstruation et la grossesse, *Ann. Gynécol.* **24**, 257-273.

Michalek, J.E., Akhtar, F.Z., Ketchum, N.S. and Jackson, W.G. (2001) The Air Force health study: A summary of results, *Organohal. Com.* **54**, 396-398.

Michalek, J.E., Young, A.L., Newton, M. and Kang, H.K. (2001) Use of Agent Orange, *Organohal. Com.* **54**, 384-411.

Mocarelli, P., Gertroux, P.M., Ferrari, E., Patterson, D.G., Kieszak, S.M., Brambilla, P., Vincoli, N., Signorini, S., Tramacere, P. and Carreri, V. (2000) Paternal concentrations of dioxin and sex ratio of offspring, *Lancet* **355**, 1858-1863.

Ott, M.G., Zober, A. and Germann, C. (1994) Laboratory results for selected target organs in 138 individuals occupationally exposed to TCDD, *Chemosphere* **29**, 2423-2437.

Patandin, S., Koopman-Esseboom, C., de Ridder, A.A.J., Weisglas-Kuperus, N. and Sauer, P.J.J. (1998) Effects of environmental exposure to polychlorinated biphenyls and dioxins on birth size and growth in Dutch children, *Pediatr. Res.* **44**, 538-545.

Patandin, S., Lanting, C.I., Mulder, P.G., Boersma, E.R., Sauer, P.J. and Weisglas-Kuperus, N. (1999) Effects of environmental exposure to polychlorinated biphenyls and dioxins on cognitive abilities in Dutch children at 42 months of age, *J. Pediatr.* **134**, 33-41.

Piomelli, S. (1999) Lead poisoning, in David, N. G. and Frank A.O, (eds) *Hematology of Infancy and Childhood*, W.B.Saunders Company, 1993, pp. 472-494.

Pratt, C.B., George, S.L., O'Connor, D. and Hoffman, R.E. (1987) Adolescent colorectal cancer and dioxin exposure, *Lancet* **2**, 803.

Revich, B., Aksel, E., Ushakova, T., Ivanova, I., Zhuchenko, N., Klyuev, N., Brodsky, B. and Sotskov, Y. (2001) Dioxin exposure and public health in Chapaevsk, Russia, *Chemosphere* **43**, 951-966.

Rogan, W.J., Gladen, B.C., Hung, K.L., Koong, S.L., Shih, L.Y., Taylor, J.S., Wu, Y.C., Yang, D., Ragan, N.B. and Hsu, C.C. (1988) Congenital poisoning by polychlorinated biphenyls and their contaminants in Taiwan, *Science* **241**, 334-336.

Sassa, S., De Verneuil, H. and Kappas, A. (1984) Inhibition of uroporphyrinogen decarboxylase activity in polyhalogenated aromatic hydrocarbon poisoning, in: *Biological Mechanisms of Dioxin Action 1984*, pp. 215-224.Banbury report 18. New York Eds. A. Poland and R. Kimbrough. Cold Spring Harbor Laboratories. Cold Spring Harbor.

ten Tusscher, G.W. (2002) *Neurodevelopmental influences of perinatal dioxin exposure as assessed with magnetoencephalography, electroencephalography, psychological and neuromotor tests*, PhD thesis, University of Amsterdam, The Netherlands.

ten Tusscher, G.W. and Koppe, J.G. (2004) Perinatal dioxin exposure and later effects-a review, *Chemosphere* **54**, 1329-1336.

ten Tusscher, G.W., de Weerdt, J., Roos, C.M., Griffioen, R.W., De Jongh, F.H., Westra, M., van der Slikke, J.W., Oosting, J., Olie, K. and Koppe, J.G. (2001) Decreased lung function associated with perinatal exposure to Dutch background levels of dioxins, *Acta Paediatr.* **90**, 1292-1298.

ten Tusscher, G.W., Stam, G.A. and Koppe, J.G. (2000) Open chemical combustions resulting in a local increased incidence of orofacial clefts, *Chemosphere* **40**, 1263-1270.

ten Tusscher, G.W., Steerenberg, P.A., van Loveren, H., Vos, J.G., von dem Borne, A.E.G.K., Westra, M., van der Slikke, J.W., Olie, K., Pluim, H.J. and Koppe, J.G. (2003) Persistent hematologic and immunologic disturbances in 8-year-old Dutch children associated with perinatal dioxin exposure, *Environ. Health Perspect.* **111**, 1519-1523.

Thunberg T. (1984) *Effect of TCDD on vitamin A and its relation to TCDD-toxicity*. (A Poland and R Kimbrough, ed). New York: Cold Spring Harbor Laboratory, Cold Spring Harbor, 333-344.

Vos, J.G., Moore, J.A. and Zinkl J.G. (1974) Toxicity of 2,3,7,8-tetrachlorodibenzo-p-dioxin in Rhesus monkeys (Macaca mulatta) following a single oral dose, *Toxicol. Appl. Pharmacol.* **29**, 229-241.

Vreugdenhil, H.J., Slijper, F.M., Mulder, P.G. and Weisglas-Kuperus, N. (2002) Effects of perinatal exposure to PCBS and dioxins on play behaviour in Dutch children at school age, *Environ. Health Perspect.* **110**, A593-A598.

Warner, M, Eskenazi, B., Mocarelli, P., Gerthoux, P.M., Samuels, S., Needham, L., Patterson, D. and Brambilla, P. (2002) Serum dioxin concentrations and breast cancer risk in the Seveso women's health study, *Environ. Health Perspect.* **110**, 625-628.

Watanabe, S., Kitamura, K., Otaki, M. and Waechter, G. (2001) Health effects of chronic exposure to polychlorinated dibenzo-p-dioxins, dibenzofurans and co-planar PCBs in Japan, *Organohal. Com.* **53**, 136-140.

Zeilmaker, M.J., Houweling, D.A., Cuijpers, C.E.J., Hoogerbrugge, R. and Baumann, R.A. (2002) *Verontreiniging van moedermelk met gechloreerde koolwaterstoffen in Nederland: niveaus in 1998 en tijdtrends*. 52. 529102012/2002. Bilthoven, the Netherlands.

LINKS BETWEEN *IN UTERO* EXPOSURE TO PESTICIDES AND EFFECTS ON THE HUMAN PROGENY. DOES EUROPEAN PESTICIDE POLICY PROTECT HEALTH?

C. WATTIEZ
Consultant for Pesticides Action Network Europe
Leonard Street 56-64
London
UNITED KINGDOM
Email: catherine.wattiez@skynet.be

Summary

There are many difficulties in establishing a causal link between exposure to chemicals and human diseases or disorders, when a mixture of chemicals is the causal factor under examination, and when controls have also some degree of contamination. Despite this, significant risk increases have been demonstrated for various congenital malformations including central nervous system, cardiovascular, urogenital and limb defects, and orofacial clefts, all of which have been shown after parental exposure to several or to specific pesticides. The same has been observed for intra uterine growth retardation and neurodevelopment impairments, involving short term or longer term functional anomalies. However, so far, epidemiological research has had very little influence on the authorisation of pesticides.

The adequacy of the risk assessment, which is the basis of the authorisation of Pesticides Directives, is questioned. The shortcomings in active substance regulatory testing are highlighted, together with the lack of consideration for combined effects, potential toxicities of formulated products, the poor consideration of vulnerable groups, and pesticide exposure evaluation deficits.

In order to compensate for the shortage of information and for the potentially large societal costs of filling this deficit, arguments are presented for the adoption of precautionary exposure reduction measures. These could include exclusion criteria

P. Nicolopoulou-Stamati et al. (eds.), Congenital Diseases and the Environment, 183–206.

based on pesticide intrinsic properties, the substitution principle and mandatory national pesticide dependency reduction programmes, with targets and timetables, accompanied by a significant promotion of low input crop production systems.

These cost effective precautionary measures, to prevent or reduce health impacts of pesticides, should now be considered for adoption as the revision of the plant protection products authorisation Directive and a Thematic Strategy on the sustainable use of pesticides are on the political agenda.

1. Introduction

Pesticides are synthetic chemicals developed to kill or control certain organisms and, as a consequence, having the potential to cause adverse effects to non-target organisms. However, most of time their use involves deliberate release into the environment.

This paper reviews epidemiological findings, illustrating the degree of association between parental pesticides exposure and congenital diseases, as well as some experimental findings. At the same time, it will discuss the limitations of the tests required in the pesticides authorisation Directives which contribute, together with poor evaluation of exposure, to the inaccurate assessment of the risk, which is the basis for the authorisation of each individual pesticide. This paper also discusses key precautionary policy measures which need to be adopted, in order to reduce exposure to pesticides and to prevent pesticide contribution to diseases.

Establishing a causal link between exposure to one specific pesticide or several pesticides and health problems is difficult, as human diseases or health disorders are the result of many interacting influences, including radiation, chemicals, genetic background, lifestyle choices, and dietary habits. Only when a chemical or a group of chemicals exert a very strong impact has it been possible to highlight an association. It is much more difficult to prove small but existing influences of chemicals on health. A weak association is largely attributable to the lack of specificity of association, when a mixture is the causal factor under examination. In the case of pesticides, it is often difficult to isolate the effects of specific products because, in many cases, there are simultaneous exposures to several pesticides and the type of pesticides used varies with the season. Moreover, the power of epidemiology is further weakened, as there are no true control groups. Everyone has some degree of contamination with relevant pollutants. Studies focusing on gestational exposure are, however, of particular interest, as they consider the most important potential link between endocrine disrupters and human health.

The difficulties in assessing real risk and determining a causal link are underlined by an increasing number of scientists, including health professionals (Sanborn *et al.*, 2004; Paris Appeal, 2004; Prague Declaration, 2005), who call for precautionary exposure reduction. These calls are timely, as modifications to the pesticides policy are now being prepared at European level. The European Commission is soon to publish a proposal for revision of the Directive 91/414/EEC of 15 July 1991 concerning the placing on the market of plant protection products. Moreover, a Commission's proposal for a Thematic Strategy 'on the sustainable use of pesticides' will address the use phase of pesticides, which up to now has not been regulated at all, as only the authorisation phase was addressed These proposals will have to be discussed and finally adopted by the European Parliament and the Council during 2006-2007.

2. Congenital disorders

Congenital disorders are, of course, only one of the numerous health outcomes associated with pesticides exposure, as shown in the 'Systematic Review of Pesticides Human Health Effects' coordinated by Sanborn *et al.*, 2004. Congenital disorders include congenital malformations but also functional impairments expressed later in life and likely to result from embryo/fetal exposure during key windows of sensitivity.

As far as embryos, fetuses and infants are concerned, many pesticides are transferred across the placenta and via breast milk. Moreover, since in the female, ova are formed in the fetal stage of development, and environmental contaminants have been found in follicular fluid, the next generation of children may be affected by their grandmother's exposure (Chance and Harmsen, 1998). Mothers can be directly exposed through food, water and other drinks, occupational use, gardening and household use (including by handling of treated pets), use by professional applicators, and if the house is located near sprayed field or in an intensive pesticide use area. Mothers can also be indirectly exposed by their partner's professional or amateur use. Indeed, pesticides and dust with adsorbed pesticides can be brought back home on clothing, shoes, vehicles…

For most of the epidemiological studies mentioned, more information on the methodological aspects, such as population description; pesticide types and exposure assessment; covariates; statistical analysis; as well as on the quality rating of the studies, can be found in the review of Sanborn *et al.* (2005).

2.1. *Congenital malformations*

The prevalence of birth defects is usually derived from data on fetuses that survive until birth. However, such an approach ignores malformations associated with syndromes incompatible with fetal life or those borne by fetuses electively aborted due to prenatal screenings. Despite those limitations, increases in risks associated with pesticides exposure have been shown for various birth defects.

2.1.1. *Several birth defects*

A case control study conducted in Spain (Garcia *et al.,* 1998) showed an association between paternal agricultural work and various congenital malformations, such as nervous system, cardiovascular, musculoskeletal, other multiple and unspecified defects, oral clefts and hypospadias/epispadias. Some increased risks were shown for aliphatic hydrocarbons (adjusted Odds Ratio (OR) = 2.05 (0.62-6.80), inorganic compounds (adjusted OR = 2.02 (0.53-7.72) and glufosinate (adjusted OR = 2.45 (0.78-7.70) as well as a significant association for pyridil derivatives (adjusted OR = 2.77(1.19-6.44). Garcia *et al.* (1998, 1999) subsequently defined two exposure periods: the acute risk period (father three months prior to conception and/or first trimester of pregnancy and mother one month prior to conception and/or first trimester of pregnancy) and a non acute risk period. The authors showed a significant association between maternal agricultural work in the acute risk period and the birth defects mentioned in the 1998 publication (adjusted OR = 3.16 (1.1-9.0)). Fathers who reported ever handling pesticides had an adjusted OR = 1.49 (0.94-2.35), mainly related to an increased risk for nervous system and musculoskeletal defects.

Another case control study conducted in Finland (Nurminen *et al.,* 1995) showed an association between parental agricultural pesticides exposure and pooled birth defects (orofacial clefts, central nervous system and skeletal defects). The adjusted OR for agricultural work in the first trimester of pregnancy was 1.4 (0.9-2.00). Pesticides used include 2,4-D, MCPA and dimethoate.

In a retrospective cohort study, Crisostomo and Molino (2002) recorded exposures three months before conception up to the first trimester of pregnancy and showed that conventional pesticide users in the Philippines were four times more at risk for birth defects (spontaneous abortion, various birth defects and preterm delivery), OR = 4.56 (1.21-17.09), than farmers using integrated pest management (IPM).

One Minnesota study (Garry *et al.,* 1996) showed the highest rate of a variety of birth anomalies per 1,000 live births for pesticide applicators: 30.0 versus 26.9 for the general population, the lower rates occurring in non-crop regions. Adjusted OR for all anomalies was 1.41 (1.18-1.69) comparing with general population.

Sherman (1995) reports data on four cases involving an unusual pattern of birth defects affecting the eye, ear, palate, teeth, heart, feet, nipples, genitalia and brain. Each of the children was exposed *in utero* to a pesticide product containing chlorpyrifos. Romero *et al.,* (1989) observed multiple cardiac defects, bilateral optic nerve colobomas, microphthalmia, cerebral and cerebella atrophy and facial anomalies at birth after a mother's contamination, due to entering a cauliflower field contaminated with residues of oxydemeton methyl, mevinphos and methomyl, when the fetus was 4 weeks old.

2.1.2. Central nervous system defects

A case control study (Shaw *et al.,* 1999) in the USA showed, for mothers exposed one month before and three months after conception, a significant increased risk for neural tube defect with an OR = 1.6 (1.1-2.5) when professionals applied pesticides to their homes and an OR = 1.5 (1.1-2.1) when they lived within 0.25 miles of an agricultural crop.

A Norwegian case control study (Kristensen *et al.,* 1997) showed an OR = 3.49 (1.34-9.09) for hydrocephaly associated with parental agricultural work exposure in orchards and greenhouses and an OR of 2.76 (1.07-7.13) for spina bifida.

2.1.3. Cardiovascular defects

A case control study (Loffredo *et al.,* 2001) in Maryland and Virginia (USA) resulted in an association between mothers exposed to any pesticides during the first trimester of pregnancy as well as during the preceding three months and transposition of the great arteries: OR = 2.0 (1.2-3.3). The OR was 2.8 (1.3-7.2) for maternal exposure to herbicides, 4.7 (1.4-12.1) for exposure to rodenticides in the form of pellets, powder or food imitators and 1.5 (0.9-2.6) for exposure to insecticides.

Bivariate analysis of total anomalous pulmonary venous return and exposure in life style, hobbies and work showed association for exposure to pesticides (OR = 2.7 (1.2-6.4) in the Baltimore – Washington Infant Study (Correa-Villasenor *et al.,* 1991).

2.1.4. *Orofacial clefts*

In a case control study (Nurminen *et al.,* 1995), oral clefts were associated with parental agricultural work: adjusted OR = 1.9 (1.1-3.5).

In another case control study (Shaw *et al.,* 1999), multiple cleft lip with/without cleft palate was associated with paternal occupational and household exposure one month before and three months after conception : OR = 1.7 (0.9-3.5).

2.1.5. *Urogenital defects*

In an ecological study (Garcia *et al.,* 1996) in Spain, provinces were categorized by four levels of exposure to pesticides. Cases of orchidopexy were associated with the three highest levels of exposure (OR = 2.32 (1.26-4.29)) and the increase of risk was positively correlated with the exposure level: OR = 2.54, 4.29 and 5.74 for levels of exposure 1, 2, 3, respectively.

In a Norwegian retrospective cohort study (Kristensen *et al.,* 1997), parents' pesticides exposure, based on money spend on the farm to buy pesticide and on tractor spraying equipment, was significantly associated with cryptorchidism and hypospadias: OR = 1.7 (1.16-2.50).

Weidner *et al.* (1998) in Denmark investigated, in a registered-based case control study, parental occupation in the farming and gardening industry, among cases of cryptorchidism and hypospadias. They showed that the risk of cryptorchidism but not hypospadias was significantly increased in the sons of women working in farming or gardening (OR = 1.38 (1.10-1.73)). The risk was more pronounced in the sons of women working in gardening (OR = 1.67 (1.14-2.47).

A Dutch case control study (Pierik *et al.,* 2004) related to cases of cryptorchidism and hypospadias confirmed the previous findings of Weidner and found that paternal pesticide exposure was associated with cryptorchidism (OR = 3.8 (1.1-13.4).

2.1.6. *Limb defects*

Engel *et al.* (2000), in a retrospective cohort study in Washington State, compared infants of mothers professionally exposed to pesticides with those of mothers not exposed, as well as with those of mothers whose husbands were professionally exposed. They found an increased level of limb defects for mothers agriculturally exposed (OR = 2.6 (1.1- 5.8)).

Table 10. Congenital malformations epidemiology

Authors	Exposure	Pesticide type	Malformations	OR
Garcia *et al.*, 1998	Paternal agricultural work	• Aliphatic hydrocarbons	Various	2.05 (0.62-6.80)
		• Inorganic compounds		2.02 (0.53-7.72)
		• Glufosinate		2.45 (0.78-7.70)
		• Pyridil derivatives		2.77 (1.19-6.44)
Garcia *et al.*, 1999	Parental agricultural work during acute risk period (1-3 months prior to conception and first trimester of . pregnancy)		Various	
	• maternal			3.16 (1.10-9.0)
	• paternal			1.49 (0.94-2.35)
Nurminen *et al.*, 1995	Parental agricultural work during first trimester of pregnancy	• includes 2,4-D, MCPA, dimethoate	Various	1.40 (0.90-2.00)
			Oral clefts	1.90 (1.10-3.50)
Crisostoma *et al.*, 2002	Parental agricultural work during 3 months before and first trimester pregnancy		Various	
	Conventional pesticide users compared to IPM users			4.56 (1.21-17.09)
Garry *et al.*, 1996	Parental agricultural work		Various	1.41 (1.18-1.69)

Table 10. (Continued)

Authors	Exposure	Pesticide type	Malformations	OR
Kristensen *et al.*, 1997	Parental agricultural work in orchards and greenhouses		Spina bifida	3.49 (1.34-9.09) 2.76 (1.07-7.13)
	Parental agricultural work		Chryptorchidism and hypospadias	1.70 (1.60-2.50)
	Parental agricultural work in grain farming		limb reduction	2.50 (1.06-5.90)
Shaw *et al.*, 1999	Mothers exposed 3 months before and 3 months after conception		Neural tube defects	
	• Indoor use			1.60 (1.10-2.50)
	• Proximity sprayed fields			1.50 (1.10-2.10)
	Fathers' occupational and household exposure 1 month before and 3 after conception		Multiple cleft lip (with/without cleft palate)	1.70 (0.90-3.40)
	Mothers exposed after professional application at home 1 month before and 3 months after conception		Limb anomalies	1.60 (1.00-2.70)
Loffredo *et al.*, 2001	Mothers exposed 3 months before and 3 months after conception	• All pesticides • Herbicides • Rodenticides • Insecticides	Transposition of great arteries	2.00 (1.20-3.30) 2.80 (1.30-7.20) 4.70 (1.40-12.10) 1.50 (0.90-2.60)
Correa-Villasnor *et al.*, 1991	Parental exposure		Total anomalous venous return	2.70 (1.20-6.40)

Table 10. (Continued)

Authors	Exposure	Pesticide type	Malformations	OR
Garcia *et al.*, 1996	Provinces characterized by 4 levels of exposure		Orchidopexy	
	• For the 3 highest levels of exposure			2.32 (1.26-4.29)
Weidner *et al.*, 1998	Mothers in farming/gardening		Cryptorchidism	1.38 (1.10-1.73)
	Mothers in gardening		Cryptorchidism	1.67 (1.14-2.47)
Pierik *et al.*, 2004	Maternal agricultural work		Cryptorchidism	3.80 (1.10-13.40)
Engel *et al.*, 2000	Maternal agricultural work		Limb defects	2.60 (1.10-5.80)
Schwartz and Logerfo, 1988	Mothers in counties of:			(RR)
	• high agricultural productivity		Limb reduction	1.70 (1.10-2.70)
			Limb reduction + 1 other anomaly	2.40 (1.20-4.70)
	• high pesticide use		Limb reduction	1.90 (1.20-3.10)
			Limb reduction + 1 other anomaly	3.10 (1.50-6.50)
Kricker *et al.*, 1986	Maternal professional/house - hold exposure during first trimester pregnancy		Limb reduction	(RR)
	• for 1 exposure			3.10 (1.80-5.30)
	• for more than 1 exposure			7.00 (2.80-17.50)

Kristensen *et al.* (1997) found an association between parents in grain farming and limb reduction defects (OR = 2.50 (1.06-5.90)) and Shaw *et al.* (1999) established that women exposed during one month before and three months after conception who reported that a professional applied pesticides to their home had increased risk for limb anomalies (OR = 1.6 (1.0-2.7)).

A case control study (Schwartz and Logerfo, 1988) using California birth records from 1982, 1983 and 1984, compared cases with limb reduction defects and randomly selected controls with regard to parental occupation and maternal county of residence. The relative risk (RR) among mothers who resided in a county of high agricultural productivity, as compared with minimal agricultural productivity, was 1.7 (1.1-2.7), while the RR associated with residence in a county with high pesticide use, compared with minimal pesticide use, was 1.9 (1.2-3.1). When they limited the cases to children with limb reduction defects who had at least one additional anomaly, and compared them to the control births, the corresponding RR were 1.6 (0.7-3.6) for parents involvement in agricultural work, 2.4 (1.2-4.7) for a county with high agricultural productivity and 3.1 (1.5-6.5) for a county with high pesticide use.

Lin *et al.* (1994), in a case-referent study utilizing New York State congenital malformation register data, did not find significant positive association between parental farming occupations and potentially pesticide-exposed occupations and overall limb reduction defects. However, when considering the group of limb reduction defects associated with other congenital defects, a weak but consistent positive association was shown with parental involvement in either pesticide-exposed work or agricultural occupations.

An Australian study (Kricker *et al.,* 1986), showed limb reduction associated with maternal professional/household exposure: RR = 3.1 (1.8-5.3) for one exposure during the first trimester of pregnancy and 7.0 (2.8-17.5) for more than one exposure during that particular period.

2.2. Other congenital disorders

Other congenital disorders associated to pesticide exposure can be the result of short or longer term functional anomalies and include those linked to intrauterine growth retardation (IUGR) or neurodevelopmental impairment.

2.2.1. *Intrauterine growth retardation*

IUGR, usually defined as birth weight below the 10th percentile of a reference standard for a given gestational age, could be considered as a functional congenital disorder. It is indeed associated not only with poor neonatal health but with considerable chronic problems in adulthood, such as hypertension, type 2 diabetes, atherosclerosis and cardiovascular disease (Barker *et al.,* 2002). Sanborn *et al.* (2004) reviewed ten papers from Europe, the United States, Canada, Mexico and the Philippines, to examine the association between pesticide exposure and birth weight, IUGR, "small for gestational age" and preterm delivery. The results of this review suggest that there may be a possible association between occupational exposure to agricultural chemicals and IUGR but, according to the reviewers, there is a need for more advanced study design that include precise measurements of the date of the last menstrual period.

2.2.2. *Neurodevelopmental impairments*

As underlined by Schettler *et al.* (2000), it is widely recognised that neurodevelopmental disorders result from complex interactions among genetic, environmental and social factors, that impact progeny during vulnerable periods of development. However, the toxic exposures deserve special scrutiny because they are preventable causes of harm. Those impairments include attention deficit hyperactivity disorder (ADHD), learning disabilities, mental retardation, autism and autism-like disorders, memory losses, aggressive behaviour and dyslexia. They also could include reading skills, intro/extraversion, social adjustment, attention span and impulsivity.

Schettler *et al.* (2000) report on the contribution of chemical contamination, including pesticides to an epidemic of developmental, learning and behavioural disabilities affecting America's children today. The authors of the 2003 Commission baseline report on neurodevelopmental disorders (Commission Baseline Report on neurodevelopmental disorders, 2003) stated that recent data confirm, for the European Union, the apparent increase in ADHD, autism and neurobehavioral problems but were of the opinion that it is hard to say whether such increase is due to broader diagnostic criteria or to environmental factors. These authors, however, acknowledged that the rise in hypospadias could be correlated with a rise in neurodevelopmental disorders, as the same hormone disruption mechanism could be involved.

Extensive information on neurodevelopmental properties of toxicants is only available for a few chemicals. Regulatory toxicity testing for pesticides does not yet allow the detection of toxicities to a human's developing brain. Moreover, as new research is contemplated, a question can be raised about the degree to which animal data can predict neurological consequences of exposure in humans. Rice *et al.* (1996) concluded that rodent studies often vastly underestimate the sensitivity of the human brain.

Brain development begins early in the uterus and continues into adolescence. According to Laessing *et al.* (1999), it is known to be subject to environmental influences at all phase of development (cellular division, migration, differentiation, formation and pruning of synapses, apoptosis and myelination). Interference with any stage of this cascade of events may alter normal progression of subsequent stages, so that even short-term disruptions may have long-term effects later in life. Each of these events is subject to disruption by environmental agents, and brain and behaviour are likely to be the most sensitive endpoints vulnerable to endocrine disruption. Thyroid hormones play a major role. Sex steroids contribute to, among other factors, sexual differentiation of the brain centres, and thereby, to the development of sexual identity and sexual behaviour. It is therefore perfectly possible that subtle neurodevelopmental effects arise from low-level exposures to pesticides, some of which are by design neurotoxic.

Human studies relating to pesticide impacts on neurodevelopment are very scarce. Neurodevelopmental effects, such as impaired stamina coordination, memory and capacity to represent familiar subjects in drawing, were found in pre-school children in pervasive pesticide exposure situations in Mexican valley agriculture, and likely resulted from maternal, *in utero*, and early childhood exposures (Guillette *et al.*, 1998). A variety of organochlorine pesticides were measured in the umbilical chord, blood and breast milk of individuals in the pesticide-exposed community but additional exposure to other pesticides were also likely. Garry *et al.* (2002) showed, in Minnesota, adverse neurological and neurobehavioral developmental effects among the children born to users of glyphosate (OR = 3.6 (1.3-9.6)).

If human studies are scarce, specifically designed animal studies show neurodevelopmental impairments caused by some organophosphate, carbamate, organochlorine and pyrethroid pesticides (Ahlbom *et al.*, 1995; Spiker and Avery, 1977; vom Saal *et al.*, 1995; Malaviya *et al.*, 1993; Commission Baseline Report on neurodevelopmental disorders, 2003). For instance, offspring of rats given the organophosphate chlorpyriphos, at 6.25, 12.5 and 25 mg/kg/day by injection at days 12-19 of gestation, had fewer muscarinic cholinergic receptors in their brains and markedly altered righting reflex and cliff avoidance tests (Chanda and Pope, 1996). In addition, when rats are treated by gavage with 5 mg chlorpyriphos /kg/day at gestational day 6 until postnatal day 11, offspring have a decreased brain weight and a decreased auditory startle response (Makris *et al.,* 1998). Other tests on rats showed that chlorpyriphos also decreases DNA synthesis, independent of its cholinergic mechanisms, resulting in deficits in numbers of cells in the developing brain. The low concentration of chlorpyriphos necessary to impair DNA synthesis and cell division are actually lower than exposure levels of children under some pesticide home-use conditions (Slotkin, 1999; Campbell *et al.,* 1998). According to Slotkin (2004), these effects occur at exposure levels that did not elicit the traditional signs of cholinergic hyperstimulation and that were devoid of apparent systemic toxicity. Accordingly, the ability of chlorpyriphos to elicit developmental systemic toxicity became disconnected from the accepted methods of detection and hence from the existing regulatory guidelines. Many recent publications describe further the neurodevelopmental effects of chlorpyriphos. However, despite these findings, chlorpyriphos has just been reviewed and accepted on the EU positive list of the authorisation directive 91/414/EEC. Doses of 0.5 mg/kg of the organochlorine DDT were given on day 3, 10 and 19 of life, to mice which were examined at 4 months of age. The young animals exposed at day 10 of age were the only ones to show significant increases in activity at adult age and decreases in muscarinic cholinergic receptor levels (Eriksson *et al.,* 1992). These results show a short window of vulnerability. Mice given small doses such as 0.21, 0.42, 0.70 and 42 mg/kg of the pyrethroid bioallethrin I on day 10 of life show reduction in muscarinic cholinergic receptor levels and hyperactivity (Eriksson and Fredriksson, 1991).

Table 11. Other congenital disorders epidemiology

Authors	Exposure	Pesticide type	Other disorders	OR
Garry *et al.,* 2002	Parental agricultural work	Glyphosate	Neurological and behavioural effects	3.60 (1.30-9.60)
Guillette *et al.,* 1998	Maternal exposure	Organochlorines + other pesticides	Lower stamina, ability to catch balls, memory, fine eye-hand coordination, ability to draw a person.	
Sanborn *et al.,* 2004 review of 10 papers	Parental occupational exposure		Intrauterine growth retardation	Possible association but need for more advanced study design

The hyperactivity of the adult mice increased with increasing level of exposure through the 0.70 mg/kg dose but then fell sharply with the 42 mg/kg dose. This means that testing at higher doses of exposure may fail to identify an adverse effect only seen at lower doses of exposure. Current methods for dose selection for pesticide regulatory testing may miss this effect and should be re-examined. Porter *et al.* (1999) exposed adult mice to mixtures of aldicarb, atrazine and nitrates in their drinking water, at concentrations within the range of exposures regularly encountered in human drinking water in mid-West agricultural regions of the United States. They found, among other effects, that mice became significantly more aggressive after exposure.

3. Does European pesticide policy protect our health?

3.1. *The plant protection product authorisation directive*

The risk assessment is the basis for inclusion on a positive list of active substances accepted at European Union level and hence for their authorisation in pesticide products. However, the risk is not properly assessed, as both toxicity and exposure are poorly evaluated.

3.1.1. *Regulatory testing shortcomings*

In Directive 91/414/EEC concerning the placing on the market of plant protection products, there are still no specific tests for the identification of possible endocrine disrupting properties of the active substance. The 90-day oral toxicity study in rodents gives only an indication of immunological effects (OECD Test Guideline (TG) 408, incorporated in Directive 2001/59/EC, 28th adaptation to technical progress of directive 67/548/EEC, since end of August 2001 and hence of application in Directive 91/414/EEC). Neurotoxicity testing in adults has only been mandatory for some neurotoxic pesticides, such as organophosphates and carbamates.

Core tests required for the active substance, such as the multigenerational teratology short term toxicity and long term toxicity/carcinogenicity and eventually neurotoxicity tests, may indicate a potential impact on the developing nervous, immune, reproductive or endocrine systems. However, some pesticides may have effects on these systems, in the absence of any sign from the results of these core tests (Tirado, 2002; EU SCF, 1998).

Developmental neurotoxicity tests are rarely conducted on the active substance of pesticides. The OECD has prepared a draft Test Guideline, TG 426 on developmental neurotoxicity studies, but this test is not yet a requirement under the authorisation Directive. Of critical concern is the possibility that developmental exposure to neurotoxicants may result in an acceleration of age-related decline in function (Reprosafe Conference, 2003)

Some active substances may interfere with the developing immune system and give rise to persistent adverse effects, such as reduced ability to respond to immune challenge. Therefore, the EU SCF 1998 noted the need to consider the criteria that might trigger a requirement for immunotoxicity studies in developing animals.

As far as developmental endocrine and reproductive toxicities are concerned, the multigenerational study (usually two generations) and the teratology study designs (respectively following OECD TG 416 and 414, as since 5 July 2004 those testing requirements are incorporated in Directive 2004/73/EEC, 29th adaptation to technical progress of Directive 67/546/EEC and hence of application in Directive 91/414/EEC) do not adequately detect all endocrine disrupting effects of active substances. Therefore, they are in the process of being updated. All vulnerable periods of development are covered only in the two-generation study design. Late effects are partly covered in young adults, especially in relation to reproductive function and developmental neurotoxicity, but potentially important late effects are not assessed (Reprosafe Conference, 2003). Effects becoming manifest during ageing, including cardiovascular and various metabolic diseases, are not included in the guidelines for reproductive toxicity. Besides these limitations, it has to be remembered that, before 5 July 2004, tests in Directive 91/414/EEC for two-generation and teratology studies were done according to Directive 88/302/EEC of 18 November 1987, adapting to technical progress for the 9th time Council Directive 67/564/EEC, and were, consequently, even much less developed. This implies that normally all existing and new active substances accepted on the European positive list, but still tested according to Directive 88/302/EEC test requirements, may be even less screened for potential developmental toxicity properties and that pesticides presently on the market in Europe are very unequally tested.

3.1.2. No consideration for combined effects

Even the known cumulative effects of multiple pesticides that have a common mechanism of toxicity are not considered. Moreover, recent studies have shown, in humans (Zeliger, 2003) as *in vitro* (Seralini, 2003; Howard, 2003; Richard *et al.*,

2005), that pesticides mixture effects can occur, even when each component is present at a dose that individually is not supposed to produce effects, and can even produce unpredictable new effects. These observations undermine the belief that threshold doses can be applied meaningfully for the risk assessment.

3.1.3. Toxic properties of the formulated products are not properly evaluated

According to annex III of Directive 91/414/EEC, for formulated products those toxicological studies which are required are limited to acute toxicity. So-called inert ingredients added to the formulated product are not tested but only the available toxicity data relating to those non-active substances are considered. The new authorisation legislation in preparation could, however, require some testing for some formulants such as safeners and synergists. Those substances would also have to be accepted, or not, onto a EU positive list. Some *in vitro* studies show that combined effects can occur, between the active ingredients in the product and its inert ingredients (Howard, 2003; Seralini, 2003). For instance, new findings (Richard *et al.,* 2005) show that Roundup can have endocrine disrupting properties and that its formulation products amplify the effects of the active ingredient glyphosate.

3.1.4. No systematic review of the scientific literature is required

Data in the open literature, generated from studies that do not comply with the current test guidelines, should be taken more into account in risk assessment. This view was also expressed in its Opinion of November 2003 by the Commission's Scientific Committee on Toxicity, Ecotoxicity and the Environment (CSTEE). The CSTEE noted that available data on endocrine disrupter effects, especially for pesticides, have not been used to any great extent in risk assessment under Directive 91/414/EEC and considered that this information should be assessed (Opinion of the Scientific Committee on Toxicity, 2003)

3.1.5. No exclusion criteria are defined for active substances, based on intrinsic properties

As the authorisation of each pesticide is based on risk assessment, pesticides showing toxic properties of concern in regulatory tests can continue to be marketed if their individual exposure is estimated to be too low to result in an unacceptable risk for humans or the environment.

A fundamental element of risk assessment is the assumption of a threshold dose below which there are no effects. For instance, as declared by Kortenkamp *et al.*

(2005), co-signatories of the Prague Declaration, this may not be tenable when dealing with endocrine disrupters, because certain hormonally active chemicals act in concert with natural hormones already present in exposed animals.

Additional and more stringent exclusion criteria for persistence and bioaccumulation are also needed. At present, according to annex VI of Directive 91/414/EEC, a plant protection product cannot be authorised if the half life of the active substance or of relevant residues in soil exceeds 3 months. No cut-off criteria is given for persistence in other media. Where there is a possibility of aquatic organisms being exposed, a bioconcentration threshold factor of the active substance in those organisms of 1,000 or 100 has been defined, beyond which no plant protection product can be authorised, being 1,000 or 100 according to the degree of persistence of the active substance. When there is a possibility of birds and other non target terrestrial vertebrates being exposed, the product cannot be authorised if the bioconcentration factor related to their fat tissue exceeds 1, after normal use of the product.

3.1.6. Exposure evaluation deficits

To assess the risk of each pesticide, exposure of organisms is calculated using mathematical models, assuming normal conditions of use for the product. Pesticide use data are rare and incomplete in Europe. Until now, sales data have usually been used as a proxy for use data, which is rather imprecise. According to Eurostat, Member States usage data are shown to be different to that of the pesticides industry. Therefore, Eurostat is developing a Regulation on the collection and reporting of data on the sales and use of pesticides. Farmers will, as of January 2006, have to make a record of all their pesticides applications, according to Regulation EC/852/2004 concerning the hygiene of foodstuffs. This will help Member States to report usage data to Eurostat, which will publish regular reports. A publicly accessible geographical mapping of each pesticide use would indicate hot spots of exposure and meet citizens' right to know, but there is no guarantee that such reporting will be requested. Moreover, aggregate exposure, the total exposure to a pesticide, occurring via all routes and pathways such as inhalation, dietary and non-dietary ingestion and dermal contact, are not considered in the risk assessment. No data of human biomonitoring are available for risk assessment at European level.

3.1.7. The substitution principle and comparative assessment are not considered

If a substance is of concern and if there are one or several other substances for the same product type, or other non-chemical methods or practices of crop management which present less risk to health and environment, this substance of concern should be replaced. This substitution principle will probably be integrated in the review of the authorisation Directive, but its relevance will depend on the criteria used to define a substance of concern and on whether non-chemical practices of crop protection are considered.

3.2. The thematic strategy on the sustainable use of pesticides

According to the Commission Communication COM (2001) 31 of 24 January 2001 on the Sixth Environmental Action Programme, there is a sufficient evidence to suggest that the scale and trends of problems caused by pesticides are serious and growing. Therefore, in order to guarantee a high level of health and environmental protection, the Decision No 1600/2002 of 22 July 2002, laying down the Community Environmental Action Programme 2001-2010, calls to elaborate a Thematic Strategy on the sustainable use of pesticide to address the use phase of pesticides and to achieve (art. 7, 1, 5th indent):

> *'... a significant overall reduction in risks and of the use of pesticides consistent with the necessary crop protection.'*

Recent consultations launched by the Commission during the elaboration of the Thematic Strategy made clearer the intentions of the Commission. . It is expected that a part of the Thematic Strategy proposal, to be published by the Commission during 2006, will consist of a Framework Directive. It is highly probable that, according to that Directive proposal, Member States will be required to implement national action plans to reduce risks from pesticide use. Some minimal measures might be required for these national reduction plans, such as: training and certification of users, certification and monitoring of spraying equipment, pest forecasting systems, specific requirements but no ban for aerial spraying, some protection for the aquatic environment, designation of zones where the use of pesticides has to be reduced or banned, systems for collection of packaging and obsolete pesticides, promotion of organic farming and integrated pest and crop management.

However, the Commission seems to intend to reduce risks from pesticide usage, mostly by reduction of the use of 'unintended' pesticides. In theory, the use of 'unintended' pesticides can be avoided by using the authorised products in the correct way, according to good farming practice, but without aiming at pesticide dependency reduction. The Commission is of the opinion that there is no direct link between the overall reduction of the quantities of pesticides used and the risks involved, as some new low dose pesticides can result in higher risks than some high dose pesticides. Therefore, the Commission does not propose to set out legally binding reduction targets and does not presently intend to adopt precautionary usage reduction measures. However, this absence of link between usage reduction and risk reduction remains to be proven, when overall use reduction is the result of pesticides dependency reduction measures which also address new low dose pesticides. Precautionary pesticide dependency/use reduction, as measured by a treatment frequency index, seems indeed to be the best way to cope with the difficulties in determining the real risk associated with pesticide use.

Besides the absence of targets and timetables for national risk reduction plans, there is an alarming real shortage of financial incentives or other support for farmers to convert towards integrated crop management (ICM) and organic farming, whose development is essential for pesticide dependency reduction. No pesticide tax will be requested to finance measures in the national reduction plans such as research, training and advice on integrated crop management, aimed at pesticide dependency reduction and further promotion of organic farming.

4. Conclusions

Establishing a causal link between exposure to pesticides and health problems is difficult. Human diseases are indeed the result of many interacting influences. Weak associations are largely attributable to lack of specificity of association, when a mixture of chemicals is the causal factor under examination, and when controls have also some degree of contamination with relevant pollutants. Despite those difficulties, significant associations have been shown between *in utero* exposures to pesticides and various congenital malformations or congenital functional diseases or disorders. Associations have also been shown for specific active substances. However, so far, epidemiological research has had very little influence on pesticide authorisation and use. New toxicological findings increasingly show chemicals' impacts at every level of exposure. Therefore, the epidemiological link is even more difficult to highlight. These factors contribute to justify the fact that negative epidemiological results are not reported here. The assumption is that epidemiology tends to produce more false negatives than false positives.

At the same time, the risk assessment of pesticides, which is the basis of the authorisation of each individual pesticide, lacks reliability, as both toxicity and exposure are poorly evaluated. Regulatory testing has to be strongly improved in order to integrate new scientific findings, including those related to endocrine disrupters, to low dose effects, to combination effects, to particular windows of vulnerability during the embryo/fetal life and to effects of inert ingredients. Exposure evaluation of each individual pesticide has to be based on field pesticide usage data and aggregate exposure has to be considered.

Congenital diseases are not the only health impacts of pesticides and all classes of pesticides include substances of potential health concern. In this context, it seems relevant to remember the opinion of the Ontario College of Family Physicians about the need for exposure reduction:

> *'The results of the systematic review [of Pesticide Human Health Effects] do not help indicate which pesticides are particularly harmful. Exposure to all the commonly used pesticides — phenoxyherbicides, organophosphates, carbamates, and pyrethrins — has shown positive associations with adverse health effects. The literature does not support the concept that some pesticides are safer than others; ; it simply points to different health effects with different latency periods for the different classes.' and that '...our message to patients should focus on reduction of exposure to all pesticides, rather than targeting specific pesticides or classes.'* (Sanborn *et al.*, 2004, ch.11)

It also seems relevant to remember elements of the conclusions of the 2003 Reprosafe conference on reproductive toxicology and chemicals, in favour of precautionary measures.

> *'Estimates should be made of the very large gap between the current toxicity testing guidelines and those required to adequately assess all risks from all important and relevant impacts on humans, wildlife and eco-systems; from all relevant exposure opportunities; across all relevant time windows; with all relevant dose regimes; supported by all necessary monitoring and modelling of likely exposures; and with sufficient statistical power to detect all unacceptable impacts. The large costs of filling that information gap should then be estimated and widely publicised. Society could then agree on the more cost effective precautionary and proxy measures that are needed to compensate for the absence of adequate information in order to strike a better balance between economic activity and the hazards arising from it.' (*Reprosafe, 2003, p.12)

Cost effective precautionary measures to reduce health impacts of pesticide use, such as

- meaningful exclusion criteria for active substances, based on intrinsic properties,
- substitution of substances of concern, considering not only the replacement of a chemical by another but also alternative crop production practises,
- inclusion of integrated crop management (ICM) in the definition of 'proper use' of pesticides,

should now be considered for adoption in the context of the revision of the authorisation of plant protection products Directive 91/414/EEC.

Cost effective precautionary measures, aimed at pesticide dependency reduction and organic farming should be considered for adoption as well, such as

- mandatory national pesticide dependency / use reduction programmes with targets and timetables,
- accurate and publicly accessible pesticide use reporting,
- sufficient support to farmers to convert to ICM,

in the context of the elaboration of the Thematic Strategy on the sustainable use of pesticides.

References

Ahlbom, J., Fredriksson, A. and Eriksson, P. (1995) Exposure to an organophosphate (DFP) during a defined period in neonatal life induces permanent changes in brain muscarinic receptors and behaviour in adult mice, *Br. Res.* **677**, 13-19.

Barker, D., Eriksson, JG., Forsen, T. and Osmone, D. (2002) Fetal origins of adult disease: strenght of effect and biological basis, *Int. J. Epidemiol.* **31**, 1235-1239.

Campbell, C.G., Seidler, F.J. and Slotkin, T.A. (1998) Chlorpyriphos interferes with cell development in rat brain regions, *Brain Res. Bull.* **43**, 179-189.

Chance, G.W. and Harmsen, E. (1998) Children are different: environmental contaminants and chidren's health, Can, *J. Public Health* **89** suppl 1, S9-S19.

Chanda, S.M. and Pope, C.N. (1996) Neurochemical and neurobehavioral effects of repeated gestational exposure to chlorpyriphos in maternal and developing rats, *Pharmacol. Biochem. Behavior.* **53**, 771-776.

Commission Baseline report on neurodevelopmental disorders in the framework of the European Environment and Health Strategy (2003) (COM (2003) 338 final) 2004, (online) http://www.europa.eu.int/comm/environment/health/pdf/neurodevelopmental_disorders.pdf

Correa–Villasenor, A., Ferencz, C., Boughman, J.A., Neil, C.A. (1991) Total anomalous pulmonary venous return: familial and environmental factors. The Baltimore – Washington Infant Study Group, *Teratology* **44**, 415-428.

Crisostomo, L. and Molina, V.V. (2002) Pregnancy outcomes among farming households of Nueva Ecija with conventional pesticide use versus integrated pest management, *Int J Ocur. Environ. Health* **8**, 232-242.

Engel, L.S., O'Meara, E.S.and Schwarts, S.M. (2000) Maternal occupation in agriculture and risk of limb defects in Washington State, 1980-1993, *Scand. J. Work Environ. Health* **26**, 193-198.

Eriksson, P. and Fredriksson, A. (1991) Neurotoxic effects of two different pyrethroïds, bioallethrin and deltamethrin, on immature and adult mice: changes in behavioural and muscarinic receptor variables, *Toxicol. Appl. Pharmacol.* **108**, 78-85.

Eriksson, P., Ahlbom, J. and Fredriksson, A. (1992) Exposure to DDT during a defined period in neonatal life induces permanent changes in brain muscarinic receptors and behaviour in adult mice, *Br. Res.* **582**, 277-281.

EU SCF (1998) *Further advice on the opinion of the Scientific Committee for Food expressed on the 19 September 1997 on a maximum residue limit (MRL) of 0,01 mg/kg for pesticides in foods intended for infants and young children*, Adopted by the SCF 4 June 1998.

Garcia, A.M., Benavides, F.G., Fletcher, T. and Orts, E. (1998) Paternal exposure to pesticides and congenital malformations, *Scand. J. Work Environ. Health* **24**, 473-480.

Garcia, A.M., Benavides, F.G., Fletcher, T. and Orts, E. (1999) Parental agricultural work and selected congenital malformations, *Am. J. Epidemiol.* **149**, 64-74.

Garcia-Rodriguez, J., Garcia-Martin, M., Noguerasa-Ocana, M., de Dios Luna-del-Castillo, J., Espigares-Garcia, M. and Olea, N. (1996) Exposure to pesticides and cryptorchidism: geographical evidence and a possible association, *Environ. Health Perspect.* **104**, 1090-1095.

Garry, V.F., Harkins, M.E., Erickson, L.L., Long-Simpson, L.K., Holland, S.E. and Burroughs, B.L. (2002), Birth defects, season of conception and sex of children born to pesticide applicators living in the Red River Valley of Minnesota, USA, *Environ. Health Perspect.* **110**, 441-449.

Garry, V.F., Schreinemackers, D., Harkins M.E. and Griffith, J. (1996) Pesticide appliers, biocides, and birth defects in rural Minnesota, *Environ. Health Perspect.* **104**, 394-99.

Gordon, J.E. and Shy, C.M. (1981) Agricultural chemical use and congenital cleft lip and/or palate, Arch, *Envron. Health Perspect.* **36**, 213-220.

Guillette, E.A., Meza, M.M., Aquilar, M.G., Soto, A.D. and Enedina, I. (1998) An antropological approach to the evaluation of preschool children exposed to pesticides in Mexico, *Environ. Health Perpect.* **106**, 347-353.

Howard, C.V. (2003) *The inadequacies of the current licensing system for pesticides*, in Pesticides Action Network (PAN) Europe (ed) Proceedings of the PAN Europe 20 November 2003 conference on Reducing Pesticide Dependency in Europe to Protect Health, Environment and Biodiversity, pp 11-14, (online) http://www.pan-europe.info.

Kricker, A., Mc Credie, Elliott, J. and Forrest, J. (1986) Women and the environment: A study of congenital limb anomalies, Comm, *Health Stud.* **10**, 1-11.

Kristensen, O., Irgens, L.M., Andersen, A., Bye, A.S. and Sundheim, L. (1997) Birth Defects among offspring of Norwegian farmers, 1967-1991, *Epidemiology* **8**, 537-544.

Laessing, S.A., McCarthy, M.M. and Silbergeld, E.K. (1999) Neurotoxic effects of endocrine disrupters, *Curr. Opinion. Neurol.* **12**, 745-751.

Lin, S., Marshall, E.G. and Davidson, G.K.(1994) Potential parental exposure to pesticides and limb reduction defects, *Scand. J. Work. Env. Health* **20**, 166-179.

Loffredo, C.A., Silbergeld, E.K., Ferencz, C. and Zhang, J. (2001) Association of transposition of the great arteries (TGA) in infants with materanl exposures to herbicides and rodenticides, *Am. J. Epidemiol.* **153**, 529-536.

Makris, S., Raffaele, K., Sette, W. and Seed J. (1998) *A retrospective analysis of twelve developmental neurotoxicity.* Studies submitted to the US EPA Office of Prevention, Pesticides, and Toxic Substances (OPPTS), Wasington DC, USA.

Malaviya, M, Husain, R., Seth, PK. and Husain R. (1993) Perinatal effects of two pyrethroid insecticides on brain neurotransmitter function in the neonatal rat, *Vet. Hum.Toxicol.* **35**, 109-122.

Nurminen, T., Rantala, K., Kurppa, K. and Holnberg, P.C. (1995) Agricultural work during pregnancy and selected structural malformations in Finland, *Epidemiology* **6**, 23-30.

Opinion of the Scientific Committee on Toxicity, Ecotoxicity and the Environment (CSTEE) on two study reports on endocrine disrupters by WRc-NSF and BKH Consulting Enginheers (*WRc-NSF Ref: UC 6052; BKH Ref: M0355037)* adopted by the CSTEE during the 40th plenary meeting of 12-13 November 2003.

Paris Appeal (2004) International Declaration on diseases due to chemical, pollution, coordinated by the French Association for Research on Treatments Against Cancer (ARTAC), 57-59 rue de la Convention, 75015 Paris, France, (online) http://www.artac.info.

Pierik, F.H., Burdorf, A., James, Deddens, J.A., Juttman, R.E. and Weber, R.F.A. (2004) Maternal and paternal risk factors for cryptorchidism and hypospadias: a case-control study in newborn boys, *Environ. Health Perspect.* **112**, 1570-1576.

Porter, W.P., Jaeger, J.W. and Carlson, I.H. (1999) Endocrine, immune and behavioral effects of aldicarb (carbamate), atrazine (triazine) and nitrate (fertilizer) mixtures at groundwater concentrations, *Toxicol. Ind. Health* **15**, 133-150.

Prague Declaration on Hormone Disruption (2005) Centre for Toxicology, University of London School of Pharmacy, 29/39 Brunswick Square, London WC1N 1AX, United Kingdom, (online) *http://www.comprendo-project.org/_files/Prague%20Declaration%2017%20June%202005.pdf.*

Reprosafe Conference (2003) *Reproductive toxicology and chemicals: a matter of timing?* Synopsis Copenhagen 2-3 October 2003, A conference hosted by the European Environment Agency and organised by Reprosafe, a research programme supported by the Swedish EPA.

Rice, D., Evangelista de Duffard, A., Duffard, R., Iregren, A., Satoh, H. and Watanabe, C. (1996) Lessons for neurotoxicology from selected model compounds: SGOMSEC joint report, *Environ. Health Perspect*. **104** (suppl. 2), 205-215.

Richard, S., Moslemi, S., Sipahutar, H., Benachour, N. and Seralini, G.E. (2005) Differential effects of glyphosate on human placental cells and aromatase, *Environ. Health Perspect.* **113**, 716-720.

Romero, B., Barnett, P.G. and Midtling, J.E. (1989) Congenital anomalies associated with maternal exposure to oxydemeton methyl, *Environ. Res.* **50**, 256-261.

Sanborn, M., Cole, D., Kerr, K., Vakil, C., Sanin, L.H. and Bassil, K. (2004) *Systematic review of pesticide human health effects, ontario college of family physicians*, Toronto, Canada.

Schettler, T., Stein, J., Reich, F., Valenti, M. and Wallinga, D (2000) in In Harm's Way: *Toxic threats to child development*, Greater Boston physicians for social responsibility.

Schwartz, D.A. and LoGerfo, J.P. (1988) Congenital limb reduction defects in the agricultural setting, *Am. J. Pub. Health* **78**, 654-659.

Seralini, G.E. (2003) *A new concept useful for pesticide assessment : ecogenetics, in Pesticides Action Network (PAN) Europe* (ed) Proceedings of the PAN Europe 20 November 2003 conference on Reducing Pesticide Dependency in Europe to Protect Health, Environment and Biodiversity, 15-16, (online) http://www.pan-europe.info.

Shaw, G.M., Wasserman, C.R., O'Malley, C.D., Nelson, V. and Jackson, R.J. (1999) Maternal pesticide exposure from multiple sources and selected congenital anomalies, *Epidemiology* **10**, 60-66.

Sherman, J. (1995) Chlorpyrifos (Dursban) associated birth defects: a proposed syndrome, report of four cases and discussion of the toxicology, *Int. J. Occur. Med. Toxicol.* **4**, 1-13.

Slotkin, T.A. (1999) Developmental cholinotoxicants: nicotine and chlorpyriphos, *Environ. Health Perspect.* **107** (suppl 1), 71-80.

Slotkin, T.A. (2004) Guidelines for developmental neurotoxicity and their impact on organophosphate pesticides: a personal view from an academic perspective, *Neuro Toxicol.* **25**, 631-640.

Spyker, J.M., Avery, D.L. (1977) Neurobehavioral effects of prenatal exposure to the organophosphate diazinon in mice, *J. Toxicol. Environ. Health* **3**, 989-1002.

Tirado, C. (2002) *Pesticides*, Chapter 11, in Tamburlini, G., von Ehrenstein, O.S., Bertollini, R.(eds) (2002) Children's health and environment: a review of evidence, A joint report from the European Environment Agency and the WHO Regional Office for Europe.

vom Sall, F.S., Nagel, S., Palanza, P., Boechler, M., Parmigiani, S. and Welshons, W. (1995) Estrogenic pesticides: binding relative to estradiol in MCF-7 cells and effects of exposure during fetal life on subsequent territorial behaviour, *Toxicol. Lett.* **77**, 343-350.

Weidner, I.S., Moller, H., Jensen, T.K. and Skakkebaek, N. (1998) Chrytorchidism and hypospadias in sons of gardeners and farmers, *Environ. Health Perspect.* **106**, 793-796.

Zeliger, H.I. (2003) Toxic effects of chemical mixtures, *Arch. Environ. Health* **58**, 23-59.

ASSOCIATION OF *INTRA-UTERINE* EXPOSURE TO DRUGS WITH CONGENITAL DEFECTS: THE THALIDOMIDE EFFECT

M. CLEMENTI[1*], K. LUDWIG[1], AND A. ANDRISANI[2]
[1] *CEPIG, Genetica Clinica ed Epidemiologica*
Dipartimento di Pediatria
Università di Padova
Via Giustiniani 3
Padova
ITALY
[2] *Dipartimento Di Scienze Ginecologiche E Della Riproduzione Umana, Università di Padova*
Via Giustiniani 3
Padova
ITALY
(author for correspondence, Email: maurizio.clementi@unipd.it)*

Summary

Thalidomide was first synthesised in 1954 and subsequently marketed in 1956. In pre-clinical studies, thalidomide was found to have no apparent side effects. Consequently, it was distributed as an "over-the-counter" drug and used by pregnant women in order to combat morning sickness. In 1961, a significant increase of congenital limb defects in prenatally-exposed newborns became evident. As a result, thalidomide was withdrawn from the market, due to its severe teratogenic effects. It has been estimated that about 6,000 children were affected with thalidomide embryopathy. The clinical findings of the embryopathy are unusual and characteristic.

Despite the fact that scientists have studied thalidomide over the years, its pharmacokinetics, metabolism and teratogenic mechanisms have never been fully understood.

P. Nicolopoulou-Stamati et al. (eds.), Congenital Diseases and the Environment, 207–221.

In recent years, thalidomide has been rediscovered and used for the treatment of several immunological diseases and inflammatory dermatoses, especially lepromatous leprosy. A specific programme (STEPS) to support the appropriate use of thalidomide is being implemented in the United States, in order to avoid thalidomide related birth defects; whereas, on the other hand, an increase of thalidomide embryopathy in newborns from under developed countries has been noted.

1. Introduction

Thalidomide ((alpha)-(N-phthalmido)glutarimide) was synthesised in 1954, in the former West Germany, by Chemie Grunenthal GmbH. Thalidomide was studied in adult rodents for various kinds of toxicities, and even very high dosages were not found to have apparent side effects. Clinical trials noted the drug to be an effective sedative, helpful in treating anxiety, insomnia, gastritis and tension. An overdose apparently seemed to be impossible in humans, since huge amounts were only noted to prolong sleep without provoking further side effects (Mellin and Katzenstein, 1962), and thus confirming the low toxicity which had been described in animals. The drug, thought safe, was distributed to employees of the company. In this period, one newborn was affected by anotia, but the malformation was not considered to be related to the prenatal exposure to thalidomide.

At first, thalidomide was marketed in Germany in 1956 for the treatment of influenza under the name Grippex. After the discovery of its sedative properties, it was promoted as a remedy for sleeplessness and sold in 46 countries (with the trade name Contergan in Germany in 1957, Distavil in UK and Kevadon in Canada in 1958, etc.). It was supposed to be a "no-side-effect" treatment and distributed as an "over-the-counter" drug.

Because of its effectiveness in treating morning sickness, it was used frequently by women in order to treat their nausea during pregnancy (Ances, 2002). The production of thalidomide increased in the late 1950s and reached about 15 tons in Germany by 1960. In the United States, however, following concerns about the development of peripheral neuropathies by the FDA primary reviewer of the thalidomide application, Dr. Frances Kelsey, the drug was never approved.

Soon after, in the early 1960s, an increasing number of newborns with congenital limb defects were noted by some "alert" physicians all over Australia (Mc Bride, 1961), Germany (Lenz, 1962) and England (Smithells, 1962).

Subsequently, the drug was withdrawn from commercial sale in 1961 (Public Affair Committee, 2000). It has been estimated that worldwide there have been more than 6,000 children affected by the "thalidomide embryopathy" which is characterized by variable degrees of reduction deformities of the limbs and many other non-skeletal malformations (table 12). This number does not include spontaneous abortions and fetal deaths, in which diagnosis was not made. *Ad hoc* studies, performed after the thalidomide tragedy, have demonstrated that the specific pattern of malformations observed in humans is presented only in rabbits and nonhuman primates, while other species (chicken, dogs, rats, mice) do not seem to be susceptible.

2. Pharmacokinetics

Thalidomide ((alpha)-(N-phthalmido) glutarimide) is a derivative of glutamic acid. Thalidomide is off-white to white, odourless, crystalline powder that is soluble at 25°C in dimethyl sulfoxide and sparingly soluble in water and ethanol. The empirical formula for thalidomide is $C_{13}H_{10}\ N_2O_4$ and the gram molecular weight is 258.2. It consists of a left-sided phthalimide ring and a right-sided glutarimide ring. The right-sided glutarimide ring is similar in structure to other hypnotic drugs. It is believed that this portion of thalidomide mediates its sedative properties. Thalidomide is slowly absorbed from the gastrointestinal tract. Peak plasma concentrations occur approximately 4 hours after the dose is administered (Tseng *et al.,* 1996). Food does not affect peak concentrations although absorption of thalidomide increases after a high fat meal in healthy subjects (Teo *et al.,* 2000).

The absolute bioavailability of thalidomide has not been determined in humans; however, in animal studies bioavailability ranged from 67% to 93% (Chen *et al.,* 1989).

The drug is distributed extensively throughout the body fluids and tissues. It has been found also to accumulate in semen; thus male patients on thalidomide should use a contraceptive if engaging in sexual activity with a female partner of childbearing potential (see paragraph STEPS). The exact metabolic route and sequence of thalidomide has not been determined in humans so far (Micromedex, 2006). The drug is mainly metabolised through a nonenzymatic pathway, which leads to spontaneous hydrolysis in plasma (Product Information: Thalomid(R), 2001; Schumaker *et al.,* 1965). Results of *in vitro* studies have suggested that pharmacological effects of thalidomide might be secondary to those of metabolites formed via hepatic metabolism by the hepatic cytochrome P450 system (Wood and Proctor, 1990; Ando *et al.,* 2002), similar to those already described in animals

(Czejka and Koch, 1987). However, in the only well-conducted pharmacokinetical study, no metabolites of the drug could be detected in plasma or urine samples (Chen *et al.,* 1989). The mean elimination time is 5-7 hours. Less than 0.7% of the drug can normally be found in urine samples, confirming that there is no renal excretion.

3. Mechanism of action

The exact mechanism of action of thalidomide has not yet been clearly defined, and the mechanisms leading to teratogenicity have remained elusive.

During the last 40 years, at least 30 hypotheses about its mechanism of action have been suggested. Some of them have been proved to be false. Many others still have to be analysed adequately (table 12).

Table 12. Thalidomide proposed mechanisms of action (modified from Stephens and Fillmore, 2000)

Mechanisms supported by data
Angiogenesis inhibitor
Axial limb artery degeneration
B vitamin antagonism (folic acid)
Cell death
Cell-cell interactions
Decreased mesonephric induction of chondrogenesis
Direct effect on the limb bud
Distalization without outgrowth
Down regulation of certain integrins
Faulty chondrification and calcification
IGF-I and FGF-2 antagonist
Intercalation into DNA
Interference with glutamic acid metabolism
Nucleic acid synthesis
Oxidative DNA damage
Pteroylglutamic acid antagonism
TNF-alpha antagonist

The anti-inflammatory, immunomodulatory and anti-angiogenic effects of thalidomide have been studied extensively.

Inhibition of the monocyte-derived TNF-alpha

Investigations have shown that various anti-inflammatory and immunomodulating effects are due to the suppression of excessive tumour necrosis factor alpha (TNF-alpha) production (Sanpaio *et al.,* 1991; Sanpaio *et al.,* 1993) and down-modulation of selected cell surface adhesion molecules involved in leukocyte migration. However, this mechanism is still unclear and the inhibition incomplete and selective. Increased plasma TNF-alpha concentrations have been observed in some patient groups (HIV-seropositive patients) (Jacobson *et al.,* 1997).

Inhibition of leukocyte chemotaxis

Other immunomodulatory and anti-inflammatory properties include the inhibition of leukocyte chemotaxis into areas of inflammation and subsequently a reduction of phagocytosis by polymorphonuclear leucocytes (Klausner *et al.,* 1996).

Immunomodulation

Thalidomide also seems to modulate interleukins, even if results of the effects on specific interleukins and interferon gamma have so far been equivocal. Effects on CD4+ cells and variable effects on other mediators of intercellular pathways have also been implicated.

Angiogenesis

Thalidomide inhibits angiogenesis (the formation of new vessels), which is a fundamental process in the growth of organs in the embryo and tumours/metastases. The angiogenesis effect has been hypothesized to explain the limb malformations in the embryopathy and the effect has been demonstrated in laboratory experiments (D'Amato *et al.,* 1994; Tabin, 1998). The angiogenesis inhibition may be of special importance regarding the treatment of patients with solid tumours and some other diseases. A very interesting hypothesis of the thalidomide mechanism of action in the inhibition of angiogenesis (Stephens and Fillmore, 2000). The authors suggest that thalidomide (or its metabolic products), by binding to the GC (GGGCGG) promoter region of the aV and b3 integrin, interferes with the transcription and suppresses angiogenesis.

It is noteworthy that this mechanism of action, although proven in laboratory and animal tests, has not been demonstrated to be the cause of the thalidomide teratogenicity.

4. Thalidomide side-effects

Teratogenicity is thalidomide's most severe side effect, and therefore thalidomide is absolutely contraindicated in pregnancy.

From the epidemiological data obtained in the 1960s, it was possible to establish the period of greatest sensitivity, which lies between 20 and 36 days after fertilization or 34-50 days after the last menstrual period (Lenz and Knapp, 1962).

Table 13. Major clinical findings in thalidomide embryopathy (Smithells and Newman, 1992)

Ear	Absent ears
	Microtit
	Absent auditory canal
	Narrow/atretic auditory canal
	Deafness (conductive/neurosensorial)
	Abnormal auditory ossicles
Eyes	Anophthalmia
	Microphthalmia
	Coloboma of iris/retina/choroid
Nose	Choanal atresia
Heart	Fallot tetralogy
	Patent ductus arteriosus
	Ventricular septal
	Other congenital cardiac malformations
Abdomen	Bowel atresia/obstruction
	Inguinal hernia
	Anal atresia
Genital	Abnormal labia
	Female general abnormalities
	Recto-vaginal fistula
	Hypospadias
	Hypoplastic scrotum
	Cryptorchid testes
Kidney	Renal agenesis
	Ectopic kidneys
	Hydronephrosis
Skeletal	Reduction deformity of arms/legs
	Absent/hypoplastic humerus
	Radio-humeral synostosis
	Absent thumbs/digits
	Hemivertebrae

The risk of a fetus exposed to thalidomide in this sensitive period could not be clearly defined but might well be 50% or more. No dose-effect has been observed and no safe dose of thalidomide has been established.

The clinical spectrum of the thalidomide embryopathy is very unusual, specific and characteristic (see table 13)

Limb reduction defect is the most common reported birth malformation. It is characterized by a wide range of severity from complete absence of limbs to milder defects, such as thenar hypoplasia, absent phalanges or hypoplasia of the shoulders. In the upper limbs, the defects are usually symmetrical with the preaxial bones being more severely affected. The thumb is most often involved, then the radius, the humerus and the ulna.

External ear abnormalities, showing a spectrum from anotia to sensorineural deafness secondary to inner ear anomalies, are very common. Ocular anomalies, such as coloboma, glaucoma and microphthalmia (Miller and Stromland, 1999) have been reported as well as facial palsy, heart (atrial septum defect, ventricular septum defect, and Fallot tetralogy), gastrointestinal (intestinal atresias), renal and genital malformations.

Other side effects in adults include peripheral neuropathy, hypotension, neutropenia and hypersensitivity. Among these disease patterns, the peripheral neuropathy is the most common and potentially severe one. The peripheral neuropathy can be irreversible.

It has been demonstrated that thalidomide does not cause second-generation birth defects (Smithells, 1998).

5. "New" indications

Although thalidomide was withdrawn from the market in 1962 and never approved in the United States, it has been an interesting research field for a number of scientists in the last decades.

Thalidomide reappeared in 1965 when Sheskin (1965), a dermatologist from Israel, made a fortuitous discovery while treating his leprosy patients with thalidomide. He prescribed thalidomide for its sedative properties, and noticed a significant improvement or healing of the skin disease erythema nodosum leprosum (ENL) within the first 2 days of treatment. This finding caused a revival of interest and the desire to know more about the immunomodulatory properties of thalidomide.

Several studies on the mechanisms of action have determined the effectiveness of thalidomide as an immunosuppressive. In particular, the inhibiting effect of thalidomide on the synthesis of TNF-alpha and thus slowing down immune response might be an important approach to the treatment of some severe diseases. In addition, thalidomide slows down the viral replication of Human Immunodeficiency Virus (HIV) (Moreira *et al.,* 1997).

At present, the only FDA-approved indication for thalidomide is the acute treatment of the cutaneous manifestations of ENL, a systemic disorder that typically occurs as a complication of lepromatous leprosy. A specific and comprehensive program has been developed to regulate the prescription, dispensing and use of the drug: the System for Thalidomide Education and Prescribing Safety (STEPS) (Gillis, 1997; Ances, 2002). The following “new” indications (table 14) have been identified in the last decades:

Table 14. “New” indications for thalidomide prescription

FDA-approved indication:

- Moderate to severe erythema nodosum leprosum

Literature-supported oncology and AIDS indications:

- Refractory multiple myelomaRefractory chronic graft-vs-host disease AIDS-related cachexia AIDS-related mucocutaneous ulcers

Potential Oncology uses (requires additional research):

- AIDS-related Kaposi's sarcomaPlasma cell leukemiaMiscellaneous advanced solid tumours (e.g., breast, CNS, prostate)

Potential oncology uses (potential for new phase I research):

- Cancer cachexiaSevere, uncontrollable night sweatsCombination therapy with anthracyclines (antiangiogenesis and cardioprotective?)Combination therapy with chemotherapy and/ or radiation (antiangiogenesis and effect on mucositis?)

Autoimmune diseases/inflammatory dermatoses:

- Beucet disease
- Chronic discoid lupus erythematosus
- Graft-versus-host-disease (GVHD)
- Jessner's lymphocytic infiltration
- Langerhans' cell histiocytosis
- Lichen planus
- Pemphigoid
- Sarcoidosis

5.1. Leprosy

After the first report (Sheskin, 1965), numerous studies were able to confirm the effectiveness of thalidomide as a treatment for type II leprosy reaction, which is a manifestation characterized by the appearance of vasculitic nodules and severe neuritis due to destruction of the Schwann cells (Sheskin, 1980; Zwingerberger and Wendt, 1996).

Leprosy is a severe disease which frequently occurs in some parts of the world, in particular South America. Unfortunately, no specific programs and stringent controls are required in these countries where thalidomide is commercially available for the treatment of leprosy. Subsequently, some cases of thalidomide embryopathy have already been noted within the last years (Castilla *et al.,* 1996).

5.2. Human Immunodeficiency Virus (HIV)

Many studies have demonstrated that thalidomide improves the "wasting syndrome" (weight loss without clear identifiable cause) associated with AIDS (Balog *et al.,* 1998).

The reason for thalidomide treatment again lies in the inhibition of TNF-alpha synthesis.

Thalidomide has also proven to be effective in the treatment of AIDS-related Kaposi sarcoma as well as oropharyngeal, rectal, esophageal and genital ulcers in seropositive patients (Gardner-Medwin *et al.,* 1994).

5.3. Beucet disease

Beucet disease is a type of systemic vasculitis with recurrent oral and genital ulcers, skin lesions, panuveitis and arthritis. Thalidomide has been proved to be effective in the treatment of ulcers, but the therapeutical success is restricted as ulcers reform after suspension of treatment (Gardner-Medwin *et al.,* 1994).

5.4. Dermatological and autoimmune disorders

Thalidomide has more recently been rediscovered to be a powerful immunomodulatory and anti-inflammatory agent and has shown promise in the treatment of various dermatological diseases. Among these are chronic Graft Versus

Host disease (GVHD) (Rovelli *et al.,* 1998), systemic lupus erythematosus (knop *et al.,* 1983), pyoderma gangrenosum, prurigo nodularis and lychen planus. All of these are able to be treated successfully.

5.5. *Cancer complications*

Thalidomide has also demonstrated benefits in the treatment of many tumours, for example multiple myeloma and solid tumours, as well as cancer complications. These studies are still ongoing and results will be available in the future (Kumar *et al.,* 2002).

The list of cancers being treated so far with thalidomide is in table 14.

6. STEPS programme (System for thalidomide education and prescribing safety)

The STEPS program, in response to the teratogenicity of thalidomide, is the most stringent program ever issued in the United States for any kind of treatment.

STEPS includes:

- Educational material for physicians and pharmacists about the possible risks associated with thalidomide treatment,
- A video for patients undergoing thalidomide treatment,
- Monthly pregnancy tests, (urine/blood),
- The signing of an informed consent form after a detailed explanation of the risks,
- Participation in a patient survey,
- Thalidomide can only be prescribed by physicians registered to the STEPS program, and is prescribed only for the patient himself. It must not be passed on to someone else.

During the treatment, the patient is not allowed to donate blood and female patients should forgo sexual activity if not using an effective method of contraception (IUD, hormonal method, tubal ligation or partner's vasectomy) and also an additional second one (latex condom, diaphragm, cervical cap). Male patients should use latex condoms if engaging in sexual contact with pregnant women or women with childbearing ability. At the moment, thalidomide is not marketed in Europe, but used in clinical controlled trials as well as to treat selected groups of patients (i.e. HIV, cancer, leukemia). This explains why no guidelines as to where and how to prescribe thalidomide are available in any European countries. STEPS, therefore, will have to also function as something of a guideline for these countries, which don't have their own programme.

7. Lessons for environmental teratology

Malformations and teratogenic effects have been observed and described by contemporary witnesses since Antiquity. Pictures and statues from the Middle Eastern and Greek culture, as well as mythologies, show us early knowledge of malformations and birth defects. Nevertheless, until the 1950s, scientists showed little or no interest in the field of environmental teratology.

The atomic bombing of Hiroshima and Nagasaki in the summer of 1945 lead to an exposure of millions of people to radiation and subsequently a number of studies (Boice, 1990; Otake *et al.,* 1990) were conducted, in order to analyse the effects of radiation on the development of cancers and the incidence of other diseases. However, the interest was mainly focused on radiation induced mutations (Schull, 2003).

The congenital rubella syndrome, first described and published by the Australian ophthalmologist Sir Norman McAlister Gregg in 1941 (1892-1966, Sydney), was another groundbreaking incidence towards the understanding of teratology but still did not attract scientists' attention.

However both "lessons" appeared against a historical background where people were little interested in anything but the war, its consequences and the fight to survive. So, even though there was immediate world wide protest against further atomic experiments and for nuclear disarmament, it was the thalidomide tragedy which heralded a new area.

Physicians and patients, as well as public media, were aroused by the evidence that drugs could cause birth defects. The very thought that the introduction of new substances could not only effect the health of individuals but also influence the development of embryos and fetuses was terrifying. New programs were launched, with strict rules for the adoption of new treatments and drugs, as well as existing ones. New institutions were founded and new instruments created, in order to control the health of individuals and the prescription of drugs during pregnancy, in order to avoid birth defects: clinical teratology was born.

8. Conclusions

Thalidomide, which for a long time has been associated with one of the worst medical tragedies in the last century, has actually been something of a "milestone" in the further development of teratology. It provoked the desire and the need to know more about the effects of drugs taken during pregnancy, as well as the development of birth defects and the important role of environmental influences on the genesis of birth defects. Subsequently, in the 1970s and 1980s many principles of teratology were built on what physicians and scientists had learned from the malformations caused by thalidomide.

The vast amount of research work carried out on the different aspects of thalidomide at that time may have paved the way for a possible reappearance of the drug, creating the possibility of being able to treat a whole new catalogue of conditions. Thalidomide has raised great hope and expectation all over the world for a new approach to the treatment of various diseases such as HIV, multiple myeloma and many others. Nevertheless, despite the fact that great interest in its potential use for treatment of different kinds of disorders has been created, at least two problems will have to be faced. Firstly, the dark side of its history will always render it difficult for some people to accept it and to trust its abilities. Secondly, it will be difficult to issue programs as strict as STEPS, which might lead, once more, to an increase of the number of children being born with severe birth defects, such as missing or shortened limbs. This will be especially difficult in those countries where there is a great demand for thalidomide at present (under developed countries with a high incidence of HIV and leprosy).

References

Ances, B.M. (2002) New concerns about thalidomide, *Obstet. Gynec.* **99**, 125-128.

Ando, Y., Fuse, E. and Figg, W.D. (2002) Thalidomide metabolism by CYP2C subfamily, *Clin. Cancer Res.* **8**, 1964-1973.

Balog, D.L., Epstein, M.E. and Amodio-Groton, M.I. (1998) HIV wasting syndrome: treatment update, *Ann. Pharmacother.* **32**, 446-458.

Bartlett, J.B., Dredge, K. and Dalgleish, A.G. (2004) The evolution of thalidomide and its IMiD derivatives as anticancer agents. *Nat. Rev. Cancer* **4**, 314-322.

Boice Jr. J.D. (1990) Studies of atomic bomb survivors. Understanding radiation effects, *JAMA* **264**, 622-623.

Castilla, E.E., Ashton-Prolla, P., Barreda-Mejia, E., Brunoni, D., Cavalcanti, D.P., Correa-Neto, J., Delgadillo, J.L., Dutra, M.G., Felix, T., Giraldo, A., Juarez, N., Lopez-Camelo, J.S., Nazer, J., Orioli, I.M., Paz, J.E., Pessoto, M.A., Pina-Neto, J.M., Quadrelli, R., Rittler, M., Rueda, S., Saltos, M., Sanchez, O. and Schuler, L. (1996) Thalidomide: a current teratogen in South America, *Teratology* **54**, 273-277.

Chen, T.L., Vogels, G.B., Petty, B.G., Brundrett, R.B., Noe, D.A.; Santos, G.W. and Colvin, O.M. (1989) Plasma pharmacokinetics and urinary excretion of thalidomide after oral dosing in healthy male volunteers, *Drug Metab. Disp.* **17**, 402-405.

Czejka, M.J. and Koch, H.P. (1987) Determination of thalidomide and its major metabolites by high-performance liquid chromatography, *J. Chromatogr.* **413**, 181-187.

D'Amato, R.J., Loughnan, M.S., Flynn, E. and Folkman, J. (1994) Thalidomide is an inhibitor of angiogenesis, *PNAS* **91**, 4082-4085.

Gardner-Medwin, J.M., Smith, N.J. and Powel, R.J. (1994) Clinical experience with thalidomide in the management of severe oral and genital ulcerations in conditions such as Bcchcet disease: use of neurophysiological studies to detect thalidomide neuropathy, *Ann. Rheum. Dis.* **53**, 828-832.

Gillis, J. (1997) *Victims and drugmaker unlikely collaborators*. The Washington Post. Washington, DC, A2.

Jacobson, J.M., Greenspan, J.S., Spritzler, J., Ketter, N., Fahey, J.L., Jackson, J.B., Fox, L., Chernoff, M., Wu, A.W., MacPhail, L.A., Vasquez, G.J. and Wohl, D.A. (1997) Thalidomide for the treatment of oral aphthous ulcers in patients with human immunodeficiency virus infection, *N. Engl. J. Med.* **336**, 1487 –1493.

Klausenr, J.D., Freedman, V.H. and Kaplan, G. (1996) Short analytical review: thalidomide as an anti-TNF-alpha inhibitor: implications for clinical use, *Clin. Immunol. Immunoapthol.* **81**, 219-223.

Knop, J., Bonsmann, G., Happle, R., Ludolph, A., Matz, D.R., Mifsud, E.J. and Macher, E. (1983): Thalidomide in the treatment of 60 cases of chronic discoid lupus erythematous, *Br. J. Dermatol.* **108**, 461-466.

Kumar, S., Witzig, T.E. and Rajkumar, S.V. (2002) Thalidomide as an anti-cancer agent, *J. Cell. Mol. Med.* **6**,160-174.

Lenz, W. (1962) Thalidomide and congenital malformations, *Lancet* **1**, 271-272.

Lenz, W. and Knapp, K. (1962) Die thalidomide-embryopathie, *Dtsch. Med. Wochenschr.* **87**, 1232-1242.

McBride, W.G. (1961) Thalidomide and congenital malformations, *Lancet* **16**, 79-82.

Mellin, G.S. and Katzenstein, M. (1962) The saga of thalidomide. Neuropathy to embryopathy, with case reports of congenital anomalies, *N. Engl. J. Med.* **267**, 1184-1192.

Miller, M.T. and Stromland, K. (1999) Teratogen update: thalidomide: a review, with focus on ocular findings and new potential use, *Teratology* **60**, 306-321.

Moreira, A.L., Corral, L.G., Ye, W., Johnson, B., Stirling, D., Muller, G.W., Freedman, V.H. and Kaplan, G. (1997) Thalidomide and thalidomide analogs reduce HIV type 1 replication in human macrophages *in vitro*, *AIDS Res. Hum. Retroviruses* **13**, 1857-1863.

Otake, M., Schull, W.J. and Neel, J.V. (1990) Congenital-malformations, stillbirths, and early mortality among the children of atomic-bomb survivors – a reanalysis, *Radiat. Res.* **122**, 1-11.

Product Information: Thalomid(R) thalidomide (2001). *Celgene Corporation*, Warren, N.J, (PI revised 08/2001) reviewed 11/2001.

Public Affairs Committee (2000) Teratology Society Affairs Committee Position Paper, Thalidomide, *Teratology* **62**, 172-173.

Rovelli, A., Arrigo, C., Nesi, F., Balduzzi, A., Nicolini, B., Locasciulli, A., Vassallo, E., Miniero, R. and Uderzo, C. (1998) The role of thalidomide in the treatment of refractory chronic graft-versus-host disease following bone marrow transplantation in children, *Bone Mar. Trans.* **21**, 577-581.

Sampaio, E.P., Kaplan, G., Miranda, A., Nery, J.A., Miguel, C.P., Viana, S.M. and Sarno, E.N. (1993) The influence of thalidomide on the clinical and immunologic manifestation of erythema nodosum leprosum, *J. Infect. Dis.* **168**, 408-414.

Sampaio, E.P., Sarno, E.N., Galilly, R., Cohn, Z.A. and Kaplan, G. (1991) Thalidomide selectively inhibits tumor necrosis factor alpha production by stimulated human monocytes, *J. Exp. Med.* **173**, 699-703.

Schull, W.J. (2003) The children of atomic bomb survivors: a synopsis, *J. Radiol. Prot.* **23**, 369-384.

Schumacher, H., Smith, R.L. and Williams, R.T. (1965) The metabolism of thalidomide and some of its hydrolysis products in various species, *Br. J. Pharmacol.* **25**, 338-351.

Sheskin, J. (1965) Thalidomide in the treatment of lepra reactions, *Clin. Pharmacol. Ther.* **6**, 303-306.

Sheskin, J. (1980) The treatment of lepra reactions in lepramatous leprosy: Fifteen years' experience with thalidomide, *Int. J. Dermatol.* **19**, 328-332.

Smithells, D. (1998) Does thalidomide cause second generation birth defects? *Drug Saf.* **19**, 339-341.

Smithells, R.W. (1962) Thalidomide and malformations in Liverpool, *Lancet* **1**, 1270-1273.

Smithells, R.W. and Newman, C.G. (1992) Recognition of thalidomide defects, *J. Med. Genet.* **29**, 716-723.

Stephens, T.D. and Fillmore, B.J. (2000) Hypothesis: thalidomide embryopathy-proposed mechanism of action, *Teratology* **61**, 189-195.

Tabin, C.J. (1998) A developmental model for thalidomide defects, *Nature* **396**, 322-323.

Teo, S.K., Scheffler, M.R., Kook, K.A., Tracewell, W.G., Colburn, W.A., Stirling, D.I. and Thomas, S.D. (2000) Effect of a high-fat meal on thalidomide pharmacokinetics and relative bioavailability of oral formulations in healthy men and women, *Biopharm. Drug Dispos.* **21**, 33-40.

Tseng, S., Pak, G., Washenik, K., Pomeranz, M.K. and Shupack, J.L. (1996): Rediscovering thalidomide: a review of its mechanism of action, side effects, and potential uses, *J. Am. Acad. Dermatol.* **35**, 969-979.

Wood, P.M.D. and Proctor, S.J. (1990) The potential use of thalidomide in the therapy of graft-versus-host disease - a review of clinical and laboratory information, *Leuk. Res.* **14**, 395-399.

Zwingenberger, K. and Wendt, S. (1995-96) Immunomodulation by thalidomide: systematic review of the literature and of unpublished observations, *J. Inflamm.* **46**, 177-211.

SECTION 3:

CONGENITAL DISEASES

ENDOCRINE DISRUPTER EXPOSURE AND MALE CONGENITAL MALFORMATIONS

M.F. FERNÁNDEZ* AND N. OLEA
Laboratory of Medical Investigations
San Cecilio University Hospital
Granada
SPAIN
(author for correspondence, Email: marieta@ugr.es)*

Summary

Exposure of the developing animal male fetus to environmental pollutants, in particular to endocrine disrupting chemicals (EDC), is responsible for sexual maturation anomalies and reproductive malfunction in adult life. Human maternal-infant exposure during pregnancy is of special importance because it represents a very likely window of high susceptibility that can lead to severe and irreversible effects during critical developmental periods. This chapter reviews the epidemiological evidence linking EDC exposure to human male tract malformations, e.g. cryptorchidism and hypospadias, focusing on exposure to pesticides as potential EDCs. Epidemiological studies have yielded contradictory results and do not provide sufficient grounds to confirm the hypothesis that environmental estrogens are associated with these urogenital anomalies. The main reason for the conflicting data may be the difficulty of comparing studies that consider different exposure times and study populations, and do not use the same clinical definitions or diagnostic criteria for these diseases. In addition, the EDC hypothesis poses several challenges to the exposure assessment process that need to be addressed: i) classification of exposure by using direct determinations instead of crude proxies, ii) interpretation of complex dose-effect relationships based on U- or inverted U-shaped dose-response curves, iii) estimation of exposure that takes account of the highly heterogeneous chemical classes implicated, and

P. Nicolopoulou-Stamati et al. (eds.), Congenital Diseases and the Environment, 225–244.

iv) development of biomarkers that allow investigators to quantify exposure to mixtures of EDCs and to differentiate their effects from those of endogenous hormones.

1. Introduction

It has been suggested that exposure of the developing male fetus to environmental pollutants is responsible for anomalies of sexual maturation and reproductive malfunction in adult life (Sharpe and Skakkebaek, 1993; Kennedy and Snyder, 1999; Skakkebaek *et al.,* 2001). In particular, recent human data have focused attention on human exposure to endocrine disrupting chemicals (EDC) during pregnancy, because it may play a role in the development of male sexual disorders (Male, 1995; Toppari *et al.,* 1996; Oliva *et al.,* 2001; Itoh *et al.,* 2001). Male sexual differentiation and reproductive functioning are critically dependent on the ratio of androgen to estrogen, and an imbalance in this ratio may be responsible for male congenital anomalies, supporting the EDC hypothesis. Male congenital malformations include hypospadias, a urethral opening on the underside of the penis or on the perineum, and cryptorchidism, a failure of one or both testicles to descent into the scrotum. These are birth defects of exclusively prenatal origin with an easy and early diagnosis, making them a good choice as biomarkers of EDC exposure and effects.

EDCs are defined as exogenous substances or mixtures with the ability to disrupt normal endocrine homeostasis and alter endocrine system function, thereby producing adverse health effects in an intact organism or its progeny or in (sub-) populations (Colborn and Clement, 1992; European, 2001). Pesticides are well-known chemical pollutants that have been associated not only with male genital malformations but also with semen quality and testicular, prostate, ovarian, and breast cancer (Garry *et al.,* 1996; Koifman *et al.,* 2002). Because of their ubiquity and persistence in the environment, pesticides can be found in soil, water, wildlife, and in the adipose tissue of mothers, reaching children during pregnancy and lactation (Olea *et al.,* 1999). Some pesticides are EDCs, including compounds with estrogenic or antiandrogenic properties as well as those with antihormonal activities. Although most EDC pesticides have been characterized as acting in the living organisms in an estrogen-like manner (Soto *et al.,* 1995), e.g. DDT and endosulfan, they can also induce anti-androgen activity, e.g. p,p′-DDE and vinclozolin (Sohoni *et al.,* 1998; Gray *et al.,* 1999), or more complex interactions affecting endocrine homeostasis (Sultan *et al.,* 2001).

If, indeed, the environment plays a critical role in the pathogeny of male tract malformations, there is a need to determine the nature of the environmental agents responsible and to assess the exposure of susceptible populations (Sultan *et al.,* 2001). In this regard, maternal-infant exposure is of special importance because it represents a very likely window of susceptibility that can lead to severe and irreversible effects at critical moments of development. Furthermore, in the case of persistent fat-soluble chemicals, their bioaccumulation in adipose tissue may imply a risk for women throughout their lifetime, not only during pregnancy. This chapter reviews the epidemiological evidence linking EDC exposure to human male tract malformations, e.g. cryptorchidism and hypospadias, focusing on exposure to pesticides as potential EDCs.

2. Review of the epidemiological evidence

There is epidemiological evidence from many countries of an increasing adverse secular trend in male reproductive outcomes (Carlsen *et al.,* 1992, 1995; Auger *et al.,* 1995; Paulozzi *et al.,* 1997; Swan *et al.,* 2000; Preiksa *et al.,* 2005; Toft *et al.,* 2005), although data from several registries reveals wide inter-country variations (Paulozzi *et al.,* 1999; Toledano *et al.,* 2003; Dolk *et al.,* 2004; Martinez-Frias *et al.,* 2004). Human exposure to EDCs at certain times of life influences normal maturation and may result in impaired male reproductive development and function. Pregnancy is a period of rapid growth, cell differentiation, immaturity of metabolic pathways, and development of vital organ systems. Each normal developmental process occurs during a specific period of a few days or weeks, and it is in this window of particular susceptibility that exposure may exert potential adverse health effects, such as urogenital anomalies. These sensitive periods largely occur during the first trimester, usually before the pregnancy has been recognized. Embryos may be affected by chemicals with long and short biological half-life to which the mother was exposed before their conception (Dolk and Vrijheid, 2003; Swan *et al.,* 2005). The history of diethylstilboestrol (DES), a potent estrogenic drug, illustrates the biological plausibility of the EDC hypothesis in humans. DES was prescribed to women between 1947 and 1971, mostly to prevent abortions and pregnancy complications (Gill *et al.,* 1977). Although DES is the best documented endocrine disrupter in humans, it is not possible to extrapolate from DES to other endocrine-disrupting compounds because the dosage and probably the action mechanism may considerably differ. Basic and clinical studies have clearly demonstrated that exposure to DES during critical period of development is associated with disorders in the reproductive system (Gill *et al.,* 1977; Bibbo *et al.,* 1977; Stenchever *et al.,* 1981; Stillman, 1982; Toppari *et al.,* 1996; Newbold, 2001) among other adverse effects. Thus, boys exposed to DES *in utero* were found to have an increased risk of hypospadias (Klip *et al.,* 2002).

Despite the above evidence, epidemiological studies to date are not considered to provide sufficient grounds to confirm the hypothesis that environmental estrogens contribute to urogenital anomalies (e.g. cryptorchidism and hypospadias) or impaired semen quality. Discrepancies in the scientific community about adverse trends in male reproductive health could be related in part to the difficulties in comparing studies from different time periods, with distinct study populations, and with varied clinical definitions and diagnostic criteria for these diseases. Moreover, epidemiologists traditionally analyse the incidence and risk factors separately for each disorder. Skakkebaek *et al.* (2001) proposed that trends on sperm counts, demand for assisted reproduction, testicular cancer, hypospadias, and undescended testes, are components of a single underlying entity, the so-called testicular dysgenesic syndrome (TDS), and that environment and life-style factors are the most likely causes, with minor participation of the genomic background (Skakkebaek *et al.,* 2001).

Studies of occupational exposure, especially in relation to EDCs and uro-genital malformations, are not abundant. The occupational activities of parents have sometimes been associated with increased risk, as is the case of farmers and gardeners (Kristensen *et al.,* 1997; Weidner *et al.,* 1998; Garcia *et al.,* 1999), vehicle use (Irgens *et al.,* 2000), forestry, carpentry, and service station attendants (Olshan *et al.,* 1991). However, other authors found no association between male reproductive health and work in agriculture or gardening (Restrepo *et al.,* 1990; Weidner *et al.,* 1998) or other occupations (Aho *et al.,* 2003), or they associated health disorders with environmental exposure in a wider sense (Robaire and Hales, 2003).

Recently, occupational exposure of the father was strongly associated with cryptorchidism in a nested case-control study conducted by Pierik *et al.* (2004) in the city of Rotterdam. Based on a cohort of 8,695 boys recruited at child health care centres during the first days of life, the authors compared 78 cryptorchidism cases and 56 hypospadias cases with 313 age-matched controls. Their main aim was to identify risk factors for cryptorchidism and hypospadias, focusing on exposure to EDCs. A parental questionnaire, completed within a few weeks of delivery, collected data on personal characteristics, health, and occupation. Occupational exposure was assessed by generic questions to elicit the description of jobs held in the year before pregnancy, and by a checklist for self-reported exposure to ionizing radiation, physical exposures and to chemicals with endocrine activity or previously described as male reproductive toxicants, such as pesticides. They assessed occupational exposure by applying a job-exposure matrix for potential EDC (Van Tongeren *et al.,* 2002). The study concluded that paternal exposure should be included in studies on risk factors for urogenital malformations, because the pesticide exposure of fathers was associated with cryptorchidism in their progeny (expressed as odds ratio [OR]) (OR 3.8, 95% CI 1.1-13.4) and their smoking habit was associated with hypospadias (OR 3.8, 95% CI 1.8-8.2). These authors found no association between the occupational exposure of mothers and any abnormality.

A register-based case-control study in Denmark investigated parental occupation in farming or gardening in the year of conception (Weidner *et al.,* 1998). All live-born males between 1983 and 1992 with a diagnosis of cryptorchidism (6,177) or hypospadias (1,345) were compared with a random control group of 23,273 boys born in the same period. The risk of cryptorchidism, adjusted for year of birth and birth weight, was greater in sons of women working in farming or gardening (OR 1.38; 95% CI, 1.10-1.73). The effect was even stronger when the analysis was restricted to mothers working in gardening (OR 1.67; 95% CI, 1.14-2.47). Interestingly, maternal occupation in farming and gardening had no significant effect on the risk of hypospadias (OR 1.27, 95% CI, 0.81-1.99). Interestingly, the occupation of the father had no effect on the risk of either cryptorchidism or hypospadias.

In contrast to the above studies, other authors found no association between cryptorchidism and maternal exposure to pesticides (Restrepo *et al.,* 1990) or between hypospadias and occupational exposure to EDCs (Vrijheid *et al.,* 2003) during pregnancy. Restrepo and co-workers selected cases and controls from among the offspring of 8867 floriculture workers and their spouses included in a prevalence survey in 1982-1983 for adverse reproductive outcomes in Colombia. They compared 222 cases of congenital malformations with 443 randomly selected controls matched by maternal age and birth order, determining exposure to pesticides during the pregnancy by means of a questionnaire. The exposure variable was treated as categorical (working or not working in floriculture during pregnancy), and the Mantel–Haenszel test was used for statistical analysis of the different types of congenital malformations, followed by multivariate analysis to control for confounding. Although the relative risk of cryptorchidism after maternal exposure to pesticides during pregnancy was 4.6, the confidence interval (not specified) and p value (>0.05) indicate imprecision and uncertainty in the statistical estimate. The authors concluded that there was no significant association between the 16 cases of cryptorchidism and maternal exposure to pesticides during pregnancy.

The Vrijheid study (2003) analysed the relation between risk of hypospadias and maternal occupation. Hypospadias (n = 3,471) diagnosed at birth and recorded on the National Congenital Anomaly System in England between 1980 and 1996 were included. The potential exposure to EDC was measured as a surrogate for exposure, using the same job-exposure matrix as the Pierik study discussed before (Van Tongeren *et al.,* 2002), categorizing 348 individual jobs titles into three categories according to their unlikely, possible, or probable exposure to seven groups of EDCs: pesticides, polychlorinated organic compounds, phthalates, alkylphenolic compounds, biphenolic compounds, heavy metals, and other hormone disrupting chemicals. Hairdressers, cleaners, and painters were the largest occupational groups with probable exposure to EDCs. There was some indication of an increased risk of hypospadias in the offspring of hairdressers and some workers with reported exposure to phthalates. Although the job-exposure matrix developed by the authors was unable to distinguish among exposure levels, compound potencies, action mechanisms, possible interactions, or changes in exposure over time, this was a very interesting approach that deserves further exploration. Vrijheid *et al.* (2003) concluded that their results should be interpreted with caution and insisted on the need for validation of the job-exposure matrix by the assessment of real exposure in human tissues.

Non-occupational exposure to environmental estrogens can occur through domestic use of pesticides in homes and gardens, consumption of treated foodstuffs or contaminated drinking water, or residence in agricultural areas associated with contaminated air, water or soil, among other routes. Studies of residential exposure to pesticides have used proxy measures of exposure such as pesticide usage at the residence, residence in or near pesticide application areas, or residence near agricultural crops (Dolk and Vrijheid, 2003). Thus, Dolk *et al.* (1998) investigated 1089 cases of congenital anomaly and 2366 control births without malformation in a collaborative European (Eurohazcon) study on the risk of congenital anomaly among people living near hazardous-waste landfill sites. The study was set up to analyse the risk of congenital anomaly among people exposed to chemicals present in hazardous-waste landfill sites. Hypospadias was one of the outcomes examined in this study. Twenty-one such sites were included, and the distance of the mother's place of residence from the nearest one was used as a surrogate measurement of exposure to the chemical contaminants within it. An area of 7 km radius around each landfill was considered, and a zone of 3 km radius around each site was defined as the "proximate zone" within which most exposure to chemical contaminants would occur. Socioeconomic status and maternal age were recorded as confounding variables. Residence within 3 km of a landfill site was associated (borderline significance) with a higher risk of hypospadias (OR 1.96, 95% CI 0.98-3.92).

Kristensen *et al.* (1997) adopted a different approach, studying 192,417 children born to farming families in Norway from 1967 to 1991 and relating different birth defects to the purchase of pesticides and the amount of pesticide spraying equipment on the farm, with no consideration of parental occupation. They found a significant positive relationship with cryptorchidism and a non-significant association with hypospadias. The prevalence of specific defects at birth was determined retrospectively, and the purchases of pesticides in 1968 (from the 1969 census) and tractor-drawn pesticide spraying equipment present on the farm in 1979 (from the 1979 census) were used as proxies of exposure. ORs were calculated from contingency tables, and multiple logistic regression was used to control for potential confounders such as maternal age or birth order. The adjusted OR and 95% CI for cryptorchidism in association with pesticide purchase on the farm was 1.70 (1.16-2.50). The association was stronger when the analysis was limited to vegetable farms (adjusted OR 2.32; 95% CI, 1.34-4.01). The odds ratio for hypospadias was only moderately increased in association with the presence of spraying equipment (adjusted OR 1.38; 95% CI, 0.95-1.99). A small increase in statistical precision was noted when the analysis was limited to the presence of tractor spraying equipment on grain farms (adjusted OR 1.51; 95% CI, 1.00-2.26).

It has been suggested that epidemiological studies will benefit from a more accurate way of measuring exposure to EDCs, for example by determining their presence in tissues and fluids of the mother-child pair. In this regard, Hosie *et al.* (2000) quantified an important number of chemical residues in children, including DDT and metabolites, polychlorinated biphenyls, toxaphene, hexachlorocyclohexane, chlorinated cyclodienes and chlorinated benzenes, using fatty tissue of 48 boys undergoing surgical procedures, 18 with cryptorchidism and 30 controls. All of the substances were detected in all boys studied. The statistical analysis only revealed highly significant differences in presence of heptachlor epoxide (HE) and hexachlorobenzene (HCB) between controls and patients with cryptorchidism, with both HE and HCB showing higher mean concentrations in patients with cryptorchidism than in controls.

In another study along the same lines by Longnecker *et al.* (2002), maternal exposure was assessed by determining the serum levels of DDE, a well known EDC, in the third trimester of pregnancy, in a large sample of mother-child pairs. The authors used a nested case-control design with subjects selected from the Collaborative Perinatal Project, a birth cohort study conducted in the US. The project had open enrolment between 1959 and 1966 and was designed to investigate the relationship between maternal exposure to the pesticide and birth defects, including cryptorchidism and hypospadias. Approximately 75% of subjects born into the study could be followed up to the age of 7 years. Among the male offspring, 219 boys with cryptorchidism and 199 with hypospadias were eligible for the study and were matched with 599 male controls randomly selected from the same cohort. Maternal exposure was assessed by determining the serum levels of DDE in samples collected and stored during the third trimester of pregnancy. The data were categorized into five DDE concentration strata, and the odds of having a birth defect in relation to DDE level were estimated using logistic regression. After examining stratum-specific effects for several possible confounders and comparing the highest category of serum DDE to the lowest, the adjusted ORs were 1.3 (95% CI, 0.7-2.4) for cryptorchidism, and 1.2 (95% CI, 0.6-2.4) for hypospadias. The authors described their findings on an association between DDE and the risk of malformation as inconclusive.

3. The case of southern Spain

The hypothesis of an association between pesticide exposure and genital tract malformations was explored in southern Europe by Garcia-Rodriguez *et al.* (1996), suggesting a relationship between cryptorchidism and geographical variations in exposure to pesticides in Spain. This was an ecological study that aimed to correlate the incidence of cryptorchidism with the use of pesticides by geographic area in the Spanish province of Granada. For improved case definition, only males up to the age of 16 undergoing orchidopexy at the University of Granada Hospital between 1980 and 1991 were selected, 270 such cases being identified. The place of residence was recorded for all cases, and the health care district (the smallest administrative unit) was used as the basic geographical unit. For each geographical unit, the orchidopexy rate was calculated and compared to the degree of pesticide use on a 4-point scale (from lowest-0, to highest-3), as reported by the Regional Agrarian Agency. Logistic regression analysis results indicated that the frequency of orchidopexy in different geographic areas increased in parallel with the degree of pesticide use, except for the municipal area of Granada, where the use of pesticides was lower and the rate of orchidopexy highest. The authors concluded that their results are compatible with an association between exposure to pesticides and an increased risk of cryptorchidism, although several methodological limitations in data analysis make it necessary to evaluate the results with caution.

Beside the ecological study investigating the variations of orchidopexy rates in Granada province (Spain) in relation to geographical variations in pesticide use, few studies have addressed the EDC hypothesis in southern Spain. Nevertheless, a retrospective case-control study performed in the same geographical area found an association between cryptorchidism and the father's employment in agriculture (Rueda-Domingo *et al.,* 2001). Moreover, a frequent and abundant presence of pesticide residues was reported in fatty tissues of children residing in the same area, and some of the pesticides had been restricted or discontinued several years earlier (Olea *et al.,* 1999).

Because of the need for more robust data to support the endocrine disrupter hypothesis, a prospective mother-child cohort was established in Granada province (Spain) as part of a collaborative study that included several countries in Europe. A case-control study, with hospital-based sampling, was nested within this project with the aim of identifying the main risk factors for cryptorchidism and hypospadias and their possible association with environmental factors, with special emphasis on exposure to environmental chemicals with estrogenic activity (xenoestrogens). It was also intended to determine whether the combined effect of environmental estrogens, measured as the total effective xenoestrogen burden, is a risk factor for cryptorchidism and hypospadias.

The cohort excluded mothers without serious chronic diseases, such as diabetes, hypertension, or thyroid disease or those who developed any pregnancy complication that could affect fetal growth and development. The cohort finally included 706 boys out of 4455 newborn males at the San Cecilio University Hospital of Granada province between October 2000 and June 2002. Out of those, 34 (5%) were excluded because of medical complications, inability to collect biologic specimens, non-residence in hospital referral area, and/or refusal to participate. Testicular position was recorded after firm traction of the testis to the most distal position along the pathway of normal descent. Cases studied were 27 boys born with undescended testes (unilaterally and/or bilaterally), 19 boys born with hypospadias, and 2 boys born with both diseases. Controls were matched with cases by gestational age, day of birth, and parity (birth order), selecting three controls for each case, although only 114 controls met the matched criteria. The final subsample size for this study was 48 cases and 114 controls.

Information about characteristics of the mother, father, pregnancy, and birth was obtained using a structured face-to-face questionnaire conducted by trained interviewers, and by searching general hospital records from the maternity unit. The levels of 16 organochlorine compounds as well the total effective xenoestrogen burden (TEXB) were measured in placenta samples from case and control boys. Conditional and unconditional regression models were used for multivariate analysis to estimate the ORs and 95% CIs.

Placentas collected at the time of delivery were immediately frozen at –70ºC and stored until their analysis. Bioaccumulated compounds were extracted from samples (1.6 g of placenta) with hexane and eluted in a glass column filled with Alumine. The eluate obtained was concentrated and then injected into the preparative HPLC. The presence and identity of aldrin, dieldrin, endrin, lindane, methoxychlor, endosulfan I and II, mirex, *p,p'*-DDT, *o,p'*-DDT, *o,p´*-DDD, *p,p´*-DDE, endosulfan diol, sulfate, lactone, and ether was analysed by gas chromatography with electron-capture detection, using p´-dichlorobenzophenone as internal standard and mass spectrometry (GC/MS). More lipophilic xenoestrogens (alpha fraction) were separated from endogenous hormones (beta fraction) by a high performance liquid chromatography technique, using a previously developed methodology.

The statistical study was performed with conditional and unconditional logistic regression analyses. Mean age of mothers was 29.5 yrs (range, 17-43 yrs); 57% had low educational level (similar to that of fathers); 40% were multiparous, 19.25% had preterm delivery, and 21% of deliveries were caesarean. No differences in matching criteria (date of birth, gestational age, and parity) were observed between case-matched controls and overall cohort. Risk of cryptorchidism and hypospadias increased with lower birth weight, lower increase in maternal weight during pregnancy, previous history of abortion, and parental exposure to pesticides. The risk decreased with higher maternal age.

The estrogenicity of placenta extracts measured by TEXB was positive in 72.8% of alpha fractions (mean 1.82 ± 6.65 pM estradiol/g of placenta) and in 77.11% of beta fractions (mean 2.04 ± 17.3 pM estradiol/g of placenta). Higher TEXB alpha and beta levels were found in cases than in paired controls. Conditional logistical regression analysis, adjusted for mother's age at delivery and weight of the newborn, showed the TEXB of the alpha fractions to be a risk factor for cryptorchidism (p= 0.031; OR 2.82, 95% CI 1.10-7.24).

Moreover, DDT and metabolites, endosulfan and metabolites, lindane, aldrin/dieldrin/endrin, hexachlorobenzene, methoxychlor, and mirex were found in placentas. The most frequently detected pesticide was *p,p´*-DDE (84%) followed by lindane (61%). The number of residues found in placenta samples was significantly higher in cases than in controls (p = 0.002). The presence of lindane, DDTs, and endosulfan I was associated with increased risk of malformation, with ORs of 3.38 (95% CI 1.3-8.4), 2.6 (95% CI 1.2-5.7) and 2.2 (95% CI 0.99-4.8), respectively. This risk was also associated with higher concentrations of lindane ($p = 0.007$) and dieldrin ($p = 0.052$). In addition, the mother's occupation was categorized, after performing numerous more complex stratifications, into two levels (agriculture and others). An almost 3.5-fold increased risk for urogenital malformation was observed when the mother reported taking part in agricultural activities (OR 3.47; 95% CI 1.33-9.03). In contrast, no association was found when the father's work was categorized in the same way as the mothers. However, when a different approach to assessing the occupational exposure of the father was adopted (by questions on specific tasks and chemical exposure), occupational exposure was found to be a risk factor for urogenital malformations, with an almost 3-fold increased risk for the highest level. Finally, in order to explore the effect of a rural or urban setting on mother-child exposure, the place of residence was categorized into populations below (rural settings) and above (urban settings) 10,000 inhabitants, but no association with either setting was observed. No association was seen with place of residence.

Although the power of this study was limited, its findings are consistent with those of other studies that have implicated environmental factors in cryptorchidism and hypospadias, such as employment by the mother in agriculture and the bioaccumulation of pesticides in placentas and their combined hormonal effects.

4. Difficulties in exposure assessment: Implications for future research

The EDC hypothesis poses several challenges to epidemiological research (Krimsky *et al.,* 2000). Among these, a correct exposure assessment is a major barrier because of the complexity of the very heterogeneous chemical classes that are implicated. Although some authors have claimed that EDCs are not capable of inducing full effects because their levels are much lower than those of endogenous hormones, the dose-effect relationship based on a threshold model may underestimate the activity of environmental estrogens. EDCs induce an inverted–U dose-response curve, and risk assessment in this area can not rely solely on linear measurements of effect (Almstrup *et al.,* 2002; Weltje *et al.,* 2005).

In addition, the identification of pathways between the source of the chemicals and human exposure is very complex. The evaluation process is further complicated by the fact that many EDCs are persistent and accumulate in human tissues and that there can be long latency periods between exposure and manifestation of a response (e.g. *in utero* exposure may lead to developmental effects that are only evident when the offspring reaches the age of sexual maturity). Furthermore, although some EDCs (e.g. many organochlorine pesticides) are no longer used in many countries, areas that were heavily polluted in the past may serve as redistribution sources for these compounds in the present. Finally, there is evidence that endocrine disrupters may pass from countries where they are still in use to countries where they are not, via atmospheric deposition or perhaps as residues in imported foods.

Misclassification of exposure is possible when registry and census data are used and may also result from the use of crude proxies for exposure instead of direct determinations. Although some specific occupations such as farming, gardening or floriculture are known to be associated with exposure to significant pesticide levels, exposure determination by questionnaire can introduce recall bias.

Over the past few years, the scientific community developed techniques that will allow investigators to quantify exposure to EDCs and to discriminate between endogenous hormones and xenoestrogens (Sonnenschein and Soto, 1998). This methodology is critical for the future application of biomarkers in epidemiological studies. To address this issue, we developed a method to assess the total effective xenoestrogen burden (TEXB) in human samples (Soto *et al.,* 1997; Rivas *et al.,* 2001; Fernández *et al.,* 2004). TEXB measures the cumulative estrogenic effect of chemicals extracted from tissue specimens and fluids by determining the combined effect of the xenoestrogens using the E-Screen bioassay for estrogenicity. This bioassay compares the cell yield between cultures of MCF-7 human breast cancer treated with estradiol and those treated with different concentrations of xenobiotics suspected of being estrogenic (Soto *et al.,* 1995). The process ends with assessment of the proliferative effect, expressed in estradiol equivalents per gram of lipid (Eeq/g), and assignation of the total estrogenic burden to each patient. This standardized biomarker of human exposure to bioaccumulative xenoestrogens was successfully applied as a tool to measure exposure in a case-control study of breast cancer (Ibarluzea *et al.,* 2004), and recent studies in south-eastern Spain have again demonstrated the usefulness of this approach in exposure assessment.

There is adequate evidence of a link between EDCs and congenital malformation for a precautionary approach to be taken, implementing measures to reduce community exposure to these chemicals. Importantly, the planning of care for women of childbearing age should take account of these environment concerns, not only during pregnancy but also even before conception.

Acknowledgements

We are indebted to Richard Davies for editorial assistance. It was supported by grants from the Regional Health Organization (SAS 202/04), Spanish Ministry of Health (INMA network, FIS G03/176), and the European Union Commission (Environ Reprod Health QLK4-1999-01422 and QLK4-2002-00603).

Table 15. Characteristics of cited studies

Study	Design and population	Exposure	Effects	Analysis
Restrepo, 1990	Case-control study of birth defects nested in a prevalence survey among floriculture workers in Bogota, Colombia, conducted in 1982-1983	Maternal exposure during pregnancy to pesticides used in floriculture, determined by questionnaire	**Cryptorchidism** (n=16) was not significantly associated with maternal exposure	Mantel-Haenszel test **RR 4.6** 95%CI (?)
Garcia-Rodriguez, 1996	Ecological study to search for variations in orchidopexy rates in the Spanish province of Granada, from 1980 to 1991	Geographical variations in pesticide use. Each area was assigned to a level (from lowest-0 to highest-3)	**Orchidopexy** (n=270) increased when the highest level of pesticide use was compared with the lowest	Logistic regression OR 2.32 (1.26-4.29)
Kristensen, 1997	Retrospective cohort from register-based study of prevalence for specific birth defects, in Norway from 1967 to 1991	Exposure to pesticides based on parental occupation in the year of birth, determined by farm purchases of chemicals and tractor spraying equipment, from five agricultural and horticultural censuses	**Cryptorchidism** was associated with pesticide purchase (*). The effect was stronger when only vegetable farms were considered (#) Hypospadias was only moderately associated with tractor spraying equipment	Contingency tables **OR* 1.70** (1.16-2.50) **OR# 2.32** (1.34-4.01) **OR 1.38** (0.95-1.99)

Table 15. (Continued)

Study	Design and population	Exposure	Effects	Analysis
Weidner, 1998	Register-based case-control study of cryptorchidism and hypospadias, in Denmark from 1983 to 1992	Exposure to pesticides based on parental occupation in the farming and gardening industry	**Cryptorchidism** (n=6,177) was associated with mother's (*) but not with father's occupation. The effect was higher for gardening (#) **Hypospadias** (n=1,345) was not significantly associated with parental occupation	Contingency tables **OR* 1.38** (1.10-1.73) **OR# 1.67** (1.14-2.47) **OR 1.27** (0.81-1.99)
Dolk, 1998	Multicentre case-control study, using data from seven regional registers of congenital anomalies in five countries from Europe. The study period was different for each area (at least 5 years)	Exposure to chemical contaminants based on distance of the maternal residence from the nearest hazardous-waste landfill site (21)	**Hypospadias** (n=45) was borderline associated with residence near landfill sites	Logistic regression **OR 1.96** (0.98-3.92)
Garcia, 1999	Case-control study to assess the prevalence of congenital malformations in eight public hospitals during 1993-1994 in the Autonomous Community of Valencia, Spain	Occupational exposure to pesticides, mainly as a result of agricultural work, during the month before conception and the first trimester of pregnancy, determined by questionnaire	Selected congenital defects (n = 261) were associated with mothers involved in agricultural activities (*). Fathers handling pesticides also increased the risk (#)	Logistic regression **OR 1.30** (0.70-2.40) **OR 1.20** (0.60-2.40)

Table 15. (Continued)

Study	Design and population	Exposure	Effects	Analysis
Irgens, 2000	Register-based cohort study of birth defects, comprising all births in Norway 1970 -1993	Paternal occupational exposure based on parents' job title (36 occupational groups analysed)	**Hypospadias** (n=3) was associated with vehicle mechanics, not with agricultural work.	Contingency tables **OR 5.19** (1.31-14.24)
Longnecker, 2002	Nested case-control study of a prospective cohort of children with neurologic disorders and other conditions in USA from 1956 to 1966	Maternal serum concentrations of *p,p´*-DDT and *p,p´*-DDE in the third trimester of pregnancy	**Cryptorchidism** (n=219) was modestly-to-moderately associated when boys with the highest level of DDE exposure were compared with those with the lowest. **Hypospadias** (n=199) was not associated with DDE level in serum	Logistic regression **OR* 1.30** (0.70-2.40) **OR# 1.20** (0.60-2.40)
Aho, 2003	Finnish register-based cohort study of boys born in 1970-1986 treated for hypospadias in 1970-1996 before the age of 9 years	Exposure assessment was based on residence at time of delivery and the proportion of population employed in industry or in forestry and farming in each municipality (355)	**Hypospadias** (n=1,543) was not strongly related to employment in either industry or agriculture	Poisson regression

Table 15. (Continued)

Study	Design and Population	Exposure	Effects	Analysis
Vrijheid, 2003	Register-based cohort study from the National Congenital Anomaly System of all birth anomalies in England and Wales, 1980-96	Occupation of mothers (348 individual job titles) recorded from the ONS register, with special regard to exposure to potential endocrine disrupting chemicals (EDCs)	**Hypospadias** (n=3471) increased in the offspring of hairdressers and occupations exposed to phthalates before social class adjustment, (they were the largest occupation group with "possible" or "probable" exposure to EDCs)	Logistic regression **OR"possible"** 1.52 (1.05-2.20) **OR"probable"** 1.45 (0.99-2.12)
Pierik, 2004	Case-control study nested within a cohort of 8,698 male births, in the city of Rotterdam (the Netherlands) from October 1999 to December 2001.	Parental occupational exposure determined by job title in the year before delivery, applying a job-exposure matrix for potential EDCs, and considering self-reported exposure to EDCs.	**Cryptorchidism** (n=78) was associated with paternal pesticide exposure according to the JEM **Hypospadias** (n=56) was associated with paternal smoking	Logistic regression **OR 4.50** (1.40-13.90) **OR 3.40** (1.70-7.00)

Selected defects: nervous system, cardiovascular, oral clefts, hypospadias/epispadias, musculoskeletal, and unspecified anomalies & TEXB-alpha: estrogenicity due to organohalogenated chemicals

References

Aho, M., Koivisto, A.M., Tammela, T.L. and Auviren, A.P. (2003) Geographical differences in the prevalence of hypospadias in Finland, *Environ. Res.* **92**, 118-123.

Almstrup, K., Fernandez, M.F, Petersen, J.H., Olea, N., Skakkebaek, N.E. and Leffers, H. (2002) Dual effects of phytoestrogens result in u-shaped dose-response curves, *Environ. Health Perspect.* **110**, 743-748.

Auger, J., Kunstmann, J.M., Czyglik, F. and Jouannet, P. (1995) Decline in semen quality among fertile men in Paris during the past 20 years, *N. Engl. J. Med.* **332**, 281-285.

Bibbo, M., Gill, W.B,. Azizi, F., Blough, R., Fang, V.S., Rosenfield, R.L., Schumacher, G.F., Sleeper, K., Sonek, M.G. and Wied, G.L. (1977) Follow-up study of male and female offspring of DES-exposed mothers, *Obstet. Gynecol.* **49**, 1-8.

Carlsen, E., Giwercman, A., Keiding, N. and Skakkebaek, N.E. (1992) Evidence for decreasing quality of semen during the past 50 years, *Br. Med. J.* **305**, 609-613.

Carlsen, E., Giwercman, A., Keiding, N. and Skakkebaek, N.E. (1995) Declining semen quality and increasing incidence of testicular cancer: is there a common cause? *Environ. Health Perspect.* **103** (suppl. 7), 137-139.

Colborn, T. and Clement, C. (1992) *Wingspread consensus statement. Chemically induced alterations in sexual, functional development: the wildlife/human connection.* Princeton: Princeton Scientific Publishing, pp. 1-8.

Dolk, H. and Vrijheid, M. (2003) The impact of environmental pollution on congenital anomalies, *Br. Med. Bull.* **68,** 25-45.

Dolk, H., Vrijheid, M., Armstrong, B., Abramsky, L., Bianchi, F., Garne, E., Nelen, V., Robert, E., Scott, J.E.S., Stone, D. and Tenconi, R. (1998) Risk of congenital anomalies near hazardous-waste landfill sites in Europe: the EUROHAZCON study, *Lancet* **352,** 423-427.

Dolk, H., Vrijheid, M., Scott, J.E.S, Addor, M.C., Botting, B., de Vigan, C., de Walle, H., Garne, E., Loane, M., Pierini, A., Garcia-Minaur, S., Physick, N., Tenconi, R., Wiesel, A., Calzolari, E. and Stone, D. (2004) Toward the effective surveillance of hypospadias, *Environ. Health Perspect.* **112**, 398-402.

European Commission (2001). *European Workshops on Endocrine Disrupters 2001, June 18–20*, Aronsborg, Sweden.

Fernandez, M.F., Rivas, A., Olea-Serrano, F., Cerrillo, I., Molina-Molina, J.M., Araque, P., Martinez-Vidal, J.L. and Olea, N. (1999) Assessment of total effective xenoestrogen burden in adipose tissue and identification of chemicals responsible for the combined estrogenic effect, *Anal. Bioanal. Chem.* **379**, 163-170.

Garcia, A.M., Flecher, T., Benavides, F.G. and Orts, E. (1999) Parental agricultural work and selected congenital malformations, *Am. J. Epidemiol.* **149**, 64-74.

Garcia-Rodriguez, J., Garcia-Martin, M., Nogueras-Ocaña, M., de Dios Luna-del-Castillo, J., Espigares Garcia, M., Olea, N. and Lardelli-Claret, P. (1996) Exposure to pesticides and cryptorchidism: geographical evidence of a possible association, *Environ. Health Perspect.* **104,** 1090-1095.

Garry, V.F., Schreinemachers, D., Harkins, M.E. and Griffith, J. (1996) Pesticide appliers, biocides, and birth defects in rural Minnesota, *Environ. Health Perspect.* **104**, 394-399.

Gill, W.G., Schumacher, G.F.B. and Bibbo, M. (1977) Pathological semen and anatomical abnormalities of the genital tract in human male subjects exposed to diethylstilboestrol *in utero*, *J. Urol.* **117**, 477-480.

Gray, L.E., Ostby, J., Monosson, E. and Klece, W.R. (1999) Environmental antiandrogens: low doses of the fungicide vinclozolin alter sexual differentiation of the male rat, *Toxicol. Ind. Health* **15**, 48-64.

Hosie, S., Loff, S., Witt, K., Niessen, K. and Waag, K.L. (2000) Is there a correlation between organochlorine compounds and undescended testes? *Eur. J. Pediatr. Surg.* **10**, 304-309.

Ibarluzea, J., Fernandez, M.F., Santa-Marina, L., Olea-Serrano, M.F., Rivas, A.M., Aurrekoetxea, J.J., Exposito, J., Lorenzo, M., Torne, P., Villalobos, M., Pedraza, V., Sasco, A.J. and Olea, N. (2004) Breast cancer risk and the combined effect of environmental estrogens, *Cancer Causes Control* **15**, 591-600.

Irgens, A., Kruger, K., Skorve, A.H. and Irgens, L.M. (2000) Birth defects and paternal occupational exposure. Hypotheses tested in a record linkage based data set, *Acta Obstet. Gynecol. Scand.* **79**, 465-470.

Itoh, N., Kayama, F., Tatsuki, J. and Tsukamoto, T. (2001) Have sperm counts deteriorated over the past 20 years in healthy, young Japanese men? Results from the Sapporo area, *J. Androl.* **22**, 40-44.

Kennedy, I. W.A. and Snyder, III. H.M. (1999) Paediatric andrology: the impact of environmental pollutants, *BJU Int.* **83**, 195-200.

Klip, H., Verloop, J., van Gool, J.D., Koster, M.E., Burger, C.W. and van Leeuwen, F.E. (2002) Hypospadias in sons of women exposed to diethylstilboestrol *in utero*: a cohort study, *Lancet* **359**, 1102-1107.

Koifman, S., Koifman, R.J. and Meyer, A. (2002) Human reproductive system disturbances and pesticide exposure in Brazil, *Cad. Saude Publica* **18**, 435-445.

Krimsky, S. and Hormonal Chaos. (2000) *The scientific and social origins of the environmental endocrine hypothesis*. Baltimore, MD: Johns Hopkins University Press, USA.

Kristensen, P., Irgens, L.M., Andersen, A., Bye, A.S. and Sundheim, L. (1997) Birth defects among offspring of Norwegian farmers, *Epidemiology* **8**, 537-544.

Longnecker, M.P., Klebanoff, M.A., Brock, J.W., Zhou, H., Gray, K.A., Needham, L.L. and Wilcox, A.J. (2002) Maternal serum level of 1,1-dichloro-2,2-bis(p-chlorophenyl)ethylene and risk of cryptorchidism, hypospadias, and polythelia among male offspring, *Am. J. Epidemiol.* **155,** 313-322.

Martinez-Frias, M.L., Prieto, D., Prieto, L., Bermejo, E., Rodriguez-Pinilla, E. and Cuevas, L. (2004) Secular decreasing trend of the frequency of hypospadias among newborn male infants in Spain. Birth Defects Research (Part A), *Clin. Mol. Teratol.* **70**, 75-81.

Newbold, R.R. (2001) Effects of developmental exposure to diethylstilboestrol (DES) in rodents clues for other environmental estrogens, *APMIS* **109**, 261-271.

Olea, N., Olea-Serrano, F., Lardelli-Claret, P., Rivas, A. and Barba-Navarro, A. (1999) Inadvertent exposure to xenoestrogens in children, *Toxicol. Ind. Health* **15**, 151-158.

Oliva, A., Spira, A.and Multigner, L. (2001) Contribution of environmental factors to the risk of male infertility, *Hum. Reprod.* **16**, 1768-1776.

Olmos, B., Fernandez, M.F,. Granada, A., Lopez-Espinosa, M.J., Molina-Molina, J.M., Fernandez, J.M., Cruz, M., Olea-Serrano, F. and Olea, N. (2006) Human exposure to endocrine disrupting chemicals and prenatal risk factors for cryptorchidism and hypospadias: a nested case-control study, *Environ. Health Perspect.* (submitted).

Olshan, A.F., Teschke, K. and Baird, P.A. (1991) Paternal occupation and congenital anomalies in offspring, *Am. J. Ind. Med.* **20**, 447-475.

Paulozzi, L.J. (1999) International trends in rates of hypospadias and cryptorchidism, *Environ. Health Perspect.* **107**, 297-302.

Paulozzi, L.J., Erickson, J.D. and Jackson, R.J. (1997) Hypospadias trends in two American surveillance systems, *Pediatrics* **100,** 831-834.

Pierik, F.H., Burdorf, A., Deddens, J.A., Juttmann, R.E. and Weber, R.F.A. (2004) Maternal and paternal risk factors for cryptorchidism and hypospadias: a case-control study in newborn boys, *Environ. Health Perspect.* **112**, 1570-1576.

Preiksa, R.T., Zilaitien, B., Matulevicius, V., Skakkebaek, N.E., Petersen, J.H., Jorgensen, N. and Toppari, J. (2005) Higher than expected prevalence of congenital cryptorchidism in Lithuania: a study of 1204 boys at birth and 1 year follow-up, *Hum. Reprod.* **20**, 1928-1932.

Restrepo, M., Muñoz , N., Day, N., Parra, J.E., Hernandez, C., Blettner, M. and Giraldo, A. (1990) Birth defects among children born to a population occupationally exposed to pesticides in Colombia, *Scand. J. Work. Environ. Health* **16**, 239-246.

Rivas, A.M., Fernandez, M.F., Cerrillo, I., Ibarluzea, J., Olea-Serrano, M.F., Pedraza, V. and Olea, N. (2001) Human exposure to endocrine disrupters: Standarization of a marker of estrogenic exposure in adipose tissue, *APMIS* **109**, 1-13.

Robaire, B. and Hales, B.F. (2003) Mechanism of action of cyclophosphamide as a male-mediated developmental toxicant, *Adv. Exp. Med. Biol.* **518**, 169-180.

Rueda-Domingo, M.T., Lopez-Navarrete, E., Nogueras-Ocaña, M. and Lardelli-Claret, P. (2001) Factores de riesgo de criptorquidia, *Gaceta Sanitaria* **15**, 398-405.

Sharpe, R.M. and Skakkebaek, N.E. (1993) Are oestrogens involved in falling sperm counts and disorders of the male reproductive tract? *Lancet* **1**, 1392-1395.

Skakkebaek, N.E., Rajpert-De Meyts, E. and Main, K.M. (2001) Testicular dysgenesis syndrome: an increasingly common developmental disorder with environmental aspects, *Hum. Reprod.* **16**, 972-978.

Sohoni, P. and Sumpter, J.P. (1998) Several environmental oestrogens are also anti-androgens, *J. Endocrinol.* **158,** 327-339.

Sonnenschein, C. and Soto, A.M. (1998) An updated review of environmental estrogen and androgen mimics and antagonists, *J. Steroid. Biochem. Mol. Biol.* **65**, 143-150.

Soto, A.M., Fernandez, M.F., Luizzi, M.F., Oles, Karasko, A.S. and Sonnenschein, C. (1997) Developing a marker of exposure to xenoestrogen mixtures in human serum, *Environ. Health Perspect.* **105**, 647-654.

Soto, A.M., Sonnenschein, C., Chung, K.L., Fernandez, M.F., Olea, N. and Olea-Serrano, M.F. (1995) The E-Screen as a tool to identify estrogens: An update on estrogenic environmental pollutants, *Environ. Health Perspect.* **103**, 113-122.

Stenchever, M.A., Williamson, R.A., Leonard, J., Karp, L.E., Ley, B., Shy, K. and Smith, D. (1981) Possible relationship between *in utero* diethylstilboestrol exposure and male fertility, *Am. J. Obstet. Gynecol.* **140**, 186-193.

Stilman, R.J. (1982) *In utero* exposure to diethylstilboestrol: adverse effects on the reproductive tract and reproductive performance and male and female offspring, *Am. J. Obstet. Gynecol.* **142**, 905-921.

Sultan, C., Balaguer, P., Terouanne, B., Georget, V., Paris, F., Jeandel, C., Lumbroso, S. and Nicolas, J. (2001) Environmental xenoestrogens, antiandrogens and disorders of male sexual differentiation, *Mol. Cel. Endocrinol.* **178**, 99-105.

Swan, S.H. (2000) Intrauterine exposure to diethylstilboestrol: long-term effects in human, *APMIS* **108**, 793-804.

Swan, S.H., Elkin, E.P. and Fenster, L. (2000) The question of declining sperm density revisited: an analysis of 101 studies published, *Environ. Health Perspect.* **108**, 961-966.

Swan, S.H., Main, K.M., Liu, F., Stewart, S.L., Kruse, R.L., Calafat, A.M., Mao, C.S., Redmon, J.B., Ternand, C.L., Sullivan, S., Teague, J.L. and the Study for Future Families Research Team. (2005) Decrease in anogenital distance among male infants with prenatal phthalate exposure, *Environ. Health Perspect.* **113**, 1056-1061.

Toft, G., Hagmar, L., Giwercman, A. and Bonde, J.P. (2004) Epidemiological evidence on reproductive effects of persistent organochlorines in humans, *Reprod. Toxicol.* **19**, 5-26.

Toledano, M.B., Hansell, A.L., Jarup, L., Quinn, M., Jick, S. and Elliot, P. (2003) Temporal trends in orchidopexy, Great Britain, 1992–1998, *Environ. Health Perspect.* **111**, 129-132.

Toppari, J., Larsen, J.C., Christiansen, P., Giwercman, A., Grandjean, P., Guillette, L.J Jr, Jegou, B., Jensen, T.K., Jouannet, P., Keiding, N., Leffers, H., McLachlan, J.A., Meyer, O., Muller, J., Rajpert-De Meyts, E., Scheike, T., Sharpe, R., Sumpter, J. and Skakkebaek, N.E. (1996) Male reproductive health and environmental xenoestrogens, *Environ. Health Perspect.* **104**, 741-803.

Van Tongeren, M., Nieuwenhuijsen, M.J., Gardiner, K., Armstrong, B., Vrijheid, M., Dolk, H. and Botting, B. (2002) A job-exposure matrix for potential endocrine-disrupting chemicals developed for a study into the association between maternal occupation and hypospadias, *Ann. Occup. Hyg.* **46**, 465-477.

Vrijheid, M., Armstrong, B., Dolk, H., van Tongeren, M. and Botting, B. (2003) Risk of hypospadias in relation to maternal occupational exposure to potential endocrine disrupting chemicals, *Occup. Environ. Med.* **60**, 543-550.

Weidner, I.S., Moller, H., Jensen, T.K. and Skakkebaek, N. (1998) Cryptorchidism and hypospadias in sons of gardeners and farmers, *Environ. Health Perspect.* **106**, 793-796.

Weltje, L., vom Saal, F.S. and Oehlmann, J. (2005) Reproductive stimulation by low doses of xenoestrogens contrasts with the view of hormesis as an adaptive response, *Hum. Exp. Toxicol.* **24**, 431-437.

TESTICULAR DYSGENESIS SYNDROME AS A CONGENITAL DISEASE

H.E. VIRTANEN* AND J. TOPPARI
Departments of Physiology and Paediatrics,
University of Turku
Turku
FINLAND
(author for correspondence, Email: helvir@utu.fi)*

Summary

Sex chromosome mosaicism (45,XO/46XY karyotype) is often associated with an intersex phenotype and testicular dysgenesis. Undescended testes (cryptorchidism), hypospadias, and later on, infertility or subfertility and testicular cancer can be signs of testicular dysgenesis syndrome (TDS). The underlying reason is rarely a chromosomal abnormality, and for most cases, the etiology remains unknown. It has been hypothesized that genetic and environmental (including life style) factors that disturb gonadal development during pregnancy cause TDS, and the outcome depends on the timing and extent of the disruption. In mild cases, only spermatogenesis may be affected during adulthood, whereas in severe cases, the child may have hypospadias or develop testicular cancer at young age. The hypothesis is supported by clinical and epidemiological evidence that shows a strong association of the risks for each of the TDS signs, i.e. a boy with cryptorchidism has an increased risk of infertility and testicular cancer. Experimentally, we can induce TDS-like changes in animals by treating them with endocrine disrupting compounds. The challenge for the physicians treating TDS patients is to unravel the developmental mechanisms and causative agents.

P. Nicolopoulou-Stamati et al. (eds.), Congenital Diseases and the Environment, 245–253.

1. Introduction

Adverse trends in the incidence of cryptorchidism, hypospadias, and testicular cancer and in sperm counts have been reported during the last decade (Carlsen *et al.,* 1992; Group, 1992; Adami *et al.,* 1994; Forman and Moller, 1994; Toppari *et al.,* 1996; Paulozzi *et al.,* 1997; Swan *et al.,* 1997). In addition to the fact that they all show an increasing incidence, other associations have also been described between cryptorchidism, hypospadias, testicular cancer and low semen quality. A birth cohort effect has been observed, both in the incidence of testicular cancer and in the decrease of the sperm quality, indicating that recent generations have more problems related to reproductive health (Irvine, 1994; Bergstrom *et al.,* 1996). All four disorders (testicular cancer, cryptorchidism, hypospadias, and reduced semen quality) share the same risk factors, e.g. low birth weight and impaired androgen production/action during fetal life, and they are also risk factors to each other (Damgaard *et al.,* 2002; Sharpe, 2003). Testicular germ cell cancer arises from carcinoma *in situ* (CIS) cells, presumably deriving from germ cells that have escaped normal fetal differentiation (Skakkebaek *et al.,* 1987; Rajpert-De Meyts *et al.,* 1998). Furthermore, subfertility, and cryptorchidism have also been described as being associated with dysgenetic changes in testicular histology (Sohval, 1954; Huff *et al.,* 1993; Skakkebaek *et al.,* 2003). On the other hand, in line with the above mentioned associations, patients with rare genetic abnormalities causing testicular dysgenesis (like 45X0/46XY) have an increased risk for testicular cancer, often combined with cryptorchidism and hypospadias (Aarskog, 1970; Scully, 1981; Savage and Lowe, 1990). Based on the observed epidemiological and clinical connections, it has been proposed that all these disorders, reflecting disturbed function of Leydig or Sertoli cells or impaired germ cell differentiation, are symptoms of a testicular dysgenesis syndrome (TDS) originating from the fetal period (Skakkebaek *et al.,* 2001). Nowadays, there is comparable data available on the incidence of all four TDS-linked disorders in Denmark and Finland, and that will be presented in the following.

2. Semen quality and testicular cancer

Geographical differences in the incidence of testicular cancer and low semen quality have previously been described between Denmark and Finland. A study based on cancer registers (which in Nordic countries have almost perfect coverage of the incidence of cancers) showed that the incidence of testicular cancer has been increasing in both countries since the 1950s, and that the rates have been higher in Denmark as compared to Finland (Adami *et al.,* 1994). The latest register-based figures suggest that a similar increasing tendency and a geographical difference still exist, since the incidences were 7.8 and 1.3 per 100 000 person years in 1980 in Denmark and Finland, respectively (Adami *et al.,* 1994), whereas the current rates are 10.3 and 3.2 per 100 000 person years, respectively (Ferlay *et al.,* 2004)). The current Danish rate is one of the highest reported in the whole world (Ferlay *et al.,* 2004).

Co-ordinated cross-sectional studies have shown that both fertile Finnish men and young Finnish men not selected due to fertility have higher sperm concentration than comparable Danish men (Jorgensen *et al.,* 2001; Jorgensen *et al.,* 2002). Furthermore, up to 40% of young Danish men from the general population have been shown to have such a low sperm concentration, below 40 x 106/ml, that it may compromise their fertility (Bonde *et al.,* 1998; Andersen *et al.,* 2000). French and Scottish fertile men have been reported to have higher sperm concentration than the Danish men but lower than the Finnish men (Jorgensen *et al.,* 2001). Although a retrospective analysis suggests that the sperm counts of infertile Finnish men have been high and have not changed significantly since the 1960s (Vierula *et al.,* 1996), reliable time trend analyses concerning changes in the semen quality are possible only after the performance of comparable prospective studies in the future.

3. Birth rates of cryptorchidism and hypospadias

To explore whether the birth rates of cryptorchidism and hypospadias also show a similar difference between Denmark and Finland, prospective cohort studies were simultaneously performed in the two countries. These studies indicated that the prevalence of congenital cryptorchidism at birth is higher in Denmark as compared to Finland (9.0% and 2.4% of boys, respectively) (Boisen *et al.,* 2004). A difference between countries existed still at the age of three months, though the rates had declined in both countries due to spontaneous testicular descent (1.9% and 1.0%, respectively). The birth rate of cryptorchidism has significantly increased in Denmark as compared to the rate reported in the 1960s (Buemann *et al.,* 1961). In Finland, no previous clinical studies have been performed in this area, and therefore a temporal trend analysis is not possible. The current Finnish rate is lower, and the current Danish rate is higher, than recently reported in two cohort studies performed in Malaysia (4.8%) and Lithuania (5.7%) (Thong *et al.,* 1998; Preiksa *et al.,* 2005). Danish boys also showed a higher birth rate of hypospadias as compared to the Finnish boys (1.03% vs. 0.27%) (Virtanen *et al.,* 2001; Boisen *et al.,* 2005). A recent cross-sectional study also showed a higher birth rate of hypospadias in the Netherlands (0.73%) than in Finland (Pierik *et al.,* 2002). Comparison of the current

Finnish rate to previous data, concerning boys born 1970-1986 and operated for hypospadias by the age of eight years, indicated no increase in the birth rate of hypospadias in Finland (Aho *et al.,* 2000). Instead, comparison of the current Danish figure with previous figures concerning hospital discharge data on newborn boys suggests that the birth rate of hypospadias has also increased in Denmark (Sorensen, 1953; Boisen *et al.,* 2005).

4. Discussion

Consistently higher rates of all four described disorders in Denmark, as compared to Finland, supports the TDS hypothesis, suggesting that these disorders are interrelated via a common origin in the fetal period. Although genetic differences may partly explain the observed differences between Denmark and Finland, when taken together with the increasing incidence of cryptorchidism, testicular cancer, and hypospadias, it is more likely to be explained by non-genetic factors, such as factors related to the environment and life-style. The so-called oestrogen hypothesis originally stated that the increasing rates of male reproductive disorders would be due to increased *in utero* exposure to estrogens (Sharpe and Skakkebaek, 1993). Later on, this hypothesis has been expanded to also include environmental chemicals having estrogenic or anti-androgenic effects and thus disturbing the function of the endocrine system (Toppari *et al.,* 1996). More recently, an imbalance between fetal estrogen and androgen action has been proposed to have a role in the development of TDS-linked disorders (Sharpe, 2003).

In experimental animals, disorders linked to the human testicular dysgenesis syndrome, except for germ cell cancer, have been observed after exposure to different chemicals having an endocrine disrupting activity (table 16) (Skakkebaek *et al.,* 2001; Damgaard *et al.,* 2002; Fisher, 2004; Virtanen *et al.,* 2005). Rats exposed *in utero* to dibutyl phthalate (DBP) show abnormalities of Leydig cells, Sertoli cells, and germ cells, and it has been proposed to be a model of human TDS (Fisher *et al.,* 2003). Exposure of fetal rats to DBP has been shown to reduce fetal testicular Insl3 mRNA and testosterone production, which have been proposed to be due to the delayed maturation of the Leydig cells (Wilson *et al.,* 2004). Furthermore, such exposure has also been associated with reduced testicular Insl3 immunoexpression in adulthood, both in cryptorchid and non-cryptorchid testes (McKinnell *et al.,* 2005). This may in turn affect spermatogenesis, since INSL3 has been shown to have an anti-apoptotic effect on rat germ cells (Kawamura *et al.,* 2004). Thus, similar exposure during the fetal period may result in different outcomes. This is consistent with the theory of human TDS manifesting with a varying number of symptoms, i.e. mild forms perhaps manifesting only with slightly impaired spermatogenesis and the most serious forms showing all four symptoms (Skakkebaek *et al.,* 2001). In humans, cryptorchidism has been associated with increased chemical exposure level (Hosie *et al.,* 2000) and it remains to be seen whether further studies will reveal also other exposure-outcome relationships concerning the human TDS.

Table 16. Examples of chemicals that have endocrine disrupting activity and that have caused TDS-linked disorders in animal studies

Fungicides	Procymidone Vinclozolin	Ostby *et al.,* 1999; Wolf *et al.,* 2000; Gray *et al., 2000*
Industrial chemicals	Bisphenol A Benzyl butyl phthalate (BBP) Dibutyl phthalate (DBP) Diethylhexyl phthalate (DEHP)	vom Saal *et al.,* 1998; Gray *et al.,* 2000; Fisher *et al.,* 2003
Pesticide congener	1,1-dichloro-2,2-bis(p-chlorophenyl)ethylene (DDE) (congener of DDT)	Gray *et al.,* 1999
Pharmaceuticals	Diethylstilboestrol Ethinyl estradiol Flutamide	Yasuda *et al.,* 1985; Walker *et al.,* 1990; McIntyre *et al.,* 2001; McLachlan *et al.,* 2001

Acknowledgements

Data on the current incidence of testicular cancer were based on GLOBOCAN 2002 at http: //www-dep.iarc.fr/. This work was supported by funding from the Academy of Finland, Turku University Central Hospital, the Sigrid Juselius Foundation and the EU Quality of Life Programme.

References

Aarskog, D. (1970) Clinical and cytogenetic studies in hypospadias. *Acta. Paediatr. Scand.* Suppl **203**, 203-201.

Adami, H. O., Bergström, R., Möhner, M., Zatonski, W., Storm, H., Ekbom, A., Tretli, S., Teppo, L., Ziegler, H., Rahu, M., Gurevicius, R. and Stengrevics, A. (1994) Testicular cancer in nine northern European countries, *Int. J. Cancer* 59, 33-38.

Aho, M., Koivisto, A. M., Tammela, T. L. and Auvinen, A. (2000) Is the incidence of hypospadias increasing? Analysis of Finnish hospital discharge data 1970-1994, *Environ. Health Perspect*. **108**, 463-465.

Andersen, A. G., Jensen, T. K., Carlsen, E., Jorgensen, N., Andersson, A. M., Krarup, T., Keiding, N. and Skakkebaek, N. E. (2000) High frequency of sub-optimal semen quality in an unselected population of young men, *Hum. Reprod.* **15**, 366-372.

Bergstrom, R., Adami, H. O., Mohner, M., Zatonski, W., Storm, H., Ekbom, A., Tretli, S., Teppo, L., Akre, O. and Hakulinen, T. (1996) Increase in testicular cancer incidence in six European countries: a birth cohort phenomenon, *J. Natl. Cancer Inst.* **88**, 727-733.

Boisen, K. A., Chellakooty, M., Schmidt, I. M., Kai, C. M., Damgaard, I. N., Suomi, A. M., Toppari, J., Skakkebaek, N. E. and Main, K. M. (2005) Hypospadias in a cohort of 1072 Danish newborn boys: prevalence and relationship to placental weight, anthropometrical measurements at birth, and reproductive hormone levels at three months of age, *J. Clin. Endocrinol. Metab.* **90**, 4041-4046.

Boisen, K. A., Kaleva, M., Main, K. M., Virtanen, H. E., Haavisto, A. M., Schmidt, I. M., Chellakooty, M., Damgaard, I. N., Mau, C., Reunanen, M., Skakkebaek, N. E. and Toppari, J. (2004) Difference in prevalence of congenital cryptorchidism in infants between two Nordic countries, *Lancet* **363**, 1264-1269.

Bonde, J. P., Ernst, E., Jensen, T. K., Hjollund, N. H., Kolstad, H., Henriksen, T. B., Scheike, T., Giwercman, A., Olsen, J. and Skakkebaek, N. E. (1998) Relation between semen quality and fertility: a population-based study of 430 first-pregnancy planners, *Lancet* **352**, 1172-1177.

Buemann, B., Henriksen, H., Villumsen, A. L., Westh, A. and Zachau-Christiansen, B. (1961) Incidence of undescended testis in the newborn, *Acta. Chir. Scand. Suppl.* **283,** 289-293.

Carlsen, E., Giwercman, A., Keiding, N. and Skakkebaek, N. E. (1992) Evidence for decreasing quality of semen during past 50 years, *Br. Med. J.* **305**, 609-613.

Damgaard, I. N., Main, K., Toppari, J. and Skakkebaek, N. E. (2002) Impact of exposure to endocrine disrupters *in utero* and in childhood on adult reproduction, *Best Practise and Res. Clin. Endocrinol. Metabol.* **16**, 289-309.

Ferlay, J., Bray, F., Pisani P. and Parkin D.M.(2004) *GLOBObCAN 2002: Cancer Incidence, Mortality and Prevalence Worldwide IARC CancernBase.* No. 5. version 2.0, IARC*Press*, Lyon.

Fisher, J. S. (2004). Environmental anti-androgens and male reproductive health: focus on phthalates and testicular dysgenesis syndrome, *Reproduction* **127**, 305-315.

Fisher, J. S., Macpherson, S., Marchetti, N. and Sharpe, R. M. (2003) Human 'testicular dysgenesis syndrome': a possible model using in-utero exposure of the rat to dibutyl phthalate, *Hum. Reprod.* **18**, 1383-1394.

Forman, D. and Moller, H. (1994). Testicular cance, *Cancer Surv.* **19-20**, 323-341.

Gray, L. E., Jr., Ostby, J., Furr, J., Price, M., Veeramachaneni, D. N. and Parks, L. (2000) Perinatal exposure to the phthalates DEHP, BBP, and DINP, but not DEP, DMP, or DOTP, alters sexual differentiation of the male rat, *Toxicol. Sci.* **58**, 350-365.

Gray, L. E., Jr., Wolf, C., Lambright, C., Mann, P., Price, M., Cooper, R. L. and Ostby, J. (1999) Administration of potentially antiandrogenic pesticides (procymidone, linuron, iprodione, chlozolinate, p,p'-DDE, and ketoconazole) and toxic substances (dibutyl- and diethylhexyl phthalate, PCB 169, and ethane dimethane sulphonate) during sexual differentiation produces diverse profiles of reproductive malformations in the male rat, *Toxicol. Ind. Health* **15**, 94-118.

Gray, L. E., Ostby, J., Furr, J., Wolf, C. J., Lambright, C., Parks, L., Veeramachaneni, D. N., Wilson, V., Price, M., Hotchkiss, A., Orlando, E. and Guillette, L. (2001) Effects of environmental antiandrogens on reproductive development in experimental animals, *Hum. Reprod. Update* **7**, 248-264.

Group, J. R. H. C. S. (1992) Cryptorchidism: a prospective study of 7500 consecutive male births, 1984-8. John Radcliffe Hospital Cryptorchidism Study Group, *Arch. Dis. Child.* 67, 892-899.

Hosie, S., Loff, S., Witt, K., Niessen, K. and Waag, K. L. (2000) Is there a correlation between organochlorine compounds and undescended testes? *Eur. J. Pediatr. Surg.* **10**, 304-309.

Huff, D. S., Hadziselimovic, F., Snyder, H. M., 3rd, Blythe, B. and Ducket, J. W. (1993). Histologic maldevelopment of unilaterally cryptorchid testes and their descended partners, *Eur. J. Pediatr.* **152** (Suppl 2), S11-14.

Irvine, D. S. (1994) Falling sperm quality, *Br. Med. J.* **309**, 476.

Jorgensen, N., Andersen, A. G., Eustache, F., Irvine, D. S., Suominen, J., Petersen, J. H., Andersen, A. N., Auger, J., Cawood, E. H., Horte, A., Jensen, T. K., Jouannet, P., Keiding, N., Vierula, M., Toppari, J. and Skakkebaek, N. E. (2001) Regional differences in semen quality in Europe, *Hum. Reprod.* **16**, 1012-1019.

Jorgensen, N., Carlsen, E., Nermoen, I., Punab, M., Suominen, J., Andersen, A. G., Andersson, A. M., Haugen, T. B., Horte, A., Jensen, T. K., Magnus, O., Petersen, J. H., Vierula, M., Toppari, J. and Skakkebaek, N. E. (2002) East-West gradient in semen quality in the Nordic-Baltic area: a study of men from the general population in Denmark, Norway, Estonia and Finland, *Hum. Reprod.* **17**, 2199-2208.

Kawamura, K., Kumagai, J., Sudo, S., Chun, S. Y., Pisarska, M., Morita, H., Toppari, J., Fu, P., Wade, J. D., Bathgate, R. A. and Hsueh, A. J. (2004) Paracrine regulation of mammalian oocyte maturation and male germ cell survival, *PNAS* **101**, 7323-7328.

McIntyre, B. S., Barlow, N. J. and Foster, P. M. (2001) Androgen-mediated development in male rat offspring exposed to flutamide *in utero*: permanence and correlation of early postnatal changes in anogenital distance and nipple retention with malformations in androgen-dependent tissues, *Toxicol. Sci.* **62**, 236-249.

McKinnell, C., Sharpe, R. M., Mahood, K., Hallmark, N., Scott, H., Ivell, R., Staub, C., Jegou, B., Haag, F., Koch-Nolte, F. and Hartung, S. (2005) Expression of insulin-like factor 3 protein in the rat testis during fetal and postnatal development and in relation to cryptorchidism induced by *in utero* exposure to di (n-Butyl) phthalate, *Endocrinology* **146**, 4536-4544.

McLachlan, J. A., Newbold, R. R., Burow, M. E. and Li, S. F. (2001) From malformations to molecular mechanisms in the male: three decades of research on endocrine disrupters, *APMIS* **109**, 263-272.

Ostby, J., Kelce, W. R., Lambright, C., Wolf, C. J., Mann, P. and Gray, L. E., Jr. (1999) The fungicide procymidone alters sexual differentiation in the male rat by acting as an androgen-receptor antagonist *in vivo* and *in vitro, Toxicol. Ind. Health* **15**, 80-93.

Paulozzi, L. J., Erickson, J. D. and Jackson, R. J. (1997) Hypospadias trends in two US surveillance systems, *Pediatrics* **100**, 831-834.

Pierik, F. H., Burdorf, A., Nijman, J. M., de Muinck Keizer-Schrama, S. M., Juttmann, R. E. and Weber, R. F. (2002) A high hypospadias rate in The Netherlands, *Hum. Reprod.* **17**, 1112-1115.

Preiksa, R. T., Zilaitiene, B., Matulevicius, V., Skakkebaek, N. E., Petersen, J. H., Jorgensen, N. and Toppari, J. (2005) Higher than expected prevalence of congenital cryptorchidism in Lithuania: a study of 1204 boys at birth and 1 year follow-up, *Hum. Reprod.* **20**, 1928-1932.

Rajpert-De Meyts, E., Jorgensen, N., Brondum-Nielsen, K., Muller, J. and Skakkebaek, N. E. (1998) Developmental arrest of germ cells in the pathogenesis of germ cell neoplasia, *APMIS* **106**, 198-204; discussion 204-196.

Savage, M. O. and Lowe, D. G. (1990) Gonadal neoplasia and abnormal sexual differentiation, *Clin. Endocrinol (Oxf).* **32**, 519-533.

Scully, R. E. (1981). Neoplasia associated with anomalous sexual development and abnormal sex chromosomes, *Pediat. Adolesc. Endocr.* **8**, 203-217.

Sharpe, R. M. (2003) The 'oestrogen hypothesis'- where do we stand now? *Int. J. Androl.* **26**, 2-15.

Sharpe, R. M. and Skakkebaek, N. E. (1993) Are oestrogens involved in falling sperm counts and disorders of the male reproductive tract? *Lancet* **341**, 1392-1395.

Skakkebaek, N. E., Berthelsen, J. G., Giwercman, A. and Muller, J. (1987) Carcinoma-in-situ of the testis: possible origin from gonocytes and precursor of all types of germ cell tumours except spermatocytoma, *Int. J. Androl.* **10**, 19-28.

Skakkebaek, N. E., Holm, M., Hoei-Hansen, C., Jorgensen, N. and Rajpert-De Meyts, E. (2003). Association between testicular dysgenesis syndrome (TDS) and testicular neoplasia: evidence from 20 adult patients with signs of maldevelopment of the testis, *APMIS* **111**, 1-9.

Skakkebaek, N. E., Rajpert-De Meyts, E. and Main, K. M. (2001) Testicular dysgenesis syndrome: an increasingly common developmental disorder with environmental aspects, *Hum. Reprod.* **16**, 972-978.

Sohval, A. R. (1954) Testicular dysgenesis as an etiologic factor in cryptorchidism, *J. Urol.* **72**, 693-702.

Sorensen, H. R. (1953) *Hypospadias.* With special reference to aetiology. Copenhagen: Munksgaard.

Swan, S. H., Elkin, E. P. and Fenster, L. (1997) Have sperm densities declined? A reanalysis of global trend data, *Environ. Health Perspect.* **105**, 1228-1232.

Thong, M., Lim, C. and Fatimah, H. (1998). Undescended testes: incidence in 1,002 consecutive male infants and outcome at 1 year of age, *Pediatr. Surg. Int.* **13**, 37-41.

Toppari, J., Larsen, J. C., Christiansen, P., Giwercman, A., Grandjean, P., Guillette, L. J., Jr., Jegou, B., Jensen, T. K., Jouannet, P., Keiding, N., Leffers, H., McLachlan, J. A., Meyer, O., Muller, J., Rajpert-De Meyts, E., Scheike, T., Sharpe, R., Sumpter, J. and Skakkebaek, N. E. (1996) Male reproductive health and environmental xenoestrogens, *Environ. Health Perspect.* **104** Suppl 4, 741-803.

Vierula, M., Niemi, M., Keiski, A., Saaranen, M., Saarikoski, S. and Suominen, J. (1996) High and unchanged sperm counts of Finnish men, *Int. J. Androl.* **19**, 11-17.

Virtanen, H. E., Kaleva, M., Haavisto, A. M., Schmidt, I. M., Chellakooty, M., Main, K. M., Skakkebaek, N. E. and Toppari, J. (2001) The birth rate of hypospadias in the Turku area in Finland, *APMIS* **109**, 96-100.

Virtanen, H. E., Rajpert-De Meyts, E., Main, K. M., Skakkebaek, N. E. and Toppari, J. (2005) Testicular dysgenesis syndrome and the development and occurrence of male reproductive disorders. *Toxicol. Appl. Pharmacol.* **207** (Suppl 1), 501-505.

vom Saal, F. S., Cooke, P. S., Buchanan, D. L., Palanza, P., Thayer, K. A., Nagel, S. C., Parmigiani, S. and Welshons, W. V. (1998) A physiologically based approach to the study of bisphenol A and other estrogenic chemicals on the size of reproductive organs, daily sperm production, and behavior, *Toxicol. Ind. Health* **14**, 239-260.

Walker, A. H., Bernstein, L., Warren, D. W., Warner, N. E., Zheng, X.. and Henderson, B. E. (1990) The effect of *in utero* ethinyl oestradiol exposure on the risk of cryptorchid testis and testicular teratoma in mice, *Br. J. Cancer* **62**, 599-602.

Wilson, V. S., Lambright, C., Furr, J., Ostby, J., Wood, C., Held, G. and Gray, L. E., Jr. (2004) Phthalate ester-induced gubernacular lesions are associated with reduced insl3 gene expression in the fetal rat testis, *Toxicol. Lett.* **146**, 207-215.

Wolf, C. J., LeBlanc, G. A., Ostby, J. S. and Gray, L. E., Jr. (2000) Characterization of the period of sensitivity of fetal male sexual development to vinclozolin, *Toxicol. Sci.* **55**, 152-161.

Yasuda, Y., Kihara, T. and Tanimura, T. (1985) Effect of ethinyl estradiol on the differentiation of mouse fetal testis, *Teratology* **32**, 113-118.

ENDOCRINE DISRUPTERS, STEROIDOGENESIS AND INFLAMMATION

K. SVECHNIKOV*, V. SUPORNSILCHAI, I. SVECHNIKOVA, M. STRAND, C. ZETTERSTRÖM, A. WAHLGREN AND O. SÖDER
Department of Woman and Child Health
Pediatric Endocrinology Unit
Karolinska Institute & University Hospital
Q2:08
Stockholm
SWEDEN
(authour for correspondence, Email: Konstantin.Svechnikov@kbh.ki.se)*

Summary

During the past decade, considerable information concerning the effects of endocrine disrupters (EDCs) on animals and humans has been accumulated. These compounds of anthropogenic or natural origin mimic the action of sex hormones, thus disturbing the endocrine system. The present overview covers the different classes of EDCs, such as pesticides (that act as androgen receptor (AR) antagonists, i.e. anti-androgens), and phthalates and dioxins (which appear to inhibit fetal testosterone synthesis). The effects of these compounds on steroidogenesis by Leydig cells and reproductive development are reviewed and their possible key role in connection with increasing frequencies of abnormalities in reproductive development, such as hypospadias and cryptorchidism, is debated. Moreover, the influence of different classes of EDCs on ovarian and adrenal steroidogenesis is also described. In addition, putative interactions between EDCs and mediators of inflammation, that can potentially intensify the inflammatory process, are discussed.

P. Nicolopoulou-Stamati et al. (eds.), Congenital Diseases and the Environment, 255–280.

1. Introduction

Via the food chain, air and water, man and wildlife are exposed to numerous agents of anthropogenic or natural origin that can interact with the endocrine system. These so-called endocrine disrupters either activate or inhibit various endocrine pathways, depending on which cell-signalling receptors they bind to. The function of steroid hormone receptors such as ligand-inducible transcription factors is crucial to normal sex differentiation during embryonic development. Thus, inhibition of the action of the androgen receptor (AR) by metabolites of pesticides and fungicides during male sex differentiation and development often gives rise to abnormalities in reproductive organs. Similarly, the reduction in androgen production by fetal Leydig cells caused by exposure to phthalates leads to incomplete masculinization and various malformations in the reproductive tract of both humans and animals. Androgen deficiency during prenatal development is thought to disturb differentiation of the Wolffian duct, a process crucial to the proper development of reproductive organs.

Furthermore, different classes of EDCs exert harmful effects on hormonal function of ovaries and adrenals and can thereby affect reproductive potential and homeostasis in humans and animals.

In addition, evidence that environmental EDCs can induce the production of pro-inflammatory cytokines and thereby modulate inflammatory processes is beginning to accumulate. Thus, interaction between low-grade endotoxinemia (possibly originating from subclinical infection) and EDCs, involving pro-inflammatory cytokines as mediators, may disturb steroidogenesis during a critical window of development.

The aim of this review is to give an overview of the reported effects of EDCs on testicular, ovarian and adrenal steroidogenesis with respect to reproduction, fertility and metabolic homeostasis.

2. The role of androgens in male fetal differentiation

Development of the male reproductive tract is a dynamic process requiring the interaction of numerous factors and hormones. Among the major factors essential to the proper development of the internal and external male reproductive tract are the androgens testosterone and dihydrotestosterone (DHT) (Wilson, 1978). The testes of the human male embryo begin to secrete androgens after 9 weeks of gestation.

Development of the Wolffian duct, as well as virilization of external genitalia, occurs between 9 and 15 weeks of gestation.

Testosterone, the levels of which peaks between 11 and 18 weeks of gestation, stimulates differentiation of this Wolffian duct system into the epididymis, vas deferens and seminal vesicles. Masculinization of the external genitalia and prostate is mediated primarily by DHT, a more potent metabolite of testosterone produced by the enzyme 5α–reductase. Simultaneously, the Sertoli cells of the fetal testis secrete anti-Müllerian hormone (AMH), which induces regression of the Müllerian duct.

The actions of both testosterone and DHT are mediated by a nuclear receptor that is expressed in both the internal and the external genitalia. In the human male fetus, this androgen receptor (AR) is expressed as early as after 8 weeks of gestation, prior to the onset of androgen secretion, and at a higher level in genital than in urogenital structures. Upon high-affinity binding of its agonists, AR undergoes conformational changes and is imported into the nucleus where it dimerizes (Wong *et al.,* 1995). This dimer then binds to an androgen response element located within introns or flanking regions of responsive genes and activates their transcription (Tan *et al.,* 1992). The resulting gene products are responsible for androgen-dependent functions crucial to the sexual differentiation of developing tissues.

Binding of anti-hormones to AR is thought to induce a conformation that differs from that associated with agonist binding, thereby altering the ability of this receptor to activate gene transcription (Eckert and Katzenellenbogen, 1982). The models for this anti-hormone action that have been proposed include two mechanisms that differ with respect to the ability of AR to bind to DNA: Type I antagonists are proposed to prevent binding to DNA, whereas Type II antagonists promote this binding, but nevertheless do not allow initiation of transcription (Truss *et al.,* 1994).

The environmental anti-androgens identified to date exhibit low-to-moderate affinity for AR and act as Type I antagonists (Kelce *et al.,* 1995). It has been suggested that binding of several ligand molecules to the same receptor dimer is required for AR antagonism (Wong *et al.,* 1995), but the specific mechanism(s) by which AR binding to DNA is inhibited remains unknown. This mechanism could involve enhanced AR degradation as a consequence of an inappropriate conformation, and/or of an inability of rapidly dissociating ligands to stabilize the receptor (Kemppainen *et al.,* 1992).

Interestingly, the steroid-binding characteristics of AR isolated from the embryonic urogenital tract are identical to those of the corresponding receptor in tissues of the adult reproductive tract (Bentvelsen *et al.,* 1994). Thus, the greater sensitivity of the developing male fetus to the anti-androgenic effects of environmental chemicals may simply be a result of the lower levels of endogenous androgens present. Accordingly, any event which impairs testosterone and/or DHT formation or the normal functioning of the AR may result in deficient androgen action in a male fetus and subsequent inadequate virilization of the newborn male (Sultan *et al.,* 2001).

3. Functions of the fetal Leydig cell

During the prenatal period, the fetal Leydig cell is the primary source of androgens. During the fetal period, these cells secrete testosterone and other androgens which regulate not only the masculinization of internal and external genitalia, but also neuroendocrine function.

The development of fetal Leydig cells can be divided into three phases, idifferentiation, fetal maturation and involution (Pelliniemi and Niei, 1969). In the human fetal testis, the differentiation phase occurs at a fetal age of 8-14 weeks; fetal maturation occurs during the 14th-18th weeks of gestation; and the phase of involution extends from 18-38 weeks. Also, in humans, the maximum number of Leydig cells per pair of testes ($48x10^6$) has been reported to be present during weeks 13-16 of gestation (Codesal *et al.,* 1990).

Initial expression of the enzymes of the steroidogenic pathway and the onset of testosterone synthesis in fetal Leydig cells is independent of luteinizing hormone (LH) and its receptor; but following the appearance of this hormone and its receptor, they regulate testosterone synthesis trophically (Huhtaniemi and Pelliniemi, 1992). In the rat, the full-length mRNA encoding the extracellular domain of the LH receptor appears at a gestational age of 16.5 days and, employing in situ hybridization, has been localized to the Leydig cells (Zhang *et al.,* 1994).

The major androgen produced by fetal Leydig cells is testosterone (Murono, 1990). Androgens are synthesized from cholesterol by the sequential actions of the cytochrome P450 side-chain cleavage enzyme (P450scc), 3β–hydroxysteroid dehydrogenase (3βHSD), cytochrome P45017α–hydroxylase (P450c17) and 17β–hydroxysteroid dehydrogenase/17-ketosteroid reductase (17βHSD/17KSR), all of which are expressed in the fetal testis (Greco and Payne, 1994). Indeed, P450scç, 3βHSD, and P450c17 are expressed exclusively by the Leydig cells (Majdic *et al.,* 1998; Sha *et al.,* 1996). However, O´Shaughnessy *et al.* (2000) have shown that fetal mouse Leydig cells synthesize predominantly androstenedione and that the enzyme isoform that converts this weak androgen into the more potent testosterone, 17β-hydroxysteroid dehydrogenase type III, is expressed primarily by Sertoli cells (O'Shaughnessy *et al.*, 2000).

In addition to androgens, fetal Leydig cells secrete insulin-like factor-3 (Insl3), also referred to as relaxin-like factor, which acts on the gubernaculum of the testis and plays a key role in guiding the testis during its transabdominal descent (Sharpe, 2001).

4. Impact of endocrine disrupters on male reproductive health

Normally, descent of the testicles into the scrotum is not completed until the third trimester of gestation. In humans, initial, transabdominal phase of descent occurs at 10-15 weeks of gestational age and the fetal testicle migrates from the groin to the scrotum during weeks 26-35 (Hutson *et al.,* 1994). Inguinal descent, the phase most commonly disrupted, is dependent on testosterone. Cryptorchidism, a condition in which one or both testicles fail to descend into the scrotum is the most common ailment affecting newborn male children, exhibiting a frequency of 1-4% among live male births. Epidemiological studies have revealed that the risk for cryptorchidism is higher among the sons of women working with pesticides (Weidner *et al.,* 1998). Furthermore, significantly higher concentrations of hexachlorbenzene and heptachlorepoxide were found in the adipose tissue of boys with testicular maldescent, compared to a control group (Hosie *et al.,* 2000).

The most common birth defects caused in humans by inhibition of AR action are alterations in the development of the external male genitalia (Sweet *et al.,* 1974). During the first 12 weeks of gestation, a critical period in this development (Kalloo *et al.,* 1993), fetal androgens induce ventral folding and fusion of the urethral fold to form the penis, as well as fusion of the labioscrotal folds to form the scrotum. During the second and third trimesters of gestation, androgen-dependent growth of these structures continues.

Another clinical problem that may be related to environmental anti-androgens is the increasing incidence of hypospadias (incomplete fusion of the urethral folds of the penis, causing the urethral opening to be displaced from the tip of the penis). It has been reported that the exposure of fathers to dioxins in connection with the Seveso accident in 1976 resulted in an increased incidence of hypospadias among their sons (Baskin *et al.,* 2001). Furthermore, isolated hypospadias in the human population are seldom associated with mutations in the coding sequence of the AR gene (Allera *et al.,* 1995), with a deficiency in steroid 5α-reductase (type 2) (Wilson *et al.,* 1993) or with reduced levels of AR (Bentvelsen *et al.,* 1995). Thus, the increased incidents of human idiopathic hypospadias and cryptorchidism suggest that environmental anti-androgens may disrupt the normal development of male genitalia *in utero* (Jirasek, 1971).

5. Effects of environmental anti-androgens on the reproductive development and hormonal functions of Leydig cells

The environmental anti-androgenic compounds that have been well-characterized include procymidone, linuron, vinclozolin and p,p′-DDT and its derivatives, all of which are antagonists of AR and inhibit androgen-dependent tissue growth *in vivo* (Gray *et al.,* 1999b). None of these pesticides or their metabolites appear to display significant affinity for the oestrogen receptor or to inhibit 5α-reductase activity *in vitro* (Kelce *et al.,* 1995).

5.1. Procymidone

Procymidone(N-(3,5-dichlorophenyl)-1,2-dimethyl-cyclopropane-1,2dicarboximide) is widely used as a fungicide for the control of plant diseases. The weak anti-androgenic activity in these compound, reflected in a low affinity of its binding to the androgen receptor in the prostate, probably explains why it causes hypergonadotropism (Hosokawa *et al.,* 1993). A feasible hypothesis is that long-term administration of procymidone inhibits the negative feedback exerted by androgens on the hypothalamus and/or the pituitary, thereby causing hypergonadotropism in rats. The lesser ability of testosterone to bind to the androgen receptor in this situation suppresses the inhibitory feedback of testosterone on LH production, by preventing the hypothalamus and anterior pituitary from recognizing the presence of testosterone. This results in hypersecretion of LH with concomitant elevation of serum testosterone (see outline of proposed mechanisms in Figure 1). Long-term hypergonadotropism and hyperstimulation of Leydig cells induced by procymidone give rise to interstitial cell tumours in male rats (Murakami *et al.,* 1995).

Furthermore, in this same connection, long-term dietary administration of procymidone to rats has been reported to lead to elevated serum levels of testosterone and LH (Murakami *et al.,* 1995). Recent findings from our own laboratory reveal that dietary administration of procymidone to rats for three months not only enhances serum levels of LH and testosterone, but that the Leydig cells isolated from these animals display an enhanced capacity for producing testosterone in response to stimulation by hCG or (Bu)2cAMP, as well as elevated levels of StAR, P450scc and P450c17 (Svechnikov *et al.,* 2005).

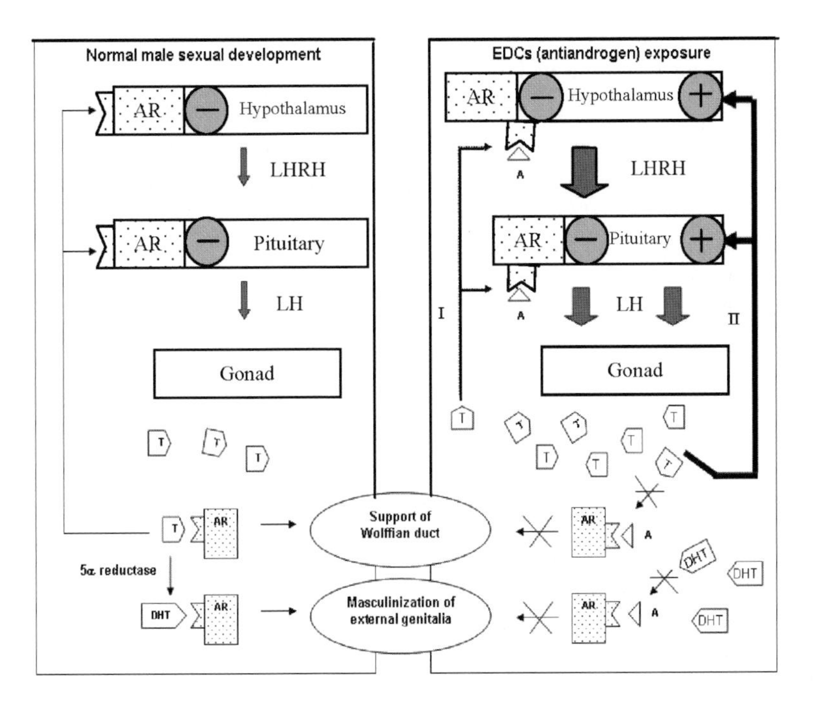

Figure 8. Schematic overview of the hypothalamo-pituitary-testis axis and the hypothesized suppressive effects of anti-androgen exposure. Anti-androgens bind to the androgen receptor (AR) disturbing the formation of the testosterone-androgen complex. The lesser ability of testosterone (T) to bind to the AR suppresses the inhibitory feed back of T on LH production by preventing the hypothalamus and anterior pituitary from recognizing the presence of T. This results in hypersecretion of LH, development of hypergonadotropism and increased production of T by Leydig cells. Blocking peripheral AR by anti-androgens inhibits androgen-mediated effects on target organs resulting in incomplete masculinization and malformations in the male reproductive tract of humans and animals

5.2. Linuron

Linuron (3-(3,4-dichlorophenyl)-1-methoxy-1-methylurea) is currently being marketed as a selective urea-based herbicide for pre- and/or post-emergence control of weeds in crops, and potatoes. This compound binds to both rat prostatic and human AR (hAR) and inhibits DHT-induced gene expression *in vitro* (Cook *et al.,* 1993). The anti-androgenic effect of linurone is observed clearly upon administration during gestation (Gray *et al.,* 1999b; Lambright *et al.,* 2000) or in a Hershberger assay (Lambright *et al.,* 2000). For example, administration of linurone (100 mg/kg daily) to rats during days 14-18 of gestation induces epispadias, decreases in the anogenital distance (AGD) and reduction of the size of androgen-dependent tissues in male offspring (Gray *et al.,* 1999b). Similarly, administration of linuron (200 mg/kg daily) to sexually immature and mature rats for 2 weeks has been shown to decrease the weights of accessory sex organs and increase serum estradiol and LH levels, without altering the serum concentration of testosterone (Cook *et al.,* 1993). Cook *et al.* concluded that this inability of linuron to increase testosterone levels may reflect the lower potency of this compound in comparison to flutamide and that the elevation of serum estradiol may be due to enhanced conversion of testosterone to estradiol via aromatase activity.

In addition, fetuses exposed *in utero* to linuron (50 mg/kg/day) demonstrated neither significant changes in their intratesticular or serum levels of testosterone nor abnormal clusters of Leydig cells (McIntyre *et al.,* 2002). In contrast, prenatal exposure to linuron (75 mg/kg/day) from gestational day 14-18 has recently been shown to decrease testosterone production by fetal Leydig cells (Hotchkiss *et al.,* 2004). This latter study suggests that linuron disrupts the development of androgen-dependent tissues primarily by acting as an AR antagonist and inhibiting testosterone synthesis by the fetal testis.

5.3. Vinclozolin

Vinclozolin is a dicarboximide fungicide that has two active metabolites, M1 and M2, which have anti-androgenic properties. These metabolites compete for AR androgen binding and inhibit DHT-induced transcriptional activation by blocking AR binding to androgen response element DNA (Wong *et al.,* 1995).

It have been shown that oral administration of vinclozolin delayed pubertal maturation, retarded sex accessory gland growth and increased serum LH, testosterone and 5α-androstane, 3α,17β–diol concentrations (Monosson *et al.,* 1999). This effect of vinclozolin is thought to result from blocking the formation of the testosterone-androgen receptor complex by this compound, as was described for procymidone. Studies *in vivo* have also revealed that vinclozolin-treated male offspring display female-like AGD at birth, retained nipples, undescended testes and small sex accessory glands (Gray *et al.,* 1999a). Further, the most sensitive period of fetal development to antiandrogenic effects of vinclozolin was found to be on gestation days 16-17 (Wolf *et al.,* 2000). Experiments *in vitro* have recently revealed that vinclozolin did not effect on basal and hCG-stimulated testosterone production by rat Leydig cells in primary culture (Murono and Derk, 2004). In addition, vinclozolin was recently reported to decrease spermatogenic capacity and to increase the incidence of male infertility in rats. These effects were transferred through the male germ line to nearly all males of all subsequent generations examined from F1 to F4 generations, suggesting an epigenetic mechanism of action (Anway *et al.*, 2005).

5.4. p,p′ DDT and its derivatives

p,p′-DDE, a metabolite of the environmentally persistent pesticide DDT, acts as an antagonist of AR both *in vivo* and *in vitro* (Kelce *et al.*, 1995). When p,p′-DDE (100 mg/kg daily) was administered to rats during development on days 14-18 of gestation, the AGD was reduced and hypospadias, retention of nipples and reduction in the weights of androgen-dependent tissues occurred (Gray *et al.*, 1999b). Moreover, long-term administration of p,p′-DDE to rat pups from the time of weaning until approximatly 50 days of age did not significantly alter the weights of accessory sex organs or serum hormone levels (Kelce *et al.*, 1995), but did produce pronounced reductions in androgen-dependent tissue weights in the Hershberger assay (Gray *et al.*, 1999b). As with rats, administration of p,p′-DDT or p,p′-DDE to rabbits during gestation induces reproductive abnormalities, including cryptorchidism, in the male offspring (Gray *et al.*, 1999b).

It has also been suggested that p,p′-DDE causes abnormalities in the reproductive development of wildlife (Facemire *et al.*, 1995). For instance, high concentrations of p,p′-DDE, together with enhanced susceptibility of the developing reproductive system to this environmental anti-androgen due to a loss of genetic diversity, may be contributing to the high incidence of cryptorchidism in the Florida panther (Facemire *et al.*, 1995).

Following the ban on DDT, methoxychlor (MC), a derivative of DDT, has found increasing use as an insecticide. Methoxychlor is converted metabolically to several monohydroxy- and dihydroxy-metabolites that exhibit estrogenic activity (Bulger *et al.*, 1978). Previous studies have shown that daily exposure of male Long-Evans hooded rats to MC (100 or 200 mg/kg, by gavage) from the day of weaning to approximately 97-100 days of postnatal age reduces both circulating levels of testosterone and the hCG-induced stimulation of testosterone production by decapsulated testes (Gray *et al.*, 1989).

HPTE, a metabolite of MC, binds to ERα (Waller *et al.,* 1996) and acts as an ERα agonist and ERβ antagonist in human HepG2 hepatoma cells (Maness *et al.,* 1998). Recent studies have revealed that HPTE inhibits both the basal and hCG-stimulated production of testosterone by cultured Leydig cells from adult rats, primarily by interfering with P450scc, which converts cholesterol to pregnenolone (Akingbemi *et al.,* 2000; Murono and Derk, 2004).

6. Effects of phthalates on Leydig cell function and reproductive development

Phthalates are widely used as plasticizers in the production of plastics, as well as solvents in inks, and are present, among other places, in food packaging and certain cosmetics (Thomas and Thomas, 1984). Although diethylhexyl phthalate (DEHP) and monoethylhexyl phthalate (MEHP) disturb reproductive development in an anti-androgenic fashion, neither of these compounds binds to AR (Parks *et al.,* 2000). It has now been established that such phthalates inhibit Leydig cell steroidogenesis at different ages of fetal development.

For example, prenatal exposure of rats to DEHP attenuates serum levels of both LH and testosterone in male offspring at 21 and 35 days of postnatal age (Akingbemi *et al.,* 2001). In contrast, treatment with DEHP for two weeks postnatally (beginning at 21 or 35 days of age) results in a significant decrease in the activity of 17βHSD and reduces testosterone production by Leydig cells by 50%. On the other hand, treatment of adult rats with DEHP does not influence their Leydig cell steroidogenesis (Akingbemi *et al.,* 2001). It has been concluded that the effects of DEHP on Leydig cell steroidogenesis are dependent on the stage of development at the time of exposure and may involve modulation of the activities of steroidogenic enzymes and/or serum levels of LH.

Furthermore, prenatal exposure of rats to di(n-butyl)phthalate (DBP) during days 12-21 of gestation gives rise to Leydig cell hyperplasia and adenomas, as well as agenesis of the epididymis. Despite the increased numbers of Leydig cells present after such exposure, testicular levels of testosterone are reduced at 18 and 21 days of gestation age. It has been suggested that the proliferation of Leydig cells observed in rat fetuses exposed to DBP may be a compensatory mechanism designed to increase testicular steroidogenesis in response to levels of testosterone that are insufficient to promote normal differentiation of the male reproductive tract (Mylchreest *et al.,* 2002).

In addition to reducing testicular levels of testosterone, inhibition of fetal Leydig cell steroidogenesis by phthalates causes numerous malformations of androgen-dependent tissues in male rats, including a decrease in the anogenital distance, hypospadias, epididymal agenesis and testicular atrophy (Gray *et al.,* 2001). These abnormalities are suggestive of sub-optimal virilisation of the Wolffian duct and urogenital sinus. Recent investigations have revealed that treatment with phthalates interferes with the transcription of several key genes involved in both cholesterol transport and the biosynthesis of testosterone (Barlow *et al.,* 2003; Thompson *et al.,* 2004). Thus, Thompson *et al.* (2004) reported that such treatment attenuates the expression of scavenger receptor B1 (SR-B1), steroidogenic acute regulatory protein (StAR), P450scc, 3βHSD and cytochrome P450c17 in the fetal rat testis. Interestingly, the suppression of testosterone synthesis induced by DBP is reversible, with recovery coinciding with the clearance of DBP and its testicular metabolite (Thompson *et al.,* 2004).

In summary, both DBP and DEHP induce a 60-85% reduction in fetal testosterone levels during the critical period of reproductive development (Akingbemi *et al.,* 2001; Mylchreest *et al.,* 2002). Thus, it is probable that exposure of fetuses to phthalates may lead to incomplete masculinization and various malformations in the male reproductive tract of humans and animals. This suggestion is supported by the recent findings of a decreased anogenital distance in boys born to mothers with high levels of phthalates in their urine. This report is one of the first directly associating prenatal exposure to phthalates at environmentally relevant levels to adverse outcome of male reproductive development in humans (Swan *et al.,* 2005).

7. Effects of dioxin on androgen production by Leydig cells and on reproductive health

Polychlorinated dibenzo-p-dioxins such as 2,3,7,8-tetrachlorodibenzo-p-dioxin (TCDD) are released into the environment as a by-product of the manufacture of chlorinated hydrocarbons and other industrial processes, and are recognized as being highly potent developmental and reproductive toxins. The toxicity of TCDD is mediated by translocation of an aryl hydrocarbon receptor (AhR) nuclear translocator complex into the nucleus, where it binds to dioxin-responsive elements on DNA, and/or by interfering with various other signalling molecules in cells (Reyes *et al.,* 1992).

TCDD suppresses steroidogenesis by Leydig cells. Thus, Leydig cells in the testis of rats exposed to this compound are reduced in size and number; contain a decreased volume of smooth endoplasmic reticulum; and carry out weak steroidogenesis

(Johnson *et al.,* 1994). Additional experiments have demonstrated that pharmacological treatment of rats with human chorionic gonadotropin (hCG) preserves the Leydig cell cytoplasm and organelles in sufficient quantities to prevent this disturbance of Leydig cell function by TCDD (Wilker *et al.,* 1995). TCDD treatment is also associated with inhibition of the mobilization of cholesterol and of the activities of cytochromes P450scc and P450c17 in Leydig cells (Mebus *et al.,* 1987; Moore *et al.,* 1991). Many or all of these alterations in steroidogenesis might be explained by the recent finding that TCDD inhibits the formation of cAMP by cultured Leydig cells (Lai *et al.,* 2005).

Interestingly, a single dose of TCDD (ranging from 50 ng to 2 μg/kg) to rats or hamsters during sexual differentiation gives rise to a number of reproductive alterations in the male, including delayed puberty and reductions in the size of the ventral prostate, seminal vesicle and testis (Gray *et al.,* 1995). The observation that these animals do not display decreased anogenital distance (AGD) or areolas suggests that TCDD may also be affecting the developing reproductive tract via pathways that do not involve the AR.

Much less is known about the potential effects of TCDD on the functions of human Leydig cells. The report that serum levels of dioxin correlated positively with serum levels of LH and negatively with testosterone levels in men exposed to dioxin in a chemical plant (Egeland *et al.,* 1994), indicates that TCDD may exert a direct toxic effect on the human Leydig cell.

8. Effects of endocrine disrupters on adrenal function

The rich supply of blood to the adrenal gland in relationship to its small tissue mass enhances the relative exposure of this endocrine organ to xenobiotics. Moreover, because of the high lipid content of the adrenal cortex, lipophilic compounds readily accumulate in this region (Szabo and Lippe, 1989).

The DDT metabolite 3-methylsulfonyl-DDE (3-MeSO2-DDE) has been reported to exert toxic effects on the mitochondria in the adrenal zona fasciculate of mice (Jonsson *et al.,* 1991). *In vitro*, this compound binds to the mitochondria-containing fraction of the rat adrenal with an affinity that is 50-fold greater than that of its binding to the post-mitochondrial supernatant fraction. Furthermore, transmission electron microscopy has revealed that a single intraperitoneal injection of MeSO2-DDE into a rat induces morphological changes in adrenal mitochondria, including the disorganization and disappearance of central cristae (Jonsson *et al.,* 1991). In addition, aryl methyl sulfone metabolites of PCBs and DDT competitively inhibit CYP11B1-catalyzed activity in the mouse adrenocortical Y1 cell line (Johansson *et al.,* 1998). Polychlorinated biphenyls (PCBs) similar in structure to dioxin have been shown to accumulate in the adrenal glands (Brandt, 1977) and to act as EDCs by disturbing adrenocortical steroidogenesis. PCB 126 was recently found to attenuate both basal and cAMP-induced androstenedione production, as well as the expression of CYP17 mRNA in the human adrenocortical H295R cell line. In contrast, high concentrations of PCB126 stimulate basal cortisol and aldosterone biosynthesis, and up-regulate the expression of mRNAs coding CYP21B, CYP11B1, and CYP11B2 in these same cells (Li and Wang, 2005). Similar observations have been made in animal studies, where treatment of male rats with Arochlor 1254 (a mixture of PCBs) by gavage for 15 weeks leads to elevation of the basal serum level of corticosterone, without affecting the increase in corticosterone level caused by stress (Miller *et al.,* 1993). It has been suggested that high concentrations of PCB126 enhance the sensitivity and responsiveness of adreno-cortical cells to ACTH by up-regulating the expression of membrane receptors (Li and Wang, 2005).

In our current studies we employ rat adrenocortical cells in primary culture (figure 9A) as a convenient model for elucidating the effects of EDCs on the hormonal functions of these cells. Recently, we have demonstrated that resveratrol, a compound that occurs naturally in grapes and red wine (Gehm *et al.,* 1997), markedly suppresses adrenocortical steroidogenesis.

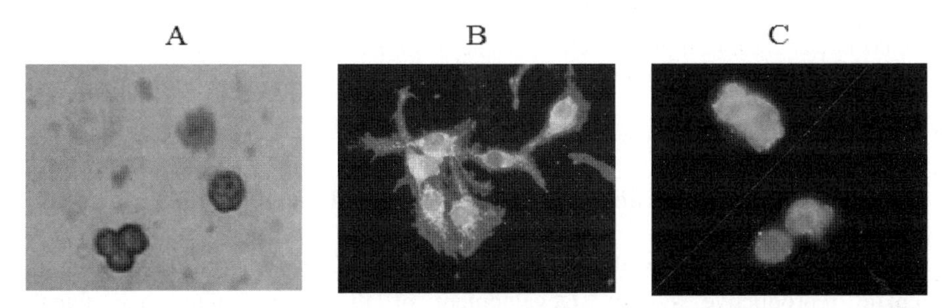

Figure 9. Light microscopic images of different types of steroidogenic cells in primary culture. These cells were isolated from rats and analyzed for the expression of various steroidogenic proteins. A) Untreated adrenocortical cells stained immunohistochemically for 11β–hydroxylase protein. B) Leydig cells stimulated with IL-1α *in vitro* and stained immunohistochemically for the StAR protein. C) Ovarian granulosa cells stained immunohistochemically for P450scc

As seen in figure 10, resveratrol decreases the ACTH-stimulated production of corticosterone by rat adrenocortical cells in culture by 40% (Supornsilchai, unpublished data). Additional experiments revealed that progesterone could not be adequately metabolized to corticosterone by these cells, due to suppression of the expression of cytochrome P450c21 (Supornsilchai *et al.,* 2005). Resveratrol is presently being used to treat several types of cancer and as a cardioprotectant, and possible alterations in the hormonal functions of the adrenal gland should be monitored during long-term therapy involving this substance.

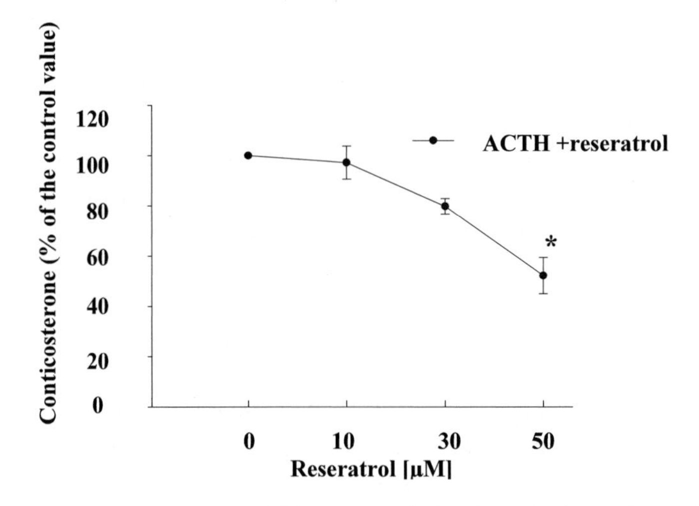

Figure 10. Resveratrol inhibits ACTH-induced activation of steroidogenesis in rat adrenocortical cells. Primary cultures of rat adrenocortical cells were pre-incubated with different concentrations of resveratrol for 24 h, following which fresh medium was added and incubation continued in the presence of ACTH (0.1 ng/ml) with or without resveratrol for an additional 24 h. Corticosterone in the medium was measured by RIA. The results presented are the means ± S.E of triplicate determinations and are expressed as percentages of control value. Each experiment was performed four times with the same results. $^{*}P<0.05$ compared to incubation with ACTH alone

9. Effects of endocrine disrupters on the hormonal functions of ovarian cells

It has been long recognized that various groups of EDCs exert harmful effects on the female reproductive system. Development of the female fetus and female reproductive system is strictly regulated by the hormones produced by the hypothalamus pituitary gonadal axis, including: gonadotropin-releasing hormone (GnRH), produced by the hypothalamus; gonadotropins such as follicle-stimulating hormone (FSH) and luteinizing hormone (LH), produced by the pituitary gland, and estrogens produced by the ovary. The signalling pathways initiated by any of these hormones may be a target for the toxic effects of endocrine-disrupting environmental chemicals, since they all involve specific hormone receptors to which xenobiotic toxicants may bind.

Exposure of primary cultures of granulosa cells to DDE, the most stable metabolite of DDT, decreases their FSH-stimulated synthesis of cAMP and progesterone and increases their rate of proliferation (Chedrese and Feyles, 2001). In contrast, lower concentrations of DDE up-regulate the expression of P450scc and enhance progesterone synthesis in these same cells (Crellin *et al.,* 2001). Clearly, the alterations in steroidogenesis by granulosa cells induced by DDE are dose-dependent. Similarly, methoxychlor, a methoxylated analogue of DDT that is more labile and readily degraded also inhibits progesterone synthesis by cultured granulosa cells, but without affecting the generation of cAMP, suggesting that the site of action of this compound is downstream from this second messenger (Chedrese and Feyles, 2001). Furthermore, it was reported recently that single exposure of a co-culture of granulosa and theca cells to p,p'-DDT and its metabolite p,p'-DDE increases estradiol secretion, suggesting an estrogenic effect (Wojtowicz *et al.,* 2004).

Phthalates are also thought to exert toxic effects on female reproduction. For example, chronic exposure of female factory workers to elevated levels of phthalates has been associated with decreased rates of pregnancy and higher rates of miscarriage (Aldyreva *et al.,* 1975). During pregnancy and delivery, both the mother and fetus may be exposed to DEHP originating from the plastic in medical devices, an exposure that can potentially disturb the normal development of the female reproductive organs.

In studies on rats, treatment with DEHP decreases serum levels of estradiol, prolongs the estrous cycle and prevents ovulation (Davis *et al.,* 1994a). The functionally significant reduction in estradiol level and the smaller follicles observed, indicate that the granulosa cells of the ovary are a target for the toxic effects of DEHP.

Primary cultures of rat granulosa cells (Figure 9C) respond to follicle-stimulating hormone (FSH) by up-regulating steroidogenic enzymes and hormone production by these cells can thus be monitored *in vitro*. Experiments with such primary cultures have been employed to elucidate the mechanism by which phthalates suppress the hormonal functions of rat granulosa cells. MEHP inhibits the FSH-stimulated production of cAMP and progesterone by these cells, effects that can be prevented by activators of adenylate cyclase or, in the latter case, by pregnenolone (Treinen *et al.,* 1990). However, MEHP also decreases estradiol production by FSH –stimulated rat granulosa cells, indicating inhibition of aromatase activity (Davis *et al.,* 1994b). Further studies in this connection revealed that MEHP decreases aromatase mRNA and protein in a dose-dependent manner (Lovekamp and Davis, 2001). In contrast, another mono-phthalate, MBP, had no influence on estradiol production by rat granulosa cells (Lovekamp-Swan and Davis, 2003).

TCDD and other dioxins are also known to adversely affect female reproductive functions in several experimental animals. A single dose of TCDD administrated to the laboratory macaque induces abortion, accompanied by a decrease in the serum level of estradiol (Guo *et al.,* 1999). Furthermore, rats exposed *in utero* to TCDD on day 8 of gestation exhibit a 30% reduction in ovarian weight at 1.5 years of age, compared to untreated animals.

Moreover, an early decline in the fertility of rats treated in this manner was also evident, since only 19% of the exposed female progeny produced a fifth litter, as compared to 61% in the control group. In addition, at 1 year of age, 47% of the rats exposed *in utero* to TCDD exhibited persistent vaginal estrus, as compared to only 16% of the control animals (Gray *et al.,* 1995). Such a state of continuous estrous can lead to infertility, as well as alterations in the cycling levels of estradiol and progesterone. It is of interest to note that in this same study, female progeny exposed to an identical dose *in utero* on day 15 of gestation displayed a lower degree of functional ovarian toxicity than those exposed on day 8, although enhanced frequencies of congenital malformations in the reproductive organs were apparent in both groups. In addition, p-dioxins and other poly halogenated aromatic hydrocarbons can increase the bio-availability of endogenous estrogens in target tissues, through the inhibition of the enzymes involved in estrogen inactivation (Kester *et al.,* 2002).

The response of primary cultures of rat granulosa cells to TCDD is dependent on the stage of sexual maturity of the animal at the time the cells are collected, the duration of exposure, and the presence or absence of FSH (Dasmahapatra *et al.,* 2000). In the case of luteinized human granulosa cells TCDD significantly attenuates the induction of P450 arom and P450scc by FSH and, in addition, exerts direct negative effect on granulosa cell steroidogenesis, by decreasing PKA activity and progesterone production (Enan *et al.,* 1996). Furthermore, a recent study found that TCDD attenuates induction of LH receptor mRNA by FSH in cultured rat granulosa cells (Minegishi *et al.,* 2004). Overall, it can be concluded that dioxin and its derivatives exert significant harmful effects on the hormonal functions of the ovary and can thereby disturb reproduction and fertility in humans and animals.

In addition, some heavy metals, such as cadmium, antimony and barium, were reported to possess estrogenic activity, as revealed by transcriptional expression and E-screen assay systems (Choe *et al.,* 2003). Further, in studies *in vivo*, exposure to cadmium was demonstrated to increase uterine wet weight, promote growth and development of the mammary glands and induced hormone-regulated genes in ovariectomized animals (Johnson *et al.,* 2003).

Additionally, some environmental chemicals such as alkylphenols, bisphenol A and several xenoestrogens may displace endogenous sex steroid hormones from human plasma sex-hormone binding globulin (SHBG) binding sites and disrupt the androgen-to-estrogen balance (Dechaud *et al.,* 1999).

10. Endocrine disrupters, inflammation and androgen production

Increasing evidence indicates that single endocrine disrupting chemicals act in synergy with other exogenous influences (including other EDCs) and endogenous host response mediators, amplifying their deleterious effects on health. In daily life, and contrary to the conditions in most experimental models, humans are never exposed to single substances alone. Instead a complex mixture of environmental xenobiotics including EDCs, microorganisms causing clinical and subclinical infections, and a multitude of other influences are continuously challenging the human organism. It is well known from chock and tissue injury models that hits from two or more factors simultaneously may result in an aggravated tissue response compared with the additive action of the single exposures alone (Colletti *et al.,* 1996; Hagberg and Mallard, 2005). Local host-defence mechanisms play an important role in these processes and may contribute to the tissue injury. Using the developing brain as model, it was found that the deleterious effects of ischemia or hypoxia at a level with little if any effect of these single factors alone were severely and synergistically aggravated in the presence of low-grade infection, experimentally mimicked by injection of endotoxin (Hagberg and Mallard, 2005). The effects of endotoxin were found to be mediated by pro-inflammatory cytokines, as cytokine inhibitors were shown to alleviate the injury.

An important finding in the present context was that endotoxin and even physiological mediators such as hCG were able to stimulate the expression of a wide array of pro-inflammatory cytokines by the testis at the transcriptional and translational levels. Thus, following systemic exposure to endotoxin (Jonsson *et al.,* 2001) or treatment with hCG (Assmus *et al.,* 2005), IL-1α and IL-6 showed de novo expression in the testis and several other pro-inflammatory cytokines including IL-1α, IL-1ra and TNFα were increased (Jonsson *et al.,* 1999). We also recently demonstrated that another member of the IL-1 super-family, interleukin-18, is expressed constitutively by the testis (Strand *et al.,* 2005).

There is also a direct link between EDCs and cytokine expression in different tissues including the testis. Thus, mono phthalates were recently shown to enhance IL-6 and IL-8 synthesis in the human epithelial A 549 cell line (Jepsen *et al.,* 2004) and acute exposure to phthalates was found to induce a rapid and transient increased production of IL-1 by the rat testis (Granholm *et al.,* 1992). From the above discussion, it might be hypothesized that exposure to EDCs during a state of chronic or acute inflammation might exacerbate the inflammatory process through the interaction with one or more pro-inflammatory cytokines, resulting in a synergistic negative effect on androgen action (see below). If such exposure coincides with hormone sensitive time-windows in fetal development malformations of reproductive organs and injury to other differentiation processes may occur.

The testicular damage caused by phthalates includes damage to the cell membrane of Sertoli cells, followed by the disruption and death of germ cells and an interstitial inflammatory reaction (Creasy *et al.,* 1987). DEHP-induced gonadotoxicity was observed to be mediated by increased production of reactive oxygen species, since antioxidant vitamins significantly accelerated regeneration of seminiferous epithelium (Ablake *et al.,* 2004). This observation is in line with the growing body of data on the role of oxidative stress in the mechanisms of toxicity of several EDCs (Abdollahi *et al.,* 2004). Thus, one common pathway by which both inflammatory cytokines and some EDCs might exert their deleterious effects on target organs is through the excessive generation of reactive oxygen species.

Chronic inflammation and systemic activation of immune responses are known to be associated with a reduction in the plasma level of testosterone (Christeff *et al.,* 1987), due to the fact that pro-inflammatory cytokines such as TNFα, IL-1α and IL-6 are potent inhibitors of the induction of testosterone production in adult Leydig cells by hCG/LH (Hales *et al.,* 1999; Saez, 1994). In contrast, IL-1 has also been reported to stimulate basal androgen production by both adult (Moore and Moger, 1991) and immature (Svechnikov *et al.,* 2001) Leydig cells in culture. We recently demonstrated that this stimulation of steroidogenesis by IL-1 reflects up-regulation of StAR in the mitochondria of these immature Leydig cells (figure 9B). Taken together, the above findings clearly demonstrate that androgen production by Leydig cells at different phases of development may constitute an important target for inflammation-induced actions of EDCs.

Exposure to anti-androgens has also been reported to be associated with inflammatory processes. The anti-androgen flutamide, which is used in the treatment of benign hyperplasia of the prostate and hirsutism, may at low but significant incidence cause an inflammation of the liver, diagnosed as hepatitis (Gomez *et al.,*

1992). Although the mechanism by which flutamide induces liver damage is still unknown, some of these cases have been associated with peripheral eosinophilia and neutropenia, suggesting involvement of the immune system and cytokines (Hart and Stricker, 1989). Furthermore, when administered to pregnant rats, another anti-androgen, a vinclozolin, induces chronic inflammation of the epididymides, prostate and seminal vesicles in male offspring (Hellwig *et al.,* 2000).

On the basis of the above discussion, we propose that exposure of pregnant women experiencing acute or chronic inflammation to EDCs activates and/or intensifies pathophysiological processes that may seriously disturb human prenatal differentiation and exert deleterious effects on the future health of their offspring. Pro-inflammatory cytokines may constitute a common pathophysiological pathway in such responses.

11. Conclusions

Anthropogenic EDC compounds present in our environment (e.g. pesticides, phthalates and other toxic substances) may alter the development of androgen-dependent tissues and reduce reproductive potential and fertility in both humans and animals. In many cases, the major origin of these alterations is a reduction in the capacity of fetal Leydig cells to produce androgens, which leads to incomplete masculinization of male fetuses and various malformations in the reproductive tract. Endocrine disrupters can disrupt Leydig cell steroidogenesis by inhibiting cholesterol delivery into mitochondria, as described for phthalates, or the activities of steroidogenic enzymes, as shown for TCDD. Furthermore, anti-androgens exert their action by competing with androgens for binding to the AR.

Moreover, environmental EDCs may act synergistically or antagonistically with various paracrine and/or endocrine factors, in such a manner as to markedly modulate responses to these disrupters. We propose that pro-inflammatory cytokines, which play a central role in the regulation of inflammation, are involved in such interactions. In addition, exposure of women experiencing chronic inflammation or subclinical infection to EDC may activate and/or intensify pathophysiological processes of inflammation and significantly harm human health. Clearly, additional studies designed to elucidate possible interactions between EDCs and endogenous mediators of inflammation, as well as the effects of such interactions on the reproductive health of humans are highly motivated.

Acknowledgements

The work discussed here was supported by grants from the European Commission (Network of Excellence "CASCADE", FOOD-CT-2004-506319; "PIONEER", FOOD-CT-2005-513991), the Swedish Research Council (project 2002-5892), the Swedish Children's Cancer Fund and the Swedish Environmental Protection Agency.

References

Abdollahi, M., Ranjbar, A., Shadnia, S., Nikfar, S.and Rezaiee, A. (2004) Pesticides and oxidative stress: a review, *Med. Sci. Monit.* **10**, RA141-147.

Ablake, M., Itoh, M., Terayama, H,. Hayashi, S., Shoji, S., Naito, M., Takahashi, K., Suna, S. and Jitsunari, F. (2004) Di-(2-ethylhexyl) phthalate induces severe aspermatogenesis in mice, however, subsequent antioxidant vitamins supplementation accelerates regeneration of the seminiferous epithelium, *Int. J. Androl.* **27**, 274-281.

Akingbemi, B.T., Ge, R.S., Klinefelter, G.R., Gunsalus, G.L. and Hardy, M.P. (2000) A metabolite of methoxychlor, 2,2-bis(p-hydroxyphenyl)-1,1, 1-trichloroethane, reduces testosterone biosynthesis in rat Leydig cells through suppression of steady-state messenger ribonucleic acid levels of the cholesterol side-chain cleavage enzyme, *Biol. Reprod.* **62**, 571-578.

Akingbemi, B.T., Youker, R.T., Sottas, C.M., Ge, R., Katz, E., Klinefeltern, G.R., Zirkin, B.R. and Hardy, M.P. (2001) Modulation of rat Leydig cell steroidogenic function by di(2-ethylhexyl) phthalate, *Biol. Reprod.* **65**, 1252-1259.

Aldyreva, M.V., Klimova, T.S., Iziumova, A.S. and Timofeevskaia, L.A. (1975) The effect of phthalate plasticizers on the generative function, *Gig. Tr. Prof. Zabol.* **2** 25-29.

Allera, A., Herbstn, M.A., Griffin, J.E., Wilson, J. D., Schweikert, H.U. and McPhaul, M.J. (1995) Mutations of the androgen receptor coding sequence are infrequent in patients with isolated hypospadias, *J. Clin. Endocrinol. Metab.* **80**, 2697-1699.

Anway, M..D., Cupp, A.S., Uzumcu, M. and Skinner, M.K (2005) Epigenetic transgenerational actions of endocrine disrupters and male fertility. *Science* **308**, 1466-1469.

Assmus, M., Svechnikov, K., von Euler, M., Setchell, B., Sultana, T., Zetterstrom, C., Holst, M., Kiess, W. and Soder, O. (2005) Single subcutaneous administration of chorionic gonadotropin to rats induces a rapid and transient increase in testicular expression of pro-inflammatory cytokines, *Pediatr. Res.* **57**, 896-901.

Barlow, N.J, Phillips, S.L., Wallace, D.G., Sar, M., Gaido, K.W. and Foster, P.M (2003) Quantitative changes in gene expression in fetal rat testes following exposure to di(n-butyl) phthalate, *Toxicol. Sci.* **73**, 431-441.

Baskin, L.S., Himes, K. and Colborn, T. (2001) Hypospadias and endocrine disruption: is there a connection? *Environ, Health Perspect.* **109**, 1175-1183.

Bentvelsen, F.M,, McPhaul, M.J., Wilson, J.D. and George, F.W. (1994) The androgen receptor of the urogenital tract of the fetal rat is regulated by androgen, *Mol. Cell. Endocrinol.* **105**, 21-26.

Bentvelsen, F.M., Brinkmann, A.O., van der Linden, J.E., Schroder FH. and Nijman J.M (1995) Decreased immunoreactive androgen receptor levels are not the cause of isolated hypospadias, *Br. J. Urol.* **76**, 384-388.

Brandt, I, (1977) Tissue localization of polychlorinated biphenyls. Chemical structure related to pattern of distribution, *Acta Pharmacol. Toxicol. (Copenh)*. **40** Suppl **2**, 1-108.

Bulger, W.H., Muccitelli, R.M. and Kupfer, D. (1978) Studies on the *in vivo* and *in vitro* estrogenic activities of methoxychlor and its metabolites. Role of hepatic mono-oxygenase in methoxychlor activation, *Biochem. Pharmacol.* **27**, 2417-2423.

Chedrese, P.J. and Feyles, F. (2001) The diverse mechanism of action of dichlorodiphenyldichloroethylene (DDE) and methoxychlor in ovarian cells *in vitro*, *Reprod. Toxicol.* **15**, 693-698.

Choe, S.Y., Kim, S.J., Kim, H.G., Lee, J.H., Choi, Y., Lee, H. and Kim, Y. (2003) Evaluation of estrogenicity of major heavy metals, *Sci. Total Environ.* **312**, 15-21.

Christeff, N., Auclair, M.C., Benassayag, C., Carli, A. and Nunez, E.A, (1987) Endotoxin-induced changes in sex steroid hormone levels in male rats, *J. Steroid. Biochem,* **26**, 67-71.

Codesal, J., Regadera, J., Nistal, M., Regadera-Sejas, J. and Paniagua, R. (1990) Involution of human fetal Leydig cells. An immunohistochemical, ultrastructural and quantitative study, *J. Anat* . **172**, 103-114.

Colletti, L.M., Kunkel, S.L., Walz, A., Burdick, M.D, Kunkel, R.G., Wilke ,C.A. and Strieter, R.M, (1996) The role of cytokine networks in the local liver injury following hepatic ischemia/reperfusion in the rat, *Hepatology* **23**, 506-514.

Cook, J.C., Mullin, L.S., Frame, S.R. and Biegel, L.B. (1993) Investigation of a mechanism for Leydig cell tumorigenesis by linuron in rats, *Toxicol. Appl. Pharmacol.* **119**, 195-204.

Creasy, D.M., Beech, L.M., Gray, T.J. and Butler, W.H, (1987) The ultrastructural effects of di-n-pentyl phthalate on the testis of the mature rat, *Exp. Mol. Pathol.* **46**, 357-371.

Crellin, N.K., Kang, H.G., Swan, C.L. and Chedrese, P.J. (2001) Inhibition of basal and stimulated progesterone synthesis by dichlorodiphenyldichloroethylene and methoxychlor in a stable pig granulosa cell line, *Reproduction* **121**, 485-492.

Dasmahapatra, A.K., Wimpee, B.A., Trewin, A.L., Wimpee, C.F., Ghorai, J.K. and Hutz, R.J. (2000) Demonstration of 2,3,7,8-tetrachlorodibenzo-p-dioxin attenuation of P450 steroidogenic enzyme mRNAs in rat granulosa cell *in vitro* by competitive reverse transcriptase-polymerase chain reaction assay, *Mol. Cell. Endocrinol.* **164**, 5-18.

Davis, B.J., Maronpot, R.R. and Heindel, J.J. (1994a) Di-(2-ethylhexyl) phthalate suppresses estradiol and ovulation in cycling rats, *Toxicol. Appl. Pharmacol.* **128**, 216-223.

Davis, B.J., Weaver, R., Gaines, L.J. and Heindel, J.J (1994b) Mono-(2-ethylhexyl) phthalate suppresses estradiol production independent of FSH-cAMP stimulation in rat granulosa cells, *Toxicol. Appl. Pharmacol.* **128**, 224-228.

Dechaud, H., Ravard, C., Claustrat, F., de la Perriere, A.B. and Pugeat, M. (1999) Xenoestrogen interaction with human sex hormone-binding globulin (hSHBG), *Steroids* **64**, 328-334.

Eckert, R.L. and Katzenellenbogen, B.S. (1982) Physical properties of estrogen receptor complexes in MCF-7 human breast cancer cells. Differences with anti-estrogen and estrogen, *J. Biol. Chem.* **257**, 8840-8846.

Egeland, G.M., Sweeney, M.H., Fingerhut, M.A., Wille, K.K., Schnorr, T.M. and Halperin, W.E, (1994) Total serum testosterone and gonadotropins in workers exposed to dioxin, *Am. J. Epidemiol.* **139**, 272-281.

Enan, E., Lasley, B., Stewart, D., Overstreet, J. and Vandevoort, C.A. (1996) 2,3,7,8-tetrachlorodibenzo-p-dioxin (TCDD) modulates function of human luteinizing granulosa cells via cAMP signaling and early reduction of glucose transporting activity, *Reprod. Toxicol.* **10**, 191-198.

Facemire, C.F., Gross, T.S. and Guillette, L.J. Jr. (1995) Reproductive impairment in the Florida panther: nature or nurture? *Environ. Health Perspect.* **103** Suppl **4**, 79-86.

Gehm, B.D,, McAndrews, J.M., Chien, P.Y. and Jameson, J.L (1997) Resveratrol, a polyphenolic compound found in grapes and wine, is an agonist for the estrogen receptor, *PNAS* **94**, 38-43.

Gomez, J.L., Dupont, A., Cusan, L., Tremblay, M., Suburu, R., Lemay, M. and Labrie, F. (1992) Incidence of liver toxicity associated with the use of flutamide in prostate cancer patients, *Am. J. Med.* **92**, 465-470.

Granholm, T., Creasy, D.M., Pollanen, P. and Soder, O. (1992) Di-n-pentyl phthalate-induced inflammatory changes in the rat testis are accompanied by local production of a novel lymphocyte activating factor, *J. Reprod. Immunol.* **21**, 1-14.

Gray, L.E., Jr., Ostby, J., Ferrell, J., Rehnberg, G., Linder, R., Cooper, R., Goldman, J., Slott, V. and Laskey, J. (1989) A dose-response analysis of methoxychlor-induced alterations of reproductive development and function in the rat, *Fundam. Appl. Toxicol.* **12**, 92-108.

Gray, L.E., Jr., Kelce, W.R., Monosson, E., Ostby, J.S. and Birnbaum, L.S. (1995) Exposure to TCDD during development permanently alters reproductive function in male Long Evans rats and hamsters: reduced ejaculated and epididymal sperm numbers and sex accessory gland weights in offspring with normal androgenic status, *Toxicol. Appl. Pharmacol.* **131**, 108-118.

Gray, L.E,, Jr., Ostby, J., Monosson, E. and Kelce, W.R, (1999a) Environmental antiandrogens: low doses of the fungicide vinclozolin alter sexual differentiation of the male rat, *Toxicol. Ind. Health* **15**, 48-64.

Gray, L.E., Jr., Wolf, C., Lambright, C., Mann, P., Price, M., Cooper, R.L. and Ostby, J. (1999b) Administration of potentially antiandrogenic pesticides (procymidone, linuron, iprodione, chlozolinate, p,p'-DDE, and ketoconazole) and toxic substances (dibutyl- and diethylhexyl phthalate, PCB 169, and ethane dimethane sulphonate) during sexual differentiation produces diverse profiles of reproductive malformations in the male rat, *Toxicol. Ind. Health* **15**, 94-118.

Gray, L.E., Ostby, J., Furr, J., Wolf, C.J., Lambright, C., Parks, L., Veeramachaneni, D.N., Wilson, V., Price, M., Hotchkiss, A., Orlando, E. and Guillette, L. (2001) Effects of environmental antiandrogens on reproductive development in experimental animals, *Hum. Reprod. Update* **7**, 248-264.

Greco, T.L. and Payne, A.H. (1994) Ontogeny of expression of the genes for steroidogenic enzymes P450 side-chain cleavage, 3 beta-hydroxysteroid dehydrogenase, P450 17 alpha-hydroxylase/C17-20 lyase, and P450 aromatase in fetal mouse gonads, *Endocrinology* **135**, 262-268.

Guo, Y., Hendrickx, A.G., Overstreet, J.W., Dieter, J., Stewart, D., Tarantal, A.F., Laughlin, L. and Lasley, B.L. (1999) Endocrine biomarkers of early fetal loss in cynomolgus macaques (Macaca fascicularis) following exposure to dioxin, *Biol. Reprod.* **60**, 707-713.

Hagberg, H. and Mallard, C. (2005) Effect of inflammation on central nervous system development and vulnerability, *Curr. Opin. Neurol.* **18**, 117-23.

Hales, D.B., Diemer. T., Hales, K.H. (1999) Role of cytokines in testicular function, *Endocrine* **10**, 201-217.

Hart, W. and Stricker, B.H. (1989) Flutamide and hepatitis, *Ann. Int. Med.* **110**, 943-944.

Hellwig, J., van Ravenzwaay, B., Mayer M. and Gembardt, C. (2000) Pre- and postnatal oral toxicity of vinclozolin in Wistar and Long-Evans rats, *Regul. Toxicol, Pharmacol.* **32**, 42-50.

Hosie, S., Loff, S., Witt, K., Niessen, K. and Waag, K.L. (2000) Is there a correlation between organochlorine compounds and undescended testes? *Eur. J. Pediatr. Surg.* **10**, 304-309.

Hosokawa, S., Murakami, M., Ineyama, M., Yamada, T., Yoshitake, A., Yamada, H. and Miyamoto, J. (1993) The affinity of procymidone to androgen receptor in rats and mice, *J. Toxicol. Sci.* **18**, 83-93.

Hotchkiss, A.K., Parks-Saldutti, L.G., Ostby, J.S., Lambright, C., Furr, J., Vandenbergh, J.G. and Gray, L.E, Jr. (2004) A mixture of the "antiandrogens" linuron and butyl benzyl phthalate alters sexual differentiation of the male rat in a cumulative fashion, *Biol. Reprod.* **71**, 1852-1861.

Huhtaniemi, I. and Pelliniemi, L.J. (1992) Fetal Leydig cells: cellular origin, morphology, life span, and special functional features, *Proc. Soc. Exp. Biol. Med.* **201**, 125-140.

Hutson, J.M., Baker, M., Terada, M., Zhou, B. and Paxton, G. (1994) Hormonal control of testicular descent and the cause of cryptorchidism, *Reprod. Fertil. Dev.* **6**, 151-156.

Jepsen, K.F., Abildtrup, A. and Larsen, S.T. (2004) Monophthalates promote IL-6 and IL-8 production in the human epithelial cell line A549, *Toxicol. In Vitro* **18**, 265-269.

Johansson, M., Larsson, C., Bergman, A. and Lund, B.O. (1998) Structure-activity relationship for inhibition of CYP11B1-dependent glucocorticoid synthesis in Y1 cells by aryl methyl sulfones, *Pharmacol. Toxicol.* **83**, 225-230.

Johnson, L., Wilker, C.E., Safe, S.H., Scott, B., Dean, D.D. and White, P.H. (1994) 2,3,7,8-Tetrachlorodibenzo-p-dioxin reduces the number, size, and organelle content of Leydig cells in adult rat testes, *Toxicology* **89**, 49-65.

Johnson, M.D., Kenney, N., Stoika, A., Hilakivi-Clarke, L., Singh, B., Chepko, G., Clarke, R., Sholler, P.F., Lirio, A.A., Foss, C., Reiter, R., Trock, B., Paik, S.and Martin, M.B., (2003) Cadmium mimics the *in vivo* effects of estrogen in the uterus and mammary gland, *Nat. Med.* **9**, 1081-1084.

Jonsson, C.J., Rodriguez-Martinez, H., Lund, B.O., Bergman, A. and Brandt, I. (1991) Adreno-cortical toxicity of 3-methylsulfonyl-DDE in mice. II. Mitochondrial changes following ecologically relevant doses, *Fundam. Appl. Toxicol.* **16**, 365-74.

Jonsson, C.K., Zetterström, R.H., Holst, M., Parvinen, M. and Söder, O. (1999) Constitutive expression of interleukin-1alpha messenger ribonucleic acid in rat Sertoli cells is dependent upon interaction with germ cells, *Endocrinology* **140**, 3755-3761.

Jonsson, C.K., Setchell, B.P., Martinelle, N., Svechnikov, K. and Söder, O. (2001) Endotoxin-induced interleukin 1 expression in testicular macrophages is accompanied by downregulation of the constitutive expression in Sertoli cells, *Cytokine* **14**, 283-288.

Kalloo, N.B., Gearhart, J.P. and Barrack, E.R. (1993) Sexually dimorphic expression of estrogen receptors, but not of androgen receptors in human fetal external genitalia, *J. Clin. Endocrinol .Metab.* **77**, 692-698.

Kelce, W.R., Stone, C.R., Laws, S.C., Gray, L.E., Kemppainen, J.A. and Wilson, E.M. (1995) Persistent DDT metabolite p,p'-DDE is a potent androgen receptor antagonist, *Nature* **375**, 581-585.

Kemppainen, J.A., Lane, M.V., Sar, M. and Wilson, E.M. (1992) Androgen receptor phosphorylation, turnover, nuclear transport, and transcriptional activation. Specificity for steroids and antihormones, *J. Biol. Chem.* **267**, 968-974.

Kester, M.H., Bulduk, S., van Toor, H., Tibboel, D., Meinl, V., Glatt, H., Falany, C.N., Coughtrie, M.W., Schuur, A.G., Brouwer, A. and Visser, T.J. (2002) Potent inhibition of estrogen sulfotransferase by hydroxylated metabolites of polyhalogenated aromatic hydrocarbons reveals alternative mechanism for estrogenic activity of endocrine disrupters, *J. Clin. Endocrinol. Metab.* **87**, 1142-1150.

Lai, K.P., Wong, M.H. and Wong, C.K. (2005) Inhibition of CYP450scc expression in dioxin-exposed rat Leydig cells, *J. Endocrinol.* **185**, 519-527.

Lambright, C., Ostby, J., Bobseine, K., Wilson, V., Hotchkiss, A.K., Mann, P.C. and Gray, L.E., Jr. (2000) Cellular and molecular mechanisms of action of linuron: an antiandrogenic herbicide that produces reproductive malformations in male rats, *Toxicol. Sci.* **56**, 389-399.

Li, L.A. and Wang. P.W., (2005) PCB126 induces differential changes in androgen, cortisol, and aldosterone biosynthesis in human adrenocortical H295R cells, *Toxicol. Sci.* **85**, 530-540.

Lovekamp, TN. and Davis, B.J. (2001) Mono-(2-ethylhexyl) phthalate suppresses aromatase transcript levels and estradiol production in cultured rat granulosa cells, *Toxicol. Appl. Pharmacol.* **172**, 217-224.

Lovekamp-Swan, T. and Davis, B.J. (2003) Mechanisms of phthalate ester toxicity in the female reproductive system, *Environ. Health Perspect.* **111**, 139-145.

Majdic, G., Saunders, P.T. and Teerds, K.J. (1998) Immunoexpression of the steroidogenic enzymes 3-beta hydroxysteroid dehydrogenase and 17 alpha-hydroxylase, C17,20 lyase and the receptor for luteinizing hormone (LH) in the fetal rat testis suggests that the onset of Leydig cell steroid production is independent of LH action, *Biol. Reprod.* **58**, 520-525.

Maness, S.C., McDonnell, D.P. and Gaido, K.W. (1998) Inhibition of androgen receptor-dependent transcriptional activity by DDT isomers and methoxychlor in HepG2 human hepatoma cells, *Toxicol. Appl. Pharmacol.* **151**, 135-142.

McIntyre, B.S., Barlow, N.J., Sar, M., Wallace, D.G. and Foster, P.M. (2002) Effects of *in utero* linuron exposure on rat Wolffian duct development, *Reprod. Toxicol.* **16**, 131-139.

Mebus, C.A., Reddy, V.R. and Piper, W.N. (1987) Depression of rat testicular 17-hydroxylase and 17,20-lyase after administration of 2,3,7,8-tetrachlorodibenzo-p-dioxin (TCDD), *Biochem. Pharmacol.* **36**, 727-731.

Miller, D.B., Gray. L.E, Jr., Andrews, J.E., Luebke, R.W. and Smialowicz, R.J. (1993) Repeated exposure to the polychlorinated biphenyl (Aroclor 1254) elevates the basal serum levels of corticosterone but does not affect the stress-induced rise, *Toxicology* **81**, 217-222.

Minegishi, T., Hirakawa, T., Abe, K., Kishi, H. and Miyamoto, K. (2004) Effect of insulin-like growth factor-1 and 2,3,7,8-tetrachlorodibenzo-p-dioxin on the expression of luteinizing hormone receptors in cultured granulosa cells, *Environ. Sci.* **11**, 57-71.

Monosson, E., Kelce, W.R., Lambright, C., Ostby, J. and Gray, L.E, Jr. (1999) Peripubertal exposure to the antiandrogenic fungicide, vinclozolin, delays puberty, inhibits the development of androgen-dependent tissues, and alters androgen receptor function in the male rat, *Toxicol. Ind. Health* **15**, 65-79.

Moore, C.and Moger, W.H. (1991) Interleukin-1 alpha-induced changes in androgen and cyclic adenosine 3',5'-monophosphate release in adult rat Leydig cells in culture, *J. Endocrinol.* **129**, 381-390.

Moore, R.W., Jefcoate, C.R. and Peterson, R.E. (1991) 2,3,7,8-Tetrachlorodibenzo-p-dioxin inhibits steroidogenesis in the rat testis by inhibiting the mobilization of cholesterol to cytochrome P450scc, *Toxicol. Appl. Pharmacol.* **109**, 85-97.

Murakami, M., Hosokawa, S., Yamada, T., Harakawa, M., Ito, M., Koyama, Y., Kimura, J., Yoshitake, A.and Yamada, H. (1995) Species-specific mechanism in rat Leydig cell tumorigenesis by procymidone, *Toxicol. Appl. Pharmacol.* **131**, 244-252.

Murono, E.P. (1990) Differential regulation of steroidogenic enzymes metabolizing testosterone or dihydrotestosterone by human chorionic gonadotropin in cultured rat neonatal interstitial cells, *Acta Endocrinol. (Copenh).* **122**, 289-295.

Murono, E.P. and Derk, R.C. (2004) The effects of the reported active metabolite of methoxychlor, 2,2-bis(p-hydroxyphenyl)-1,1,1-trichloroethane, on testosterone formation by cultured Leydig cells from young adult rats, *Reprod. Toxicol.* **19**, 135-146.

Mylchreest, E., Sar M., Wallace, D.G. and Foster, P.M. (2002) Fetal testosterone insufficiency and abnormal proliferation of Leydig cells and gonocytes in rats exposed to di(n-butyl) phthalate, *Reprod. Toxicol.* **16**, 19-28.

O'Shaughnessy, P.J., Baker, P.J., Heikkila, M., Vainio, S. and McMahon, A.P. (2000) Localization of 17beta-hydroxysteroid dehydrogenase/17-ketosteroid reductase isoform expression in the developing mouse testis--androstenedione is the major androgen secreted by fetal/neonatal Leydig cells, *Endocrinology* **141**, 2631-2637.

Parks, L.G., Ostby, J.S., Lambright, C.R., Abbott, B.D., Klinefelter, G.R., Barlow, N.J. and Gray, L.E., Jr. (2000) The plasticizer diethylhexyl phthalate induces malformations by decreasing fetal testosterone synthesis during sexual differentiation in the male rat, *Toxicol. Sci.* **58**, 339-349

Pelliniemi, L.J. and Niei, M. (1969) Fine structure of the human foetal testis. I. The interstitial tissue, *Z. Zellforsch. Mikrosk. Anat.* **99**, 507-522.

Reyes, H., Reisz-Porszasz, S. and Hankinson, O. (1992) Identification of the Ah receptor nuclear translocator protein (Arnt) as a component of the DNA binding form of the Ah receptor, *Science* **256**, 1193-1195.

Saez, J.M. (1994) Leydig cells: endocrine, paracrine, and autocrine regulation, *Endocr. Rev.* **15**, 574-626.

Sha, J., Baker, P.and O'Shaughnessy, P.J. (1996) Both reductive forms of 17 beta-hydroxysteroid dehydrogenase (types 1 and 3) are expressed during development in the mouse testis, *Biochem. Biophys. Res. Comm.* **222**, 90-94.

Sharpe, R.M. (2001) Hormones and testis development and the possible adverse effects of environmental chemicals, *Toxicol. Lett.* **120**, 221-232.

Strand, M.L., Wahlgren, A., Svechnikov, K., Zetterstrom, C., Setchell, B.P. and Soder, O. (2005) Interleukin-18 is expressed in rat testis and may promote germ cell growth, *Mol. Cell. Endocrinol.* **240**, 64-73.

Sultan, C., Paris, F., Terouanne, B., Balague,r P., Georget, V., Poujol, N., Jeandel, C., Lumbroso, S. and Nicolas, J.C. (2001) Disorders linked to insufficient androgen action in male children, *Hum. Reprod. Update* **7**, 314-322.

Supornsilchai, V., Svechnikov, K., Seidlova-Wuttke, D., Wuttke, W. and Soder, O. (2005) Phytoestrogen resveratrol suppresses steroidogenesis by rat adrenocortical cells by inhibiting cytochrome P450 c21-hydroxylase, *Horm. Res.* **64**, 280-286.

Swan, S.H., Main, K.M., Liu, F., Stewart, S.L., Kruse, R.L., Calafat, A.M., Mao, C.S., Redmon, J.B., Ternand, C.L., Sullivan, S. and Teague, J.L. (2005) Decrease in anogenital distance among male infants with prenatal phthalate exposure, *Environ. Health Perspect.* **113**, 1056-1061.

Svechnikov, K., Supornsilchai., V., Strand, M.L., Wahlgren, A., Seidlova-Wuttke, D., Wuttke, W. and Soder, O. (2005) Influence of long-term dietary administration of procymidone, a fungicide with anti-androgenic effects, or the phytoestrogen genistein to rats on the pituitary-gonadal axis and Leydig cell steroidogenesis, *J. Endocrinol.* **187**, 117-124.

Svechnikov, K.V., Sultana, T. and Söder, O. (2001) Age-dependent stimulation of Leydig cell steroidogenesis by interleukin-1 isoforms, *Mol. Cell. Endocrinol.* **182**, 193-201.

Sweet, R.A., Schrott, H.G., Kurland, R. and Culp, O.S. (1974) Study of the incidence of hypospadias in Rochester, Minnesota, 1940-1970, and a case-control comparison of possible etiologic factors, *Mayo Clin. Proc.* **49**, 52-58.

Szabo, S. and Lippe, I.T. (1989) Adrenal gland: chemically induced structural and functional changes in the cortex, *Toxicol. Pathol.* **17**, 317-329.

Tan, J., Marschke, K.B., Ho, K.C., Perry, S.T., Wilson, E.M.and French, F.S. (1992) Response elements of the androgen-regulated C3 gene, *J. Biol. Chem.* **267**, 7958.

Thomas, J.A. and Thomas, M.J. (1984) Biological effects of di-(2-ethylhexyl) phthalate and other phthalic acid esters, *Crit. Rev. Toxicol.* **13**, 283-317.

Thompson, C.J., Ross, S.M. and Gaido, K.W. (2004) Di(n-butyl) phthalate impairs cholesterol transport and steroidogenesis in the fetal rat testis through a rapid and reversible mechanism, *Endocrinology* **145**, 1227-1237.

Treinen, K.A., Dodson, W.C. and Heindel, J.J. (1990) Inhibition of FSH-stimulated cAMP accumulation and progesterone production by mono(2-ethylhexyl) phthalate in rat granulosa cell cultures, *Toxicol. Appl. Pharmacol.* **106**, 334-340.

Truss, M., Bartsch, J. and Beato, M. (1994) Antiprogestins prevent progesterone receptor binding to hormone responsive elements *in vivo*, *PNAS* **91**, 133-137.

Waller, C.L., Juma, B.W., Gray, L.E. and Kelce, W.R. (1996) Three-dimensional quantitative structure-activity relationships for androgen receptor ligands, *Toxicol. Appl. Pharmacol.* **137**, 219-227.

Weidner, I.S., Moller, H., Jensen, T.K. and Skakkebaek, N.E. (1998) Cryptorchidism and hypospadias in sons of gardeners and farmers, *Environ. Health Perspect.* **106**, 793-796.

Wilker, C.E., Welsh, T.H., Jr., Safe, S.H., Narasimhan, T.R. and Johnson, L. (1995) Human chorionic gonadotropin protects Leydig cell function against 2,3,7,8-tetrachlorodibenzo-p-dioxin in adult rats: role of Leydig cell cytoplasmic volume, *Toxicology* **95**, 93-102.

Wilson, J.D. (1978) Sexual differentiation, *Annu. Rev. Physiol.* **40**, 279-306.

Wilson, J.D., Griffin, J.E., Russell, D.W. (1993) Steroid 5 alpha-reductase 2 deficiency, *Endocr. Rev.* **14**, 577-593.

Wojtowicz, A.K., Gregoraszczuk, E.L., Ptak, A. and Falandysz, J. (2004) Effect of single and repeated *in vitro* exposure of ovarian follicles to o,p'-DDT and p,p'-DDT and their metabolites, *Pol. J. Pharmacol.* **56**, 465-472.

Wolf, C.J., LeBlanc, G.A., Ostby, J.S. and Gray, L.E., Jr. (2000) Characterization of the period of sensitivity of fetal male sexual development to vinclozolin, *Toxicol. Sci.* **55**, 152-161.

Wong, C., Kelce, W.R, Sar, M. and Wilson, E.M. (1995) Androgen receptor antagonist versus agonist activities of the fungicide vinclozolin relative to hydroxyflutamide, *J. Biol. Chem.* **270**, 1998-2003.

Zhang, F.P., Hamalainen, T., Kaipia, A., Pakarinen, P. and Huhtaniemi, I. (1994) Ontogeny of luteinizing hormone receptor gene expression in the rat testis, *Endocrinology* **134**, 2206-2213.

ENVIRONMENTAL IMPACT ON CONGENITAL DISEASES: THE CASE OF CRYPTORCHIDISM. WHERE ARE WE NOW, AND WHERE ARE WE GOING?

P.F. THONNEAU*, E. HUYGHE AND R. MIEUSSET
Human Fertility Research Group – Reproductive Health in Developing Countries; EA n°36 94
Hôpital Paule de Viguier
Aavenue de Grande Bretagne 330
Toulouse Cedex
FRANCE
(author for correspondence, Email: thonneau.p@chu-toulouse.fr)*

Summary

Cryptorchidism is the most frequent abnormality of male sexual differentiation. Recently, numerous reports have increased concerns that exposure to certain types of chemicals in the environment, including *in utero* exposure to compounds with estrogenic or antiandrogenic activities, may be linked with recently observed deleterious effects on male reproductive health, especially cryptorchidism and also decrease in sperm production and increased in incidence of testicular cancer.

In this review, we give an overview of different scientific reviews already published on cryptorchidism, its incidence rate, the potential link between various male reproductive health issues (testicular dysgenesis syndrome), the existing literature on toxic effects of anti-androgenic compounds, and some relevant data on the environmental impact of cryptorchidism in humans. Finally, we explain why cryptorchidism could be an excellent potential indicator and may potentially provide an answer to the key question of the impact of some environmental pollutants on male reproductive health.

P. Nicolopoulou-Stamati et al. (eds.), Congenital Diseases and the Environment, 281–294.

1. Background

The definition of cryptorchidism varies among authors, depending on the site of the testis and to the understanding of the testis descent. Therefore, we suggest adoption of the following definition given by Scorer (1964): '*the descent of the testis is the movement of the organ from the abdominal cavity to the bottom of a fully developed and fully relaxed scrotum*'. Several mechanisms involving estrogens, androgens, and insulin-like factor-3 are more or less inter-connected in the testicular descent. The description of the two different phases of the testis migration (a first phase of relative transabdominal migration between 10 to 15 weeks of gestation controlled by Müllerian inhibiting substance; and a second inguinoscrotal phase at 26 to 35 weeks of gestation, possibly androgen-dependent and mediated by the release of calcitonin gene-related peptide from the genitofemoral nerve) have been well described by Hutson *et al.* (1994, 1997). In the figure 11 (Sharpe, 2001) we suggest a schematic representation of the phases of testicular descent, showing hormonal-dependence, and places where hormonal disrupters may interfere.

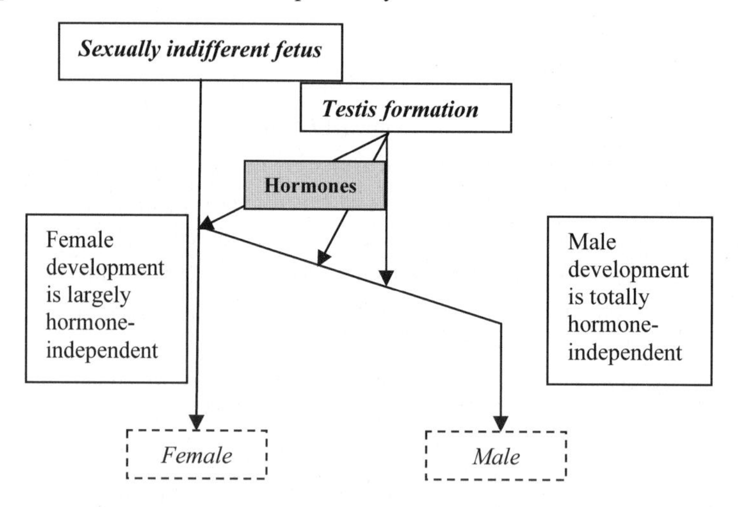

Figure 11. Sexual differentiation in mammals (adapted from Sharpe, 2001)

Cryptorchidism is the main risk factor for testicular cancer, which is currently the most frequent cancer in young men. To date, only cryptorchidism persisting until the age of 1 year has been considered a risk factor for testicular cancer. Several authors now believe that any form of cryptorchidism at birth, regardless of the outcome, should be considered a risk factor for testicular cancer. Cryptorchidism is also a major risk factor for male infertility, and orchidopexy performed in the first two years of life is considered as a useful surgical procedure to prevent future male infertility.

Over the past 10 years, a number of reports have increased concerns that exposure to certain types of chemicals in the environment, including *in utero* exposure to compounds with estrogenic or anti-androgenic activities, may be linked with recently observed deleterious effects on male reproductive health, especially a decrease in sperm production and an increased incidence of testicular cancer.

In this broad debate concerning the reliability of the environmental hypothesis, numerous studies have been published on incidence trends, risk factors for cryptorchidism, and also the possible role of environmental conditions (Carlsen *et al.,* 1992; Sharpe, 2001).

In this review, we will give an overview of different scientific papers already published on cryptorchidism, and in which directions it could be interesting to strengthen research, in order to establish links between environmental conditions and occurrence of this male malformation

2. Has the incidence of cryptorchidism increased?

First of all, and as pointed out by Toppari and more recently by Boisen, it is difficult to compare the incidence rates of cryptorchidism reported in various publications (Toppari *et al.,* 1996; Boisen *et al.,* 2004). First of all, the definition of cryptorchidism varies according to the site of the testis: for example, some authors consider the testis "normal," even when it is not located in the depth of the scrotum but in a high scrotal position. It is also often impossible to determine, in published studies, whether retractile testes (i.e., testes that are not located in the bottom of the scrotum but that can be manually pushed into this position) are classified as "cryptorchidism" or as a "normal" testicular position.

Another key point consists of clearly identifying, in published studies, the age of diagnosis of cryptorchidism, as it has been clearly established that the great majority of cases of cryptorchidism diagnosed at birth resolve spontaneously during the first years of life (natural course of cryptorchidism testes). This could explain the large range of incidence rates reported in the literature: from less than 1 to 10% in data from hospital-based or central registers (with diagnosis performed from birth to 1 year of age) and from 0.2 to 13% in surveys performed in schools and in the army (with diagnosis performed in adolescents and young adults) (Buemann *et al.,* 1961; Campbell *et al.,* 1987; Thorup and Cortes, 1990; Kaul and Roberts, 1992). Finally, very few studies have tried to analyze cryptorchidism incidence trends over recent decades based on the same diagnostic criteria and in the same areas. In the United Kingdom, by studying the Hospital Inpatient Enquiry data for England and Wales,

Chilvers reported a doubling of the incidence of cryptorchidism at birth, from 1.4% in 1952 to 2.9% in 1977 (Chilvers *et al.,* 1984).

In a meta-analysis performed in 1999, Paulozzi used data collected by the International Clearing House of Birth Defect Monitoring Systems to compare cryptorchidism incidence from several industrialized countries (Paulozzi, 1999). In the USA, a continuous increase (from 1970 to 1994) was observed, with some discordance, according to several data collection systems. Throughout Canada, the incidence of cryptorchidism has increased, with an incidence rate growing from 14 per 1,000 to 24 per 1,000, in the 20 years from 1974 to 1994. In European countries, the incidence of cryptorchidism has remained relatively stable, with about 15 cases per 1,000 from 1974 to 1996 in Norway. In France (in a specific district), the incidence has increased slightly from 8 per 1,000 in 1981 to 12 per 1,000 in 1993. In England, similar trends were observed. Unfortunately, the methodology used to assess cryptorchidism was not adequately described in all of these studies, making comparison very difficult.

The incidence of cryptorchidism was assessed in Lithuania and compared to results observed in neighbouring countries. Low birth weight, preterm delivery, small gestational weight, other congenital malformations and paternal body mass index (<20kg/m2) were the main identified risk factors. A lower incidence rate of cryptorchidism was found in Lithuania (5.7% at birth – 1.4% at 1 year) compared to 9.0% in Denmark, but higher than results noted in Finland (2.4%). Interestingly the authors observed that some male reproductive indicators (incidence rate for testis cancer and semen quality) were relatively similar between Lithuania and Finland but cryptorchidism incidence was rather different (Preiksa *et al.,* 2005).

As yet, there are no obvious reasons which could explain this phenomenon. Numerous epidemiological studies have been conducted during recent decades to identify risk factors for cryptorchidism (low birth weight, low parity, twinship, uteroplacental malfunction) and to explain the rising incidence, but without any really convincing results regarding the increase in cryptorchidism incidence (Huyghe *et al.,* 2003). Nevertheless, some key findings have recently been put before the international scientific community, leading to a better understanding of the problems surrounding cryptorchidism and also suggesting interesting directions for research.

3. Increasing evidence of a link between various male reproductive health issues

Besides the rising incidence of cryptorchidism, we have also observed a recent marked increase in hypospadias and testicular germ cell cancer rates (the occurrence of one disorder being a risk factor for the occurrence of another) and a decline in semen quality, with strong regional differences between industrialized countries (Carlsen *et al.,* 1992).

The current hypothesis, mainly developed by Skakkebaek *et al.* (2003), is that all these male reproductive disorders are interrelated, forming the testicular dysgenesis syndrome (TDS). TDS may have different clinical or biological expressions (hypospadias, cryptorchidism, testis cancer, infertility), but all arise from the same syndrome which originates during fetal development. This being so, any type of endocrine disrupter which could lead to imbalance within the male endocrine system during the period of pregnancy will have potentially damaging consequences on reproductive tract development. In figure 12, we have summarized the potential links between testis development and testicular dysgenesis syndrome.

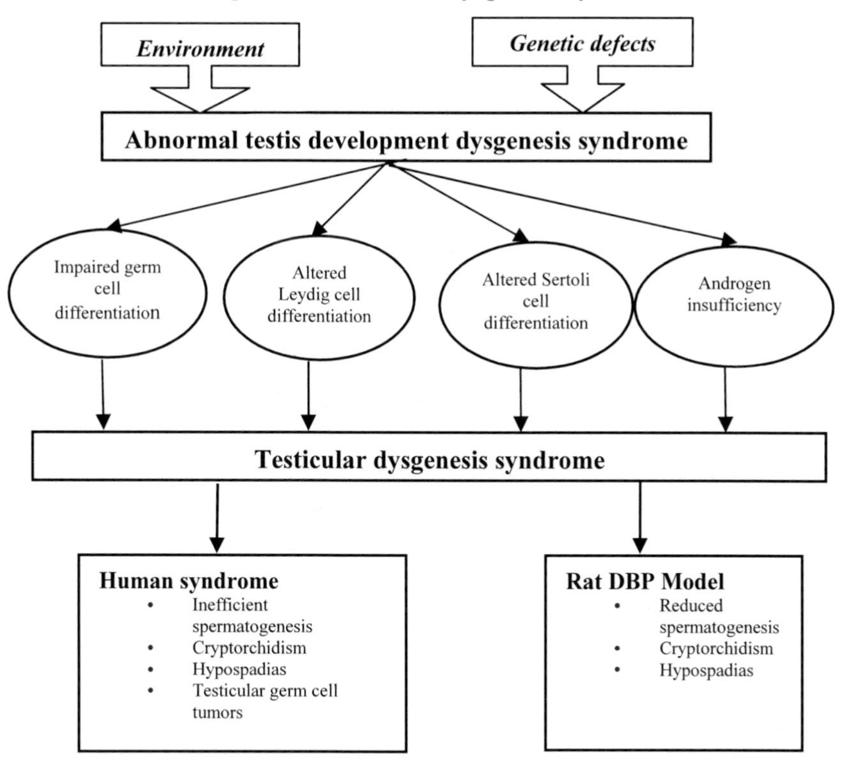

Figure 12. Links between testis development and TDS (adapted from Fisher, 2004)

The history of the use of diethylstilboestrol (DES) and its dramatic consequences on offspring are a strong argument for a potential impact of *in utero* synthetic estrogen administration on occurrence of male reproductive tract abnormalities. In animal models (rats), it has been establish that administration of DES leads to several reproductive health tract abnormalities, mainly a large reduction in testis weight, a distension and overgrowth of the rete, a distension and reduction in epithelial height of the efferent ducts, an underdevelopment of the epididymal duct epithelium, and a reduction in epithelial height in the vas deferens. Furthermore, these abnormalities (correlated to the DES doses) are associated with reduced androgen receptor and Leydig cell volume. Treatments (GnRHa, flutamide) that could interfere with androgen production were not able to obtain such abnormalities as were observed with DES. Nevertheless, large alteration of the balance between androgen and estrogen may lead to similar reproductive health tract abnormalities to those observed in animals treated with DES (McKinnel *et al.,* 2001).

In humans, several authors have demonstrated that DES is associated with undescended testes in male offspring (Gill *et al.,* 1979; Driscoll and Taylor, 1980). Furthermore, in a follow-up study, genital abnormalities were more frequent in men exposed before the 11th week of gestation, compared to those exposed later (Wilcox *et al.,* 1995). In a meta-analysis, no association was observed between genital abnormalities and exposure to sex hormones other than DES (Raman-Wilms *et al.,* 1995).

Another question which remains to be solved is the relationship between use of potent estrogens, probable decrease of androgen levels, and genital tract abnormalities.

4. Toxic effect of anti-androgenic compounds on male reproductive health

4.1. The evidence-based central role of Sertoli cells and androgen production in testis and male reproductive tract development

After activation of the sex-determining region of the Y-chromosome, the differentiation of Sertoli cells constitutes the main signal, and these cells both drive and lead testis development (Koopman *et al.,* 1990; Capel, 2000). Regression of the Müllerian duct is mediated by Sertoli-cell secretion of anti-Müllerian hormone (AMH), and the Sertoli cell is certainly also responsible for the blockage of germ cells entering the oogenic pathway. The differentiation of Leydig cells is also induced by the Sertoli cell.

As mentioned by Richard Sharpe, '*testis formation does not itself require hormones but all other aspects of masculinisation are dependent on normal testis function and on the consequent production of adequate amounts of hormones*' (Sharpe, 2001); figure 13 demonstrates this.

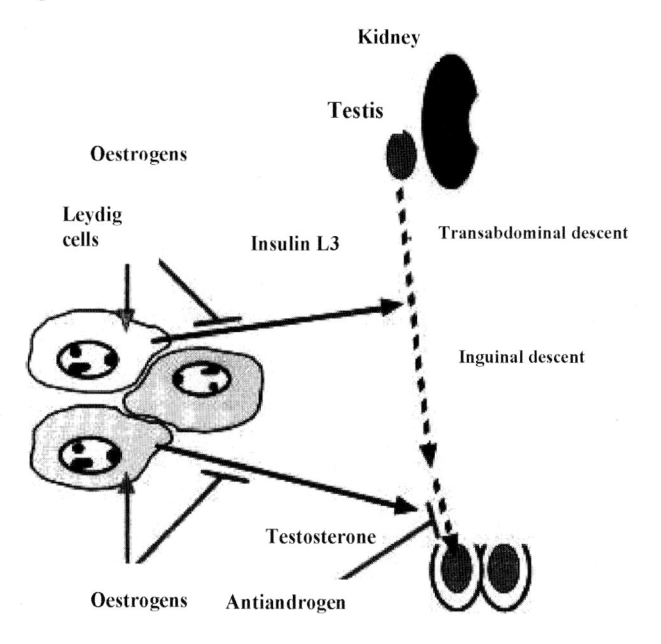

Figure 13. Representation of testicular descent in the scrotum and its hormone-dependence (modified from Sharpe, 2001)

Androgens play a key role in maintenance of the Wolffian duct, which differentiates into the epididymis, vas deferens and seminal vesicles, with masculinization mediated by testosterone. Masculinization of the external genital organs and prostate is controlled by dihydrotestosterone (DHT), a testosterone metabolite produced by the action of the 5alpha-reductase enzyme.

4.2. A potential role of Insulin-like growth factor (Insl3)

Authors have demonstrated that Insl3, a protein produced by Leydig cells, may interfere with gubernacular proliferation (using Insl3 knockout mice), and suggested an 'androgen-independent' pathway (Nef and Parada, 1999; Emmen *et al.,* 2000). Nevertheless, these results were not confirmed in similarly designed studies by other authors (Krausz *et al.,* 2000; Baker *et al.,* 2002). Furthermore, another report indicated that the calcitonin gene-related peptide (CgRP) seems not to be a major factor for cryptorchidism (Zuccarello *et al.,* 2004).

The authors demonstrated that (in rats) the Insl3 expression is suppressed in 80% of cryptorchidism and 50% of scrotal testes from rats which have been exposed to di(n-butyl) phthalate. The authors suggest that such prenatal exposure may lead to malfunction in Leydig cells (McKinell *et al.,* 2005). In humans, an interesting and recent article has demonstrated (in a series of 87 ex-cryptorchid patients and 80 controls) that Insl3-LGR8/GREAT mutations are frequently associated with cryptorchidism and are maternally inherited (Ferlin *et al.,* 2003).

4.3. A non-exhaustive list of environmental chemicals with anti-androgenic effects

First of all, we must point out that the large majority of published works in this domain have been performed in pregnant rodents, and that human data, whether epidemiological investigations or *in vitro* studies, are very scarce.

Nevertheless, the following environmental compounds, more and less widely used in industrialized countries, have demonstrated (in animals) their potential ability to inhibit the action of androgen receptors or testosterone synthesis:

- vinclozolin, a dicarboximide fungicide with anti-androgenic properties (Gray *et al.,* 2001; Wolf *et al.,* 2000),
- linuron, a herbicide and androgen receptor antagonist (McIntyre *et al.,* 2002),
- p,p'DDE and methoxychlor (derivatives of the DDT pesticide), considered as androgen and/or estrogen receptor antagonists; phthalates (mainly used in the cosmetic and manufacturing industries), which have several major impacts on androgen regulation (Fisher, 2004; Mylchreest and Foster, 2002; Mylchreest *et al.,* 2000).

The authors of one study showed higher levels of phthalate metabolites in women of reproductive age as compared with the rest of the population (Blount, 2000). It is interesting to note that relatively similar histological changes have been observed in testis after *in utero* phthalate exposure and in results of testicular biopsies performed in patients with TDS (Skakkebaek *et al.,* 2003; Hoei-Hansen *et al.,* 2003), given that, Fisher and collaborators have suggested that "*dibutyl phthalate exposure in utero may provide a useful model for defining the cellular pathways in TDS*" (Fisher *et al.,* 2003).

5. Limited data on environmental impact of cryptorchidism in humans

5.1. DES

The best documented adverse effects of estrogens concern the use of DES, a nonsteroidal estrogenic substance, in pregnant women to prevent abortion complications (Brackbill and Berendes, 1978; Stillman, 1982). DES is associated with undescended testes in male offspring (Gill *et al.,* 1979; Wilcox *et al.,* 1995).

5.2. Epidemiological studies (occupational and environmental conditions)

In a spatial ecological study undertaken in the province of Granada, Spain, the authors observed that orchidopexy tended to be more frequent in districts where the family was involved in intensive farming and pesticide spraying (Garcia-Rodriguez *et al.,* 1996). In Denmark, in a register-based case-control study of parental occupation, the authors found that sons of women working in gardening had a significantly higher risk of cryptorchidism (Weidner *et al.,* 1998). In Hungary, higher frequency of undescended testes was reported in newborns of mothers living closest to a factory producing vinyl chloride monomer and acrylonitrile (for the manufacture of stiffened plastic tubes and cartons used for the packaging of margarine) (Czeizel *et al.,* 1999).

Even these few articles are supportive of a potential impact of environmental exposures in the occurrence of cryptorchidism. Current data in this field are still scarce and relatively difficult to pool. As stated by Vidaeff, "*the results currently published in the literature do not support with certainty the view that environmental estrogens contribute to an increase in male reproductive disorders, neither do they provide sufficient grounds to reject such hypothesis*" (Vidaeff and Sever, 2005).

5.3. Bioaccumulation of chemical compounds

High-resolution gas chromatography and mass spectrometry were used to assess fat samples in patients with and without undescended testes. Heptachloroepoxide and hexachlorobenzene were present in significantly higher concentrations in patients with undescended testes (Hosie *et al.,* 2000). In a Swiss study, higher levels of estradiol were found in the placentas of neonates with cryptorchidism (Hadziselimovic *et al.,* 2000).

6. Where do we go from here, and how can we answer the question?

If we assume that the incidence of cryptorchidism is increasing in many industrialized countries (which is a plausible hypothesis), and if we are also convinced that cryptorchidism, hypospadias, germ cell testicular cancer and decreasing semen quality are on the same continuum of TDS, involving *in utero* exposure to endocrine disrupters (which is not so evident for sperm decline), we must rapidly perform appropriately designed studies to identify what kind of environmental components could be involved, how their properties or derivatives may interfere with male endocrine system balance, and certainly also to define the exact exposure window during the pregnancy period (Sharpe, 2001). As stated by Sharpe in an interesting review in the subject, several hypotheses are still 'in competition'. The estrogen hypothesis could act through diminution of testosterone production, suppression of androgen receptor expression, suppression of Innsl3, or alteration of the balance between estrogen and androgen. Nevertheless, environmental estrogens seem to have relatively weak estrogenicity, and their implication in male reproductive disorders is not evident. Some environmental chemicals have even demonstrated deleterious impact in endogenous androgens (some phthalates for example) and also in estrogens (biphenyls) (Sharpe, 2003; Williams *et al.,* 2001). So far, the molecular basis of cryptorchidism is unknown; even results regarding the Insl3 receptor and also hypothesis on some developmental transcription factors (for example with Hoxa-10 in mice) are interesting findings (Ivell and Hartung, 2003).

Hopefully, cryptorchidism is quite a good model. The relatively high frequency of cryptorchidism and the short period between *in utero* exposure and the endpoint are strong arguments for pursuing research into this malformation. Nevertheless, it would also surely be very useful to distinguish clearly between the various types of undescended testes (abdominal versus inguinoscrotal position).

From on epidemiologically pont of view, it is much more difficult and costly to undertake a prospective study and to enrol and follow several hundred thousand pregnant women (to have enough cases of cryptorchidism) than to perform appropriate case-control studies. So far, we have several different and reasonable hypotheses concerning the potential impact of some environmental compounds (phthalates for example) on cryptorchidism during pregnancy. So, the key questions are how to identify and assess maternal exposure to these compounds, and how to evaluate the anti-androgenic properties of these compounds or of their derivatives.

Finally, if we want to be able to solve the question of a potential impact of our environment on reproductive health, we definitely need close collaboration between scientists in fields as different as chemistry, toxicology, endocrinology, andrology and epidemiology! Furthermore, the link between *in vitro* and *in vivo* models must be developed, as well as collaboration with scientists working with various animal species and those involved in human studies. In conclusion, and taking all these factors into account, cryptorchidism is certainly an excellent potential indicator and may potentially provide an answer to the key question of the impact of some environmental pollutants on male reproductive health.

References

Baker, L.A., Nef S., Nguyen, M.T., Stapleton, R., Pohl, H. and Parada, L.F. (2002) The insulin-3 gene: lack of a genetic basis for human cryptorchidism, *J. Urol.* **167**, 2534-2537.

Blount, B.C., Silva, M.J., Caudill, S.P., Needham, L.L., Pirkle, J.L., Sampson, E.J., Lucier, G.W., Jackson, R.J. and Brock, J.W. (2000) Levels of seven urinary phthalate metabolites in a human reference population, *Environ. Health Perspect.* **108**, 979-982.

Boisen, K.A., Kaleva, M., Main, K.M., Virtanen, H.E., Haavisto, A.M., Schmidt, I.M., Chellakooty, M., Damgaard, I.N., Mau, C., Reumanen, M., Skakkebaek, N.E. and Toppari, J. (2004) Difference in prevalence of congenital cryptorchidism in infants between two Nordic countries, *Lancet* **283**, 1264-1269.

Brackbill, Y., and Berendes, H.W. (1978) Dangers of diethylstilboestrol : a review of a 1953 paper, *Lancet* **2**, 520.

Buemann, B., Henriksen, H., Villumsen, A.L., Westh, A. and Zachau-Christiansen, B. (1961) Incidence of undescended testes in newborn, *Acta. Chir. Scand.* **283**, 289-293.

Campbell, D.M., Webb, J.A. and Harhreave, T.B. (1987) Cryptorchidism in Scotland, *Br. Med. J.* **295**, 1235-1236.

Capel, B. (2000) The battle of the sexes, *Mecha. Dev.* **92**, 89-103.

Carlsen, E., Giwercman, A., Keiding, N. and Skakkebaek, N.E. (1992) Evidence for decreasing quality of semen during past 50 years, *Br. Med. J.* **305**, 609-613.

Chilvers, C., Pike, M.C., Forman, D. and Wadsworth, M.E. (1984) Apparent doubling of frequency of undescended testis in England and Wales in 1962-81, *Lancet* **2**, 330-332.

Czeizel, A.E., Hegedus, S. and Timar, L. (1999) Congenital abnormalities and indicators of germinal mutations in the vicinity of an acrylonitrile producing factory, *Mut. Res.* **427**, 105-123.

Driscoll, S.G. and Taylor, S.H. (1980) Effects of prenatal maternal estrogen on the male urogenital system, *Obstet. Gynecol.* **56**, 537-542.

Emmen, J.M., McLuskey, A., Adham, I.M., Engel, W., Verhoef-Post, M., Themmen, A.P., Grootegoed, J.A. and Brinkmann, A.O. (2000) Involvement of insulin-like factor 3 (Insl3) in diethylstilboestrol-induced cryptorchidism, *Endocrinology* **141**, 846-849.

Ferlin, A., Simonato, M., Bartoloni, L., Rizzo, G., Bettella, A., Dottorini, T., Dallapiccola, B. and Foresta, C. (2003) The Insl3-LGR8/GREAT ligand-receptor pair in human cryptorchidism, *J. Clin. Endocrinol. Metab.* **88**, 4273-4279.

Fisher, J.S. (2004) Environmental anti-androgens and male reproductive health: focus on phthalates and testicular dysgenesis syndrome, *Reproduction* **127**, 305-315.

Fisher, J.S., Macpherson, S., Marchetti, N. and Sharpe, R. (2003) Human 'testicular dysgenesis syndrome': a possible model in-utero exposure of the rat to dibutyl phthalate, *Hum. Reprod.* **18**, 1383-1394.

Garcia-Rodriguez, J., Garcia-Martin, M., Nogueras-Ocana, M., de Dios Lunadel-Castillo, J., Espigares Garcia, M., Olea, N. and Lardelli-Claret, P. (1996) Exposure to pesticides and cryptorchidism : geographical evidence of a possible association, *Environ. Health Perspect.* **104**, 1090-1095.

Gill, W.B., Schumacher, G.F., Bibbo, M., Straus, F.H. and Schoenberg, H.M. (1979) Association of diethylstilboestrol exposure *in utero* with cryptorchidism, testicular hypoplasia and semen abnormalities, *J. Urol.* **122**, 36-39.

Gray, L.E., Ostby, J., Furr, J., Price, M., Veeramachaneni, D.N. and Parks, L. (2001) Effects of environmental antiandrogens on reproductive development in experimental animals, *Hum. Reprod.* **7**, 248-264.

Hadziselimovic, F., Geneto, R. and Emmons, L.R. (2000) Elevated placenta oestradiol: a possible etiological factor of human cryptorchidism, *J. Urol.* **164**, 1694-1695.

Hoei-Hansen, C.E., Holm, M., Rajpert-De Meyts, E. and Skakkebaek, N.E. (2003) Histological evidence of testicular dysgenesis in controlateral biopsies from 218 patients with testicular germ cell cancer, *J. Pathol.* **200**, 370-374.

Holm, M., Rajpert-De Meyts, E., Anderson, A.M. and Skakkebaek, N.E. (2003) Leydig cell micro nodules are a common finding in testicular biopsies from men with impaired spermatogenesis and are associated with decreased testosterone, *J. Pathol.* **199**, 378-386.

Hosie, S., Loff, S., Witt, K., Niessen, K. and Waag, K.L. (2000) Is there a correlation between organochlorine compounds and undescended testes? *Euro. J. Ped. Sur.* **10**, 304-309.

Hutson, J.M., Baker, M., Terada, M., Zhou, B. and Paxton, G. (1994) Hormonal control of testicular descent and the cause of cryptorchidism, *Reprod. Fertil. Dev.* **6**, 151-156.

Hutson, J.M., Hasthorpe, S. and Heyns, C.F. (1997) Anatomical and functional aspects of testicular descent and cryptorchidism, *Endoc. Rev.* **18**, 259-280.

Huyghe, E., Matsuda, T. and Thonneau, P. (2003) Increasing incidence of testicular cancer worldwide, *J. Urol.* **170**, 5-11.

Ivell, R. and Hartung, S. (2003) The molecular basis of cryptorchidism, *Mol. Hum. Reprod.* **9**, 175-181.

Kaul, S.A. and Roberts, D.P. (1992) Preschool screening for cryptorchidism, *Br. Med. J.* **305**, 181.

Koopman, P., Munsterberg, A., Capel, B., Vivian, N. and Lovell-Badge, R. (1990) Expression of a candidate sex-determining gene during mouse testis differentiation, *Nature* **348**, 450-452.

Krausz, C., Quintana-Murci, L., Fellous, M., Siffroi, J.P. and McElreavey, K. (2000) Absence of mutations involving the Insl3 gene in human idiopathic cryptorchidism, *Mol. Hum. Reprod.* **6**, 298-302.

McIntyre, B.S., Barlow, N.J, Sar, M., Wallace, D.G. and Foster, P.M. (2002) Effects of *in utero* linuron exposure on rat Wolffian duct development, *Reprod. Toxicol.* **16**, 131-139.

McKinnell, C., Atanassova, N., Williams, K., Fisher, J.S., Walker, M., Turner, K.J., Saunders, P.T.K. and Sharpe, R.M. (2001) Suppression of androgen action and the induction of gross abnormalities of the reproductive tract in male rats treated neonatally with diethylstilboestrol, *J. Androl.* **22**, 323-338.

McKinnell, C., Sharpe, R., Mahod, K., Halmark, N., Scott, H., Ivell, R., Staub, C., Jegou, B., Haag, F., Koch-Nolte, F. and Hartung, S. (2005) Expression of insulin-like factor 3 protein in the rat testis during fetal and postnatal development and in relation to cryptorchidism induced by *in utero* exposure to Di(n-Butyl) phthalate, *Endocrinology* **146**, 4536-4544.

Mylchreest, E. and Foster, P.M. (2000) DBP exerts its antiandrogenic activity by indirectly interfering with androgen signaling pathways, *Toxicol. Appl. Pharmacol.* **168**, 174-175.

Mylchreest, E., Wallace, D.G., Cattley, R.C. and Foster, P.M. (2000) Dose-dependent alterations in androgen-regulated male reproductive development in rats exposed to di(n-butyl) phthalate during late gestation, *Toxicol. Sci.* **55**, 143-151.

Nef, S. and Parada, L.F. (1999) Cryptorchidism in mice mutant for Insl3, *Nature Genet.* **22**, 295-299.

Paulozzi, L.J. (1999) International trends in rate of hypospadias and cryptorchidism, *Environ. Health Perspect.* **107**, 297-302.

Preiksa, R.T., Zilaitiene, B., Matulevicius, V., Skakkebaek, N., Petersen, J.H., Jorgensen, N. and Toppari, J. (2005) Higher than expected prevalence of congenital cryptorchidism in Lithuania: a study of 1204 boys at birth and 1 year follow-up, *Hum. Reprod.* **20**, 1928-1932.

Raman-Wilms, L., Tseng, A.L., Wighardt, S., Einarson, T.R. and Koren, H. (1995) Fetal genital effects of first-trimester sex hormone exposure: a meta-analysis, *Obstet. Gynecol.* **85**, 141-149.

Scorer, C. (1964) The descent of the testis, *Arch. Dis. Child.* **39**, 605-609.

Sharpe, R.M. (2001) Hormones and testis development and the possible adverse effects of environmental chemicals, *Toxicol. Lett.* **120**, 221-232.

Sharpe, R.M. (2003) The 'oestrogen hypothesis' – where do we stand now? *Int. J. Androl.* **26**, 2-15.

Skakkebaek, N.E., Holm, M., Hoei-Hansen, C.E., Jorgensen, N. and Rajpert-De Meyts, E. (2003) Association between testicular dysgenesis syndrome (TDS) and testicular neoplasia: evidence from 20 adult patients with signs of maldevelopment of the testis, *Acta Patholog, Microbiol. Immunolog. Scand.* **111**, 1-9.

Skakkebaek, N.E., Rajpert-De Meyts, E. and Main, K.M. (2001) Testicular dysgenesis syndrome : and increasingly common developmental disorder with environmental aspects, *Hum. Reprod.* **16**, 972-978.

Stillman, R.J. (1982) *In utero* exposure to diethylstilboestrol: adverse effects on the reproductive tract and reproductive performance and male and female offspring, *Am. J. Obstet. Gynecol.* **142**, 905-921.

Thonneau, P.F., Gandia, P. and Mieusset, R. (2003) Cryptorchidism : incidence, risk factors and potential role of environment, *J. Androl.* **24**, 155-162.

Thorup, J. and Cortes, D. (1990) The incidence of maldescent testes in Denmark, *Pediat. Surg. Int.* **5**, 2-5.

Toppari, J., Larsen, J.C., Christiansen, P., Giwercman, A., Grandjean, P., Guillette, L.J., Jr, Jegou, B., Jensen, T.K., Jouannet, P., Keiding, N., Leffers, H., McLachlan, J.A., Meyer, O., Muller, J., Rajpert-De Meyts, E., Sceike, T., Sharpe, R., Sumpter, J. and Skakkebaek, N.E. (1998) Male reproductive health and environmental xenoestrogens, *Environ. Health Perspect.* **104**, 741-803.

Vidaeff, A.C. and Sever, L.E. (2005) *In utero* exposure to environmental estrogens and male reproductive health: a systematic review of biological and epidemiological evidence. *Reprod. Toxicol.* **20**, 5-20.

Weidner, I.S., Moller, H., Jensen, T.K. and Skakkebaeck, N.E. (1998) Cryptorchidism and hypospadias in sons of gardeners and farmers, *Environ. Health Perspect.* **106**, 793-796.

Wilcox, A.J., Baird, D.D., Weinberg, C.R., Hornsby, P.P. and Herbst, A.L. (1995) Fertility in men exposed prenatally to diethylstilboestrol, *N. Engl. J. Med.* **332**, 1411-1416.

Williams, K., McKinnell, C., Saunders, P.T.K., Walker, M., Fisher, J.S., Turner, K.J., Atanassova, N. and Sharpe, R.M. Neonatal exposure to potent and environmental estrogens and abnormalities of the male reproductive system in the rat: evidence for importance of the androgen-oestrogen balance and assessment of the relevance to man, *Hum. Reprod.* **7**, 236-247.

Wolf, C.J., LeBlanc, G.A., Ostby, J.S. and Gray, L.E., Jr. (2000) Characterization of the period of sensitivity of fetal male sexual development to vinclozolin, *Toxicol. Sci.* **55**, 152-161.

Zuccarello, D., Morini, E., Douzgou, S., Ferlin, A., Pizzuti, A., Salpietro, D.C, Foresta, C. and Dallapiccola, B. (2004) Preliminary data suggest that mutations in the CgRP pathway are not involved in human sporadic cryptorchidism, *J. Endocrinol. Invest.* **27**, 761-764.

ENVIRONMENTAL RISK AND SEX RATIO IN NEWBORNS

M. PETERKA*, Z. LIKOVSKY AND R. PETERKOVA
Department of Teratology
Institute of Experimental Medicine
Academy of Sciences CR
Videnská 1083
Prague 4
CZECH REPUBLIC
(author for correspondence, Email: peterka@biomed.cas.cz)*

Summary

Detecting intrauterine exposure to environmental pollutants, based on an increased number of malformations, has only been successful with a few teratogens. Nor does the number of spontaneous abortions represent a more reliable indicator, since a precise record of early abortions is not available. A greater vulnerability to prenatal damage leading to abortion is evident in male embryos/fetuses than in female. The newborn sex ratio (birth rate of boys/girls) is a very stable parameter in healthy populations. Its decrease has been reported after exposure to some harmful environmental factors. We document a decrease in the male birth fraction in the Czech Republic after the Chernobyl disaster in 1986. The absolute numbers of male and female births were determined in each of 600 consecutive months from 1950 to 1999. There were always more newborn boys than girls, except in November 1986, when the number of male births significantly decreased. This deficit in male births might have resulted from the spontaneous abortion of male embryos/fetuses during weeks 8-12 of pregnancy, as a consequence of their increased exposure to radiation, in particular to the radionuclide iodine131. We propose using the newborn sex ratio as a further tool for the standard evaluation of reproductive quality. Combined analyses of the incidence of newborn malformations, spontaneous abortions and stillbirths, intrauterine growth retardation and the newborn sex ratio will help to compensate for the imperfections associated with each of these parameters

P. Nicolopoulou-Stamati et al. (eds.), Congenital Diseases and the Environment, 295–319.

individually and will provide a more complete understanding of the extent of prenatal risk induced by environmental factors.

1. Introduction

Malformed children have been born since immemorial. A times figurine of dicephalic conjoined twins is thought to date from 6500 B.C. This figurine probably represents the oldest known archaeological record of a congenital defect (Warkany, 1977). It is clear that children with congenital malformations were already being born before the onset of the industrial revolution. However, we do not know whether these congenital malformations arose spontaneously or due to malnutrition, stress, infection or a genetic defect.

In the last century, several very dangerous teratogens were detected, such as the rubella virus in 1941 (Gregg, 1941), thalidomide in 1959-1961 (Lenz, 1961; McBride, 1961; Lenz and Knapp, 1962), and accutane in 1982 (Rosa, 1983). After the thalidomide affair, many countries began a careful and systematic recording of the numbers of congenital malformations. A national monitoring system of congenital malformations was started in the Czech Republic in 1961. For physicians, the reporting of congenital malformations was voluntary until 1964, then became obligatory starting in 1965 (Kučera, 1989). Many expectations were held for this national monitoring. It was believed that such monitoring would lead to the early detection of new harmful environmental factors, based on an increase in the number of congenital malformations above the basal incidence. However, although the amount of toxic chemicals in the environment has dramatically increased during the last decades, the incidence of malformations in newborns has remained low. Does this mean that thalidomide and accutane were extremely rare exceptions, or is the incidence of congenital malformations, when considered alone, not a reliable indicator of environmental risk?

In is paper, we shall first survey the basic manifestations of developmental damage and discuss the problem of dose-response dependence in teratogenesis. We shall demonstrate that relying upon only the incidence of malformed newborns (so-called residual teratogenesis) is not sufficient for monitoring environmental risk. Even the number of spontaneous abortions (prenatal deaths) does not represent a more reliable indicator, since a precise record of the number of spontaneous abortions is not available and an unknown proportion of such abortions results from purely maternal factors. The newborn sex ratio (birth rate of boys to girls) is a very stable parameter in healthy populations and has been shown to decrease due to increased environmental stress. We shall demonstrate the importance of the newborn sex ratio by documenting a change in this parameter as a consequence of the Chernobyl disaster in 1986. We propose using the newborn sex ratio as a further tool for the standard evaluation of reproductive quality. Combined analyses of the incidence of newborn malformations, the incidence of spontaneous abortions, the number of stillbirths, intrauterine growth retardation and the newborn sex ratio will help to compensate for the imperfections associated with each of these parameters individually and will provide a more complete understanding of the extent of prenatal risk induced by environmental factors.

2. Basic manifestations of developmental damage

The developmental damage of embryos or fetuses by genetic or environmental factors has three basic manifestations: lethality, growth retardation and malformation. The malformations (structural developmental defects) can be either major or minor, (Moore and Persaud, 1993). Wilson (1977) included these three basic manifestations of developmental damage under the common term "embryotoxicity". All three types of developmental damage can be seen in a single embryo/fetus.

2.1. Lethality - prenatal loss

According to estimates based on the calculated number of fertilizations, prenatal losses in humans are most severe during the first week. Leridon (1977) reported that 16% of mature oocytes are not fertilized at all. Of fertilized oocytes, 42% are lost before, during or shortly after implantation. Over the course of pregnancy, prenatal losses (11%) continue more slowly until birth (figure 14). Miller *et al.* (1980) evaluated early prenatal losses based on HCG levels and found that 43% of abortions occur shortly after implantation. Lohstroh *et al.* (2005) investigated the fate of 62 conceptions detected by HCG. Among these early pregnancies, 35% were terminated by spontaneous abortion. The risk of abortion increases with maternal age and previous abortions in anamnesis (Leridon, 1987).

The above data document that the greatest prenatal losses occur at the beginning of pregnancy. It is generally assumed that these early losses are mainly caused by chromosomal aberrations (Gardo, 1993; Brent, 2004) and impaired implantation. (Racowsky, 2002; Caglar *et al.,* 2005). Potential causes of later abortions include, for example, malformations not compatible with further prenatal life, problems arising from the placenta or umbilical cord, or a hormonal imbalance or illness of the mother. Prenatal loss is a heterogeneous phenomenon and has no uniform critical period.

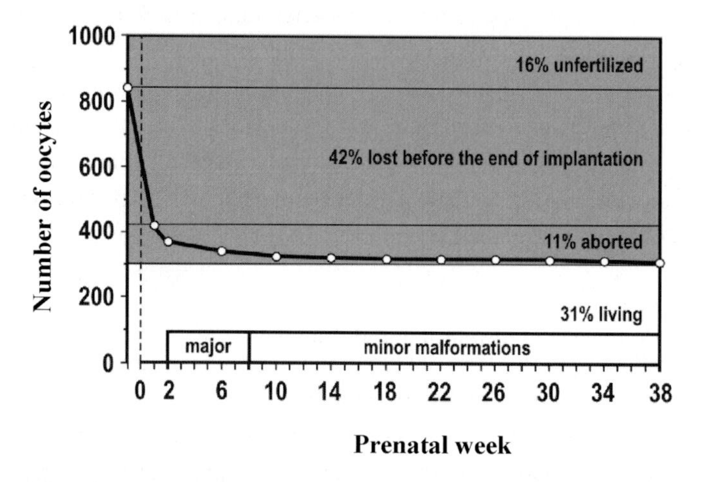

Figure 14. Estimate of prenatal losses during pregnancy (Leridon, 1977). The critical periods for the origin of major and minor malformations are indicated (modified from Moore and Persaud, 1993). Grey area – prenatal loss

The official Czech statistical monitoring system records the number of spontaneous abortions (figure 15).

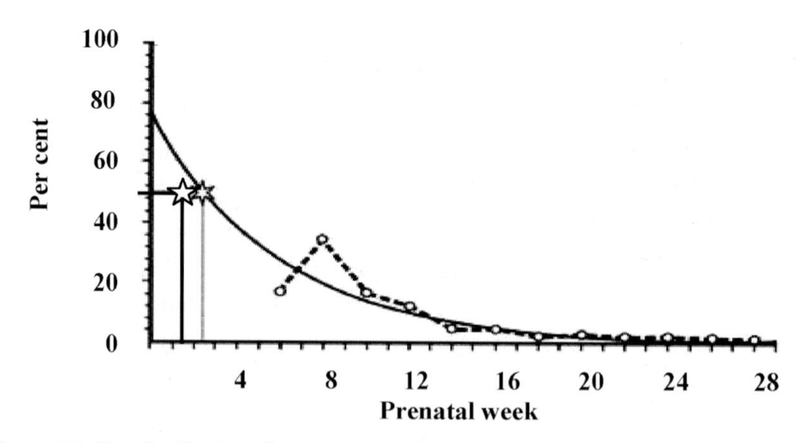

Figure 15. The distribution of spontaneous abortions over the course of pregnancy. Data by the Czech Statistical Office for prenatal weeks 5 – 28 (dashed line). The total number of abortions in the Czech Republic from 1965-1999 is taken to be 100%. The circles represent the mean percent values in particular weeks of pregnancy. The official register cannot include the complete number of spontaneous abortions during the first four weeks of pregnancy (these early abortions can go unrecognised when masked by menses, or unreported if they occur outside of hospitals). Therefore, the official data have been extrapolated mathematically using an exponential curve (full line), back to prenatal weeks 1-4. The extrapolation suggests a 50% loss during weeks 2-3 of pregnancy (asterisk) (Czech Statistical Office)

This register includes only those spontaneous abortions that took place in a hospital and so were obligatorily reported by physicians. Spontaneous abortions occurring outside of hospitals are missing from the register. The number of registered spontaneous abortions reaches a maximum at week 8 of pregnancy, then decreases to nearly zero until prenatal week 28 (figure 15). Although data are not available about the number of spontaneous abortions during the first two weeks of pregnancy, (since pregnancy is not yet confirmed at that time and early embryonic loss is masked by menstruation), we attempted to estimate the number mathematically (figure 15). The rate of spontaneous abortion during the first two-three weeks of pregnancy was estimated to be approximately 50%.

2.2. Major malformations

Each embryonic component has a critical developmental period during which its morphogenesis can be disturbed and a malformation can arise. A major malformation (monstrosity) is a structural developmental defect that is apparent at first sight (e.g. orofacial clefts, amelia, anencephaly). Major malformations originate during the embryonic and early fetal phases of prenatal development (figure 14), when the external morphology, organ systems and individual organs are established and pass through critical periods in their development. After these critical periods are over, major structural anomalies no longer arise. For example, the critical developmental period for establishing the central nervous system starts in embryonic week 3. Ordinarily, the neural tube is completely fused between embryonic days 25-27 (Kimmel and Buelke-Sam, 1981). This implies that neural tube defects (e.g. exencephaly or spina bifida) cannot arise after the end of the first prenatal month. In contrast, the critical period for the development of the external genitalia in male fetuses finishes as late as the second half of prenatal month 3; hypospadias of the male urethra can arise during prenatal weeks 9-12, but no later.

During the development of more complex structures, critical developmental periods can combine and overlap, as can be seen in the case of orofacial development. The critical period, during which different types of orofacial clefts can arise, extends from approximately prenatal day 27 until 60. Between prenatal days 27-43, the medial nasal and maxillary processes fuse; the non-fusion of these processes leads to a cleft lip. Between prenatal days 37-40, the palatal shelves appear and grow vertically down the sides of the tongue until day 55. They then rapidly elevate to a horizontal position above the dorsum of the tongue. If they are large enough, both palatal shelves fuse at the midline by day 60 of pregnancy (Ferguson, 1991). Hypoplasia of the palatal shelves is one possible reason for their non-fusion, resulting in an isolated cleft palate. However, experiments in mice have shown that an isolated cleft palate can arise even if the palatal shelves are not hypoplastic. These clefts are caused by a blockage of the horizontalisation process itself (Jelínek and Peterka, 1977). The removal of the tongue from the roof of the primitive oral cavity, and thus the creation of a free space above the tongue, is a precondition for the horizontalisation of the palatal shelves. The free space above the tongue is created by the rapid growth of the lower jaw in the forward direction, which in turn pulls the tongue down from the roof of the nasal cavity. Growth retardation of the lower jaw can, surprisingly, be an additional causal mechanism of an isolated cleft palate, because the tongue remains blocked between the palatal shelves and mechanically inhibits the horizontalisation process (Jelínek and Peterka, 1977). The origin of a free space above the tongue due to the fast growth of the mandible has also been documented in human embryos (Slípka *et al.,* 1980; Diewert, 1983). The isolated cleft palate associated with a hypoplastic mandible in Pierre-Robin syndrome (Pratt, 1966) can be explained by the above-described mechanism (Jelínek and Peterka, 1977).

2.2.1. Prenatal extinction of major malformations

It is known that not all malformed embryos survive until birth. Many malformed embryos/fetuses are spontaneously aborted (Brent and Beckman, 1994). Consequently, the number of malformations in newborns only represents a small fraction of the malformations that originated prenatally and is called "residual teratogenesis" (Kučera, 1989). In orofacial clefts, prenatal losses have been estimated to be about 90% (Kučera, 1989). Nearly 30% of early embryos (less than 35 days old) examined following induced abortions possess abnormalities (Nishimura *et al.,* 1987; Shiota *et al.,* 1987). We have studied the teratogenic effect of a higher incubation temperature on development in chick embryos and found a higher incidence of malformations in dead embryos than in surviving ones (Peterka *et al.,* 1996).

There is no doubt about the cause of death in embryos affected by a major malformation that is incompatible with further life (e.g. anencephaly). However, how does one explain prenatal extinction in embryos that are affected by a malformation that is not incompatible with further life, such as syndactyly or a cleft lip? We can speculate about additional associated and more serious developmental damage – e.g. a minor malformation that is not visible at first sight.

2.3. *Minor malformations usually manifest as functional defects*

Minor malformations are minor structural defects that arise during the fetal and early postnatal phases of development (figure 14), i.e. during the critical periods of histodifferentiation and terminal cytodifferentiation in organs. These organs need not have a major structural defect (i.e. they have been established normally). Minor structural defects can be apparent at the tissue, cellular or sub-cellular level and are usually detected due to the disturbed functioning of the affected organ (e.g. behavioural defects, blindness, deafness).

Physicians treating mothers at the end of pregnancy, during delivery and during the early postnatal period sometimes underestimate the risk of minor structural developmental defects arising in the infants. The most important period of brain growth and the differentiation of its fine structures (including myelinisation) take place during the last half of gestation and continue throughout the first two years (Levy *et al.,* 1998). During that time, for example, minor structural brain defects can arise, that can later lead to so-called "minimum brain dysfunction" ranging from hyperactivity (attention deficit hyperactivity disorder - ADHD) and learning disabilities (dyslexia, dysgraphia) to mental retardation (Levy *et al.,* 1998). Exposure to tobacco smoke *in utero* (Linnet *et al.,* 2003) and neonatal hypoxia-ischemia (Krageloh-Mann *et al.,* 1999) are suspected to be associated with ADHD. There is also a higher incidence of ADHD in boys than in girls and when the educational level of the parents is low.

2.4. *Intrauterine growth retardation and low birth weight*

Body weight is the parameter most frequently used to measure growth rate. The birth weight of a newborn reflects intrauterine growth during the prenatal period. The intrauterine growth of an embryo/fetus is a multifactorial process that is sensitive to a large variety of environmental stresses. Birth weight depends on sex (girls are always lighter) as well as ethnic and racial factors. In Sweden, the mean birth weight in both boys and girls has increased by about 500g during the last 27 years; in 2000, the mean birth weight was 3,600g in boys and 3,470g in girls (Odlind *et al.,* 2003). In comparison to Sweden, Czech boys and girls showed a lower mean birth weight in 1989-1991: 3,522g and 3,310g, respectively (Koupilova *et al.,* 1998). The mean birth weight of full term infants delivered by the Czech gypsy population was determined to be 2,939g and 2,816g in boys and girls, respectively (Bernasovsky *et al.,* 1975). Birth weight reflects the gestational age and reveals racial differences between white Americans and African-Americans (the birth weight was lower by 250g in the latter group), (Alexander *et al.,* 1999). Literature data document that birth weights differ between racial and ethnic groups (Alexander *et al.,* 1999) and that the limits of physiological and pathological birth weights should be established empirically for each individual population. When evaluating the birth weight in a specific infant, the anthropological parameters of the parents need to be taken into account as well.

A low birth weight (less than 2,500g), very low birth weight (less than 1,500g) or extremely low birth weight (less than 1,000g) have been regularly found in children born before full term (i.e. before week 40 of pregnancy), (e.g. Grandi *et al.,* 2005; Hack *et al.,* 2005; Knops *et al.,* 2005). If the attenuation of the birth weight correlates with a younger gestational age of the newborn child, it should not be classified as intrauterine growth retardation (IUGR). This term should properly be reserved for those cases in which the intrauterine growth rate is lower than would be expected from the gestational age of the embryo/fetus. IUGR is closely associated with teratogenesis and is regularly found in newborns with a major or minor structural birth defect (Brent, 1977). However, growth retardation can even be found in children without any apparent structural anomaly, e.g. after exposure to the atomic bombs in Japan (Brent, 1977). In contrast to structural defects, IUGR has no critical period and can develop at any time during pregnancy. Ergaz *et al.* (2005) have summarized the causes of IUGR as follows:

1) maternal factors (malnutrition, chronic maternal diseases, birth order, multiple birth and parental genetic factors),
2) placental pathology (mainly placental vascular damage leading to placental insufficiency),
3) infections and specific fetal syndromes including chromosomal aberrations,
4) parental life style (smoking, alcohol consumption).

Individuals affected by intrauterine growth retardation show a lower birth weight than would be expected from their gestational age.

3. Dose response in teratology

Experiments using chick embryos have demonstrated three basic types of exogenous embryotoxic factors:

1) substances inducing both teratogenic and lethal effects,

2) toxic substances that rapidly cause embryonic death without providing time for the development of a structural anomaly,
3) factors that induce neither a teratogenic nor a lethal effect, even at the highest tested doses.

3.1. Substances with teratogenic and lethal effects

These substances are usually indicated as "teratogens" when taking into account only one aspect of their effects.

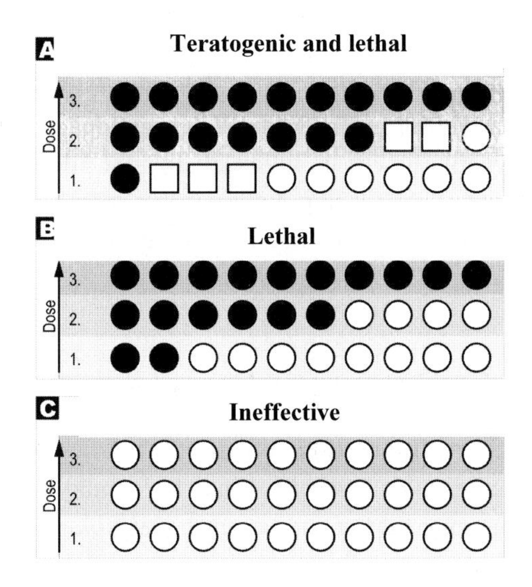

Figure 16. Comparison of three types of embryotoxic factors. A. A factor exhibiting both lethal and teratogenic effects, detected according to major malformations (e.g., hyperthermia, Peterka *et al.,* 1996). **B**. A harmful factor showing only a lethal effect (e.g., psychopharmaca, Peterka *et al.,* 1992). **C**. A factor that exhibits neither a teratogenic not a lethal effect, even if administered in the highest doses (e.g. beta–carotene, Peterka *et al.,* 1997), Note the changing numbers of dead (black circle), living normal (white circle), and living malformed (white squares) embryos after exposure to three increasing doses (1, 2, 3) (Peterka *et al.,* 1992, 1996, 1997)

However, the rate of the teratogenic effect of these substances changes with increasing dose. An increase in the dose of a teratogen leads to a marked change in the numbers of dead and living malformed embryos. Initially, increasing doses induce an increase in both teratogenic and lethal effects. However, a further increase in dose leads only to an increase in the lethal effect, while the teratogenic effect paradoxically decreases. Following the administration of the highest doses, only the lethal effect is apparent (figure 16A); the teratogenic effect completely disappears, because all embryos die before a malformation can arise (Wilson, 1977). The highest doses of a "teratogen" thus show a similar lethal effect to toxic substances (see below). We have experimentally documented changes in the ratio of living malformed embryos to dead embryos depending on increasing dose (Peterka *et al.*, 1986), (compare to figure 17).

3.2. ***Substances with a predominantly lethal effect***

These are harmful exogenous factors leading predominantly to embryonic death after the administration of their effective doses – for example, some psychopharmaceuticals. Psychopharmaceuticals have the capacity to cause arrhythmias, heart defects and cardiac arrest. The intra-amniotic injection of these drugs (e.g. maprotiline, amitriptyline, dosulepin, or imipramine) into chick embryos causes immediate heartbeat arrest followed by the death of the embryo; malformations have no time to develop. Embryos that survive exposure to effective doses of psychopharmaceuticals do not necessarily show any major structural abnormality (Peterka *et al.,* 1992) (figure 16B).

3. 3. ***Substances with neither a teratogenic nor a lethal effect***

For example, beta-carotene had no embryotoxic effect in rats and rabbits (Heywood *et al.,* 1985). In teratological experiments using chick embryos, beta-carotene or 5-aminolevulinic acid are examples of substances that exhibit no embryotoxic effects, even when they are administrated in the highest tested doses (corresponding to approximately 100g/kg of the maternal organism) (Peterka *et al.,* 1997; Peterka *et al.,* 2001), (figure 16C).

4. Residual teratogenesis and the epidemiology of malformations

What is the sumarry of the experimental and clinical data about the prenatal death of malformed embryos or fetuses and spontaneous abortions in human. The number of newborns with a malformation represents only a small portion (about 10%) of all the malformed embryos that arise prenatally. It is for this reason that this number has been called "residual teratogenesis" (Kučera, 1989). This parameter was successful used in the detection of the new teratogen thalidomide, due to three specific circumstances:

1) thousands of pregnant women consumed thalidomide. Such a sample is sufficiently large and homogeneous to manifest a teratogenic effect at the population level,

2) the typical malformations (limb deformities) induced by thalidomide apparently did not impair the viability of embryos/fetuses, which could thus survive until birth,

3) these limb deformities were so remarkable that they could not go unnoticed.

Khoury and Holtzman (1987) tried to estimate the time needed for detecting new teratogens such as thalidomide. Thanks to current programs monitoring the incidence of birth defects, it should be possible to discover such a dangerous teratogen within 2 weeks. However, detection will not be so rapid in most other cases. The detection of strong teratogens such as valproic acid and isotretinoin (accutane) required more than 20 years of monitoring, because only a relatively small population of pregnant women was treated by accutane (Khoury and Holtzman, 1987).

After prenatal exposure to very high doses of a harmful factor, all malformed embryos may die before the completion of prenatal development (Wilson, 1977). If we evaluate prenatal environmental risk based only on the number of malformed children, we might paradoxically find no increase in this parameter even in the most polluted areas. This might hypothetically explain the conclusion by Mastroiacovo *et al.* (1988), who found no malformations in the most highly dioxin-contaminated areas of Seveso. The frequencies of major defects detected in areas of low or very low contamination were 29.9/1,000 and 22.1/1,000, respectively. Since a frequency of 27.7/1,000 was found in the control area, the data collected failed to demonstrate any increased risk of birth defects associated with 2,3,7,8-tetrachlorodibenzo-p-dioxin (Mastroiacovo *et al.,* 1988). Figure 17 depicts two different situations in which the number of malformed children is the same, but the exposure to harmful environmental factors differs greatly. Exposure to the highest doses leads to prenatal death.

Using chick embryos, we have modelled experimentally this dose-response situation in humans, that is, when only data about the number of inborn developmental defects are available, while complete data about the number of spontaneous abortions are missing (figure 17).

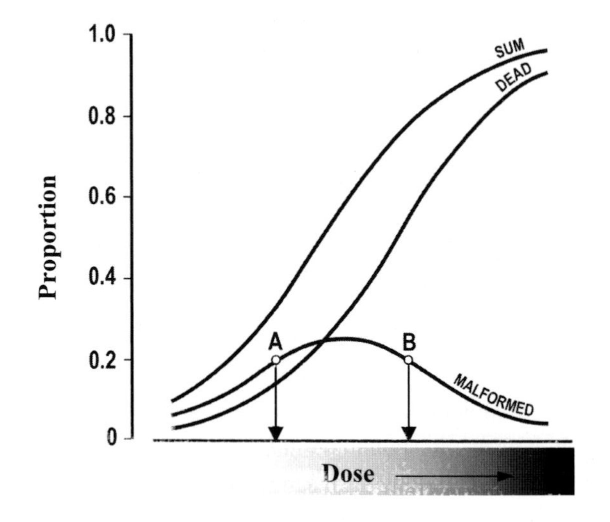

Figure 17. Changes in the numbers of malformed and dead embryos after exposure to increasing doses of an environmental factor. The number of malformed embryos is similar in locations A and B, although the dose was much higher at B than at A. This scheme is supported by experimental data collected by administering different drugs to thousands of chick embryos (Peterka *et al.,* 1986)

This analysis demonstrates why information is required not only about the number of malformed newborns, but also about the number of spontaneous abortions. Since complete epidemiological data about the number of spontaneous abortions are not available (see above and figure 17), we examined whether the sex ratio in newborns might represent a helpful indicator of prenatal losses after exposure to a harmful environmental factor.

5. Sex ratio

The sex ratio is the relative proportion of males and females and can refer to the prenatal or postnatal period. We shall document that the newborn sex ratio can be an additional tool for evaluating reproductive quality, similar to birth weight or the incidence of malformations, spontaneous abortions and stillbirths.

5.1. Sex determination

Two different sex chromosomes exist, X and Y. Normally, there are always two sex chromosomes in each cell of an individual; both are X chromosomes in women, while men have one X and one Y chromosome. The genetic sex of an individual is established at fertilization – i.e. during the fusion of an oocyte (bearing one X-chromosome) and a spermatozoon (bearing either one X or one Y chromosome). This genetic sex cannot be subsequently changed. In accordance with the genetic sex, the internal and external reproductive organs are morphologically determined during the first trimester (until prenatal week 12) (Moore and Persaud, 1993). The feminisation of male embryos (genetically XY) arises because of a defect in androgen receptors (Bulfield and Nahum, 1978), while the masculinisation of female embryos (genetically XX) is known to occur after exposure to androgens. Recently, the two genes hypothesis has been formulated with the aim of satisfactorily explaining all the known pathologies of human sex determination due to the interplay between the SRY gene, which encodes for testis-determining factor (TDF), and the Z-gene (McElreavey *et al.,* 1993).

5.2. *Newborn sex ratio*

Data from the literature document that a markedly higher number of male embryos start to develop at the beginning of pregnancy than female embryos (Wells, 2000). The original sex ratio was conservatively estimated to be 120:100 (McMillen, 1979). This "primary sex ratio" progressively decreases during prenatal development, so that the "secondary sex ratio" is actually found in newborns (Chahnazarian, 1988; Jongbloet *et al.,* 2002). A male/female sex ratio of 2.5 (861 male embryos/346 female embryos) has been determined during weeks 16-19 of pregnancy. This ratio decreases to 1.17 (268,532 male fetuses/245,862 female fetuses) in weeks 37-39 (Vatten and Skjaerven, 2004). The actual sex ratio in newborn is about 1.05 (see below). The difference between the primary and secondary sex ratios can hardly be explained by a sex reversal phenomenon (see 6.1). The most probable explanation is that male embryos and fetuses are aborted more frequently than are female ones. Indeed, a higher prenatal sex ratio (of 1.30, corresponding to 56.52% boys and 43.48% girls) has been estimated for spontaneously aborted fetuses with a normal caryotype (Hassold *et al.,* 1983; Mizuno, 2000). Data concerning differential fetal and neonatal mortality by sex in the United States from 1950 to 1972 have been analysed by McMillen (1979).

5.3. *Newborn sex ratio and the vulnerability of male embryos/fetuses*

The birth rate of newborn males is permanently higher than that of females. In the Czech Republic over the long term, on average 105 boys are born per 100 girls (Czech Statistical Office). A similar sex ratio has also been found in other populations. For example, in Sweden, Germany, Norway, Finland, the Netherlands,

Denmark, Canada and the United States, a sex ratio 1.06 has been found (Davis *et al.,* 1998) corresponding to 51.46% and 48.54% male and female births, respectively. Khoury *et al.* (1984) have reported the newborn sex ratio in white American couples to be 1.059, in black American couples 1.033, and in Native American couples 1.024. This study showed that the observed racial differences in the sex ratio are due to the effects of the father's race, not the mother's. In any event, the long-standing significantly higher male birth fraction represents a reliable indicator of reproductive stability and health (Davis *et al.,* 1998; Vartiainen *et al.,* 1999; Campbell, 2001).

Some harmful environmental factors are known to change the sex ratio in offspring (Mocarelli *et al.,* 1996; Mocarelli *et al.,* 2000). A decrease in the male birth fraction has been reported after long-term exposure to environmental chemical factors (Nicolich *et al.,* 2000; Jarrell *et al.,* 2002). A lower male birth fraction has been reported in children of fathers contaminated by TCDD – dioxin (Jongbloet *et al.,* 2002; Ryan *et al.,* 2002) and of parents who smoked more than 20 cigarettes per day (Fukuda *et al.,* 2002). One explanation of the decreased male birth fraction has been suggested by James's hypothesis (James, 1996). Genetically, gender is already determined during fertilization. Parental hormonal concentrations at the time of conception can partly control the gender of the offspring: a low concentration of testosterone and estrogens are associated with daughters (James, 1996). The anti-estrogenic and anti-androgenic properties of dioxin (Jongbloet *et al.,* 2002) and the anti-estrogenic effect of smoking are well documented (Michnovicz *et al.,* 1988). Pre- or peri-conception, both these harmful factors might reduce the number of males and change the sex ratio at the very beginning of pregnancy (James, 2002a; James, 2002b).

During pregnancy, the greater vulnerability of male embryos/fetuses to prenatal damage by environmental stress results in their abortion (Hassold *et al.,* 1983; Mizuno, 2000; Wells, 2000), thus explaining the decrease in the male birth fraction. This vulnerability results both in the predominant loss of male fetuses (Wells, 2000) and in a higher incidence of developmental malformations in newborn boys. In the Czech Republic, the long-term sex ratio of malformed newborns is 1.5 (Czech Statistical Office). Since environmental stress can affect boys more severely than girls prenatally, we examined the effect of the Chernobyl accident on the ratio of male to female births in then-Czech Republic.

6. Chernobyl

An explosion at the Chernobyl atomic power station in April 1986 caused a widespread fall-out of radionuclides, contaminating many locations around the world (Broadway *et al.,* 1988; Shizuma *et al.,* 1987). The radiation caused the death of about 100 people as well as paediatric thyroid cancers in Belarus and Ukraine (Nikiforov *et al.,* 1994; Jacob *et al.,* 2000; Leenhouts *et al.,* 2000). Increased developmental malformations (Lazjuk *et al.,* 1997; Gavyliuk *et al.,* 1992; Romanova *et al.,* 1998), spontaneous abortions (Gavyliuk *et al.,* 1992; Karakashian *et al.,* 1997) and chromosomal aberrations (Gavyliuk *et al.,* 1992) were all reported. An increase in the number of spontaneous abortions was found not only in the adjacent area (Gavyliuk *et al.,* 1992; Karakashian *et al.,* 1997), but also in Finland (Auvinen *et al.,* 2001) and Norway (Ulstein *et al.,* 1990; Irgens *et al.,* 1991). An increase in perinatal mortality (Scherb *et al.,* 2000) and trisomy 21 (Sperling *et al.,* 1994) were detected in Germany. Increased Down's syndrome and childhood leukaemia cases were reported in Sweden (Ericson and Kallen, 1994), and an increased incidence of thyroid disease was found in children in Hungary (Lukacs *et al.,* 1997).

The straight-line distance between Prague and Chernobyl is approximately 1000 km. However, the contamination of a specific place by radionuclides depends not only on distance, but also on the actual local meteorological conditions: on the winds pushing the radioactive clouds and on rain transporting the radionuclides to land (Bangert *et al.,* 1986).

6.1. Radioactive clouds and whole body radioactivity

Three waves of radioactive clouds were registered in May 1986 by Czech Republic monitoring of radionuclides in the atmospheric aerosol (Kunz, 1986). The clouds brought isotopes with different radioactive half-lives (table 17).

Table 17. Radioactive half time of some radionuclides

Radionuclide	Half time
^{140}La	1.7 days
^{132}Te	3.3 days
^{131}I	8 days
^{103}RU	1 year
^{134}Cs	2 years
^{137}Cs	30 years
^{90}Sr	30 years

In table 17, mostly tellurium (Te132), iodine (I131), ruthenium (Ru103) and caesium (Cs137 and Cs134). The maximum radiation, about 257 Bq/m3, was detected at the start of monitoring (April 30); this level could only represent the descending phase of the first wave of radioactive clouds. During the second wave (May 4), the maximum concentration of radionuclides was six times lower – only 38 Bq/m3. The third wave (about May 7) was the least radioactive – only 7 Bq/m3 at maximum.

Monitoring whole body radioactivity was begun on May 4, 1986, at the time of the peak of the second wave of radioactive clouds, when the mean value of whole body radioactivity was about 4000 Bq/s). For the period of April 30 – May 3, whole body radioactivity has been estimated to be as high as 9000 Bq/s (Kunz, 1986). The decrease in environmental radioactivity was followed, after a delay, by a gradual decrease in whole body radioactivity during the two weeks following May 4 (Kunz, 1986).

6.2. Newborn sex ratio – November 1986

The birth sex ratio is a very stable phenomenon in the Czech population. The absolute number of births has been examined separately for girls and boys in each of 600 consecutive months from 1950 - 1999. In 599 months, there was a higher number of boys born (51.42%) than girls (48.58%), by 342.1 births (2.83%) on average. The only exception was November 1986 (figure 18 A, B), when the sex ratio reversed: there were actually significantly ($p<0.05$) fewer boys born (49.35%) in this month than girls (50.65%). This month was also the only one in which the percentage of male births fell below the 50 per cent level out of 600 months from 1950-1999. This decrease cannot be explained by the premature delivery of boys, since the male birth fraction did not increase in September or October 1986 ($p > 0.05$). Sex reversal (see 6.1.) due to the feminisation of the sexual organs in male fetuses is also not a likely cause, because the number of girls born in November 1986 did not change ($p > 0.05$) compared to other November months (figure 18B). The decrease in the male birth fraction in November 1986 was absolute. The total number of missing boys in November 1986 is estimated to be 4,670 (Peterka *et al.*, 2004).

We have suggested that the negative impact of radiation on the prenatal population manifested itself as a selective loss of a portion of male fetuses (Peterka *et al.*, 2004). This is in agreement with the observed increase in the number of spontaneous abortions in Finland (Auvinen *et al.*, 2001) and Norway (Ulstein *et al.*, 1990; Irgens, 1991) after the Chernobyl explosion.

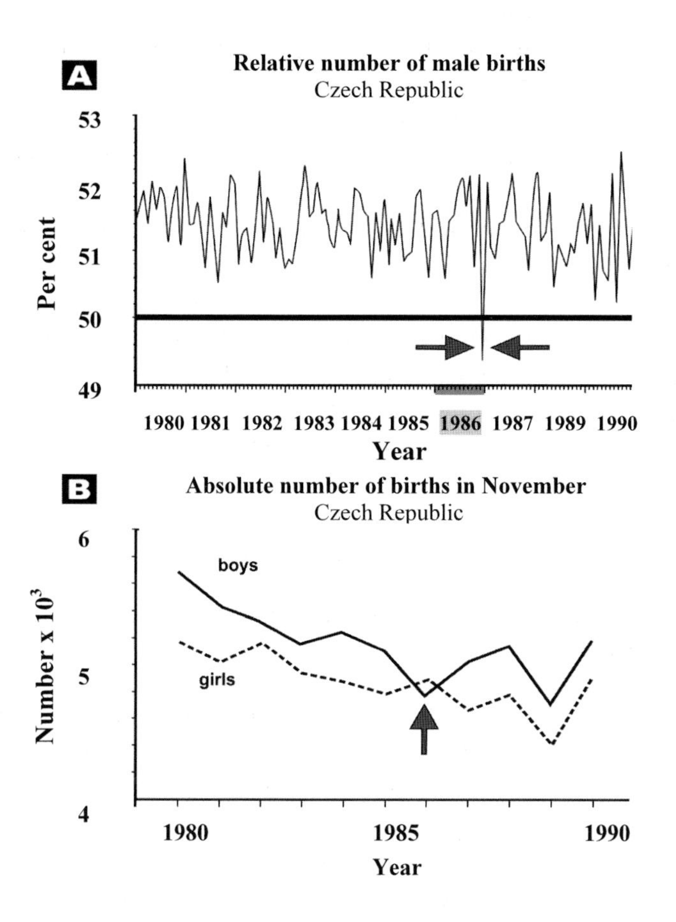

Figure 18. The sex ratio in newborn. A. Male birth rate in each month during the years 1980 – 1989. Arrows indicate the reversal in the sex ratio in November 1986. B. Absolute numbers of newborn boys (full line) and girls (dashed line) in the month of November in each year from 1980 – 1990. The curves reflect the general decline in the birth rate in Czech Republic since the end of seventies. The arrow indicates the extraordinary decline in the number of newborn boys in November 1986 (based on Czech Statistical Office)

6.3. Critical period for radiation

In the area irradiated after the Chernobyl explosion, naturally there were pregnant women in the first to the ninth month of their pregnancy. Children with a delivery date in November were conceived between February 8 and March 10, 1986. Their age was thus about 8-12 prenatal weeks during the peak of radioactivity (from April

30 to May 7). At this age, their sex was already determined and so the change in the sex ratio cannot be explained by the above-mentioned hypothesis by James (1996).

During the third month of prenatal development, the vulnerability of the developing fetal brain to radiation injury has been reported after the atomic bombings in Hiroshima and Nagasaki (Yamazaki and Schull, 1990). The radiation effect has been explained by the impaired migration of immature neurons to the brain cortex over the course of the 8th to the 15th week of prenatal development (Yamazaki and Schull, 1990). It cannot be excluded that brain development could also have been impaired in the fetuses lost after the Chernobyl explosion.

As documented in table 17, the radioactive clouds contained radionuclides with short half-lives after the Chernobyl explosion. Whole body radioactivity decreased rapidly during a few weeks after the explosion (Peterka *et al.,* 2004). This temporary exposure (end of April to beginning of May 1986), together with the above-mentioned critical period of prenatal vulnerability, can explain why the sex ratio was changed in only one month – November 1986.

6.4. Hypothesis about the reasons for the selective abortion of male fetuses after the Chernobyl disaster

We can speculate about the putative reasons for these male abortions after the Chernobyl disaster. One candidate could be the radionuclide iodine-131, whose concentration in the radioactive clouds was very high after the explosion (Broadway *et al.,* 1988; Shizuma *et al.,* 1987; Stenke *et al.,* 1987; Gembicki *et al.,* 1991). Eheman *et al.* (2003) have stressed the risk of autoimmune thyroid diseases associated with environmental radiation exposure. Radioactive iodine-131 can cause hypothyroidism and increase the maternal risk of a pre-term delivery (Tatham *et al.,* 2002) or spontaneous abortion (Grossman *et al.,* 1996). It is generally known that thyroid disturbance during pregnancy may lead to abortion, stillbirth, neonatal death or low birth weight, and the use of radioactive iodine is absolutely contraindicated during pregnancy (Bishnoi and Sachmechi, 1996; Ogris, 1997). Moreover, fetal brain development is very sensitive to an excess or deficit of thyroid hormone in the mother (Zoeller *et al.,* 2002; Poppe and Glinoer, 2003). We hypothesize that the transient increase in iodine-131 might have impaired the thyroid glands of the mothers or fetuses and thus led to the abortion of the most vulnerable part of the prenatal population.

7. Conclusion

1) The sum of all embryotoxic damage (major and minor structural malformations, prenatal death, IUGR) always shows a positive dose-response.

2) The positive dose-response relationship of a teratogen cannot be accurately detected when only malformations are taken into account. With increasing doses of a teratogen, the incidence of malformations first rises, but then declines. The highest doses induce the death of embryos before a malformation can develop.

3) When detecting a harmful prenatal factor, a thorough evaluation should focus on all types of prenatal damage (malformations, abortions, IUGR, stillbirths), including a change in the newborn sex ratio.

4) The prenatal sex ratio shows a prevalence of male fetuses from the beginning of pregnancy.

5) Male fetuses are more sensitive then female fetuses and consequently can be more easily damaged.

6) The newborn sex ratio is a very stable phenomenon. A change in the newborn sex ratio signals a harmful factor in the environment.

7) Radiation after the Chernobyl accident was followed by a transient reverse in the newborn sex ratio in November 1986 in the then-Czech Republic. The deficit in male births might reflect the spontaneous abortion of male fetuses as a consequence of the increased radiation, mostly due to the radionuclide iodine-131.

References

Alexander, G.R., Kogan, M.D., Himes, J.H., Mor, J.M. and Goldberg, R. (1999) Racial differences in birth weight for gestational age and infant mortality in extremely-low-risk US population, *Paediatr. Perinat. Epidemol.* **13**, 205-217.

Auvinen, A., Vahteristo, M., Arvela, H., Suomela, M., Rahola, T., Hakama M. and Rytomaa T. (2001) Chernobyl fallout and outcome of pregnancy in Finland, *Environ. Health Perspect.* **109**, 179-185.

Bangert, K., Blasing, C., Degener, A., Jung, A., Ratzek, R., Schennach, S., Stock, R., Urban, F.J., Reiser, W., Jonas, H., Sattler, E.E., Weingand, G., Huber, K., Barth, B., Kiefer, J.,

Elles, S., Optiz von Boberfeld, W. and Handlock, R.M. (1986) Radioactivity in air, rain, soil, plants and food after the Czernobyl incident, *Naturwissenschaften* **73**, 495-498.

Bernasovsky, I., Bernasovska. K. and Hanzelova, V. (1975) *Anthropological Characteristics of Rom new-born children in the Presov district*, Proceedings PdF UK, Prague, *Biology* **4**, 35-40.

Bishnoi, A. and Sachmechi, I. (1996) Thyroid disease during pregnancy, *Am. Fam. Physician* **53**, 215-220.

Brent, R.L. (1977) Radiations and other physical agents, chapter 5 in J.G. Wilson, F.C. Fraser (eds.), *Handbook of Teratology*, Plenum Press, New York and London, UK, pp. 153-223.

Brent, R.L. (2004) Environmental causes of human congenital malformation: The pediatrician's role in dealing with these complex clinical problems caused by a multiplicity of environmental and genetic factors, *Pediatrics* **113**, 957-968.

Brent, R.L. and Beckman, D.A. (1994) The contribution of environmental teratogens to embryonic and fetal loss, *Clin. Obstet. Gynecol.* **37**, 646-670.

Broadway, J.A., Smith, J.M., Norwood, D.L. and Porter, C.H.R. (1988) Estimates of radiation dose and health risks to the United States population following the Chernobyl nuclear plant accident, *Health Phys.* **55**, 533-539.

Bulfield, G. and Nahum A. (1978) Effect of the mouse mutants testicular feminization and sex reversal on hormone-mediated induction and repression of enzymes, *Biochem. Genet.* **16**, 743-750.

Caglar, G.S., Asimakopoulos, B., Nikolettos, N., Diedrich, K. and Al-Hasani, S. (2005) Preimplantation genetic diagnosis for aneuploidy screening in repeated implantation failure, *Reprod. Biomed. Online* **10**, 381-388.

Campbell, R.B. (2001) John Graunt, John Arbuthnott, and the human sex ratio, *Hum. Biol.* **73**, 605-610.

Chahnazarian, A. (1988) Determinants of the sex ratio at birth: review of recent literature, *Soc. Biol.* **35**, 214-235.

Davis, D.L., Gottlieb, M.B. and Stampnitzky, J.R. (1998) Reduced ratio of male to female birth in several industrial countries: A sentinel health indicator? *JAMA* **279**, 1018-1023.

Diewert, V.M. (1983) A morphometric analysis of craniofacial growth and changes in spatial relations during secondary palatal development in human embryos and fetuses, *Am. J. Anat.* **167**, 495-522.

Eheman, C.R., Garbe, P. and Tuttle, R.M. (2003) Autoimmune thyroid disease associated with environmental thyroidal irradiation, *Thyroidology* **13**, 453-464.

Ergaz, Z., Avgil, M. and Ornoy, A. (2005) Intrauterine growth restriction-etiology and consequences: What do we know about the human situation and experimental animal models? *Reprod. Toxicol.* **20**, 301-322.

Ericson, A. and Kallen, B. (1994) Pregnancy outcome in Sweden after the Chernobyl accident, *Environ. Res.* **67**, 149-159.

Ferguson, M.W.J. (1991) The orofacial region, in Wigglesworth JS, Singer DB (ed), *Textbook of Fetal and Perinatal Pathology*, Blackwell Scientific Publication, London, UK, pp. 843-880.

Fukuda, M., Fukuda, K., Shimizu, T., Andersen, C.Y. and Byskov, A.G. (2002) Parental periconceptional smoking and male: female ratio of newborn infants, *Lancet* **359**, 1407-1408.

Gardo, S. (1993) Spontaneous abortion and genetic natural selection, *Orv. Hetil.* **134**, 1459-1464.

Gavyliuk, IuI., Sozans'kyi, O.O., Akopian, G.R., Lozyns'ka, M.R., Siednieva, I.A., Glynka, P.A., Iaborivs'ka, O.M. and Grytsiuk, I.I. (1992) Genetic monitoring in connection with the Chernobyl accident, *Tsitol. Genet.* **26**, 15-29.

Gembicki, M., Sowinski, J., Ruchala, M. and Bednarek, J. (1991) Influence of radioactive contamination and iodine prophylaxis after the Czernobyl disaster on thyroid morphology and function of the Poznan region, *Endokrynol. Pol.* **42**, 273-298.

Grandi, C., Tapia, J.L. and Marshall, G. (2005) An assessment of the severity, proportionality and risk of mortality of very low birth weight infants with fetal growth restriction. A multicenter South American analysis, *J. Pediatr. (Rio J).* **81**, 198 – 204.

Gregg, N.M. (1941) Congenital cataract following German measles in the mother, *Trans. Ophthalmol. Soc. Aust.* **3**, 35-45.

Grossman, C.M., Morton, W.E. and Nussbaum, R.H. (1996) Hypothyroidism and spontaneous abortion among Hanford, Washington, downwinders, *Arch. Environ. Health* **51**, 175-176.

Hack, M., Taylor, H.G., Drotar, D., Schluchter, M., Cartar, L., Wilson-Costello, D., Klein, N., Friedman, H. Mercuri-Minich, N. and Morrow, M. (2005) Poor predictive validity of the Bayley Scales of Infant Development for cognitive function of extremely low birth weight children at school age, *Pediatrics* **116**, 333-341.

Hassold, T., Quillen, S.D. and Yamane, J.A. (1983) Sex ratio in spontaneous abortion, *Ann. Hum. Genet.* **47**, 39-47.

Heywood, R., Palmer, A.K., Gregson, R.L. and Hummler, H. (1985) The toxicity of beta-carotene, *Toxicology* **36**, 91-100.

Irgens, L.M., Lie, R.T., Ulstein, M., Skeie Jensen, T., Skjaerven, R., Sivertsen, F., Reitan, J.B., Strand, F., Strand, T. and Egil Skjeldestad, F. (1991) Pregnancy outcome in Norway after Chernobyl, *Biomed. Phar.* **45**, 233-241.

Jacob, P., Kenigsberg, Y., Goulko, G., Buglova, E., Gering, F., Golovneva, A., Kruk, J. and Demidchik, E.P. (2000) Thyroid cancer risk in Belarus after the Chernobyl accident: Comparison with external exposures, *Radiat. Environ. Biophys.* **39**, 25-31.

James, W.H. (1996) Evidence that mammalian sex ratios at birth are partially controlled by parental hormone levels at the time of conception, *J. Theor. Biol.* **180**, 271-286.

James, W.H. (2002a) Periconceptual parental smoking and sex ratio of offspring, *Lancet* **360**, 1515.

James, W.H. (2002b) Parental exposure to dioxin and offspring sex ratios, *Environ. Health Perspect.* **110**, A502.

Jarrell, J.F., Gocmen, A., Akyol, D. and Brant, R. (2002) Hexachlorobenzene exposure and the proportion of male births in Turkey 1935-1990, *Reprod. Toxicol.* **16**, 65-70.

Jelínek, R. and Peterka, M. (1977) The role of mandible in mouse palatal development revisited, *J. Cleft Palate* **14**, 211-221.

Jongbloet, P.H., Roeleveld, N. and Groenewoud, H.M. (2002) Where the boys aren't: dioxin and the sex ratio, *Environ. Health Perspect.* **110**, 1-3.

Karakashian, A.N., Chusova, V.N., Kryzhanovskaia, M.V., Lepeshkina, T.R., Martynovskaia, T.I.U., Glushchenkom S.S. and Gorbatiukm L.A. (1997) A retrospective analysis of aborted pregnancy in women engaged in agricultural production in controlled areas of Ukraine, *Lik Sprava.* **4**, 40-42.

Khoury, M.J., Erickson, J.D. and James, L.M. (1984) Paternal effects on the human sex ratio at birth: evidence from interracial crosses, *Am. J. Hum. Genet.* **36**, 1103-1111.

Khoury, M.J. and Holtzman, N.A. (1987) On the ability of birth defects monitoring to detect new teratogens, *Am. J. Epidemiol.* **126**, 136-143.

Kimmel, C.A. and Buelka-Sam, J. (1981) *Developmental toxicology*. Raven Press, N.Y, USA.

Knops, N.B., Sneeuw, K.C., Brand, R., Hille, E.T., den Ouden, L.A., Wit, J.M. and Verloove-Vanhorick, P.S. (2005) Catch-up growth up to ten years of age in children born very preterm or with very low birth weight, *BMC. Pediatr.* **5**, 26.

Koupilova, I., Vagero, D., Leon, D.A., Pikhart, H., Prikazsky, V., Holcik, J. and Bobak, M. (1998) Social variation in size at birth and preterm delivery in the Czech Republic and Sweden 1989-91, *Paediatr. Perinat. Epidemiol.* **12**, 7-24.

Krageloh-Mann, I., Toft, P., Lunding, J., Andresen, J., Pryds, O. and Lou, H.C. (1999) Brain lesions in preterms: origin, consequences and compensation, *Acta. Paediatr.* **88**, 897-908.

Kučera, J. (1989) *Population Teratology*, Avicenum, Prague, Czech Republic.

Kunz, E. (ed) (1986) *Report on Radiation Situation in CSSR at Chernobyl Accident,* Prague, Czech Republic.

Lazjuk, G.I., Nikolaev, D.L. and Novikova, I.V. (1997) Changes in registered congenital anomalies in the Republic of Belarus after the Chernobyl accident, *Stem Cells* **15**, 255-260.

Leenhouts, H.P., Brugmans, M.J.P. and Chadwick, K.H. (2000) Analysis of thyroid cancer data from the Ukraine after "Chernobyl" using a two-mutation carcinogenesis model, *Radiat. Environ. Biophys.* **39**, 89-98.

Lenz, W. (1961) Kindliche Missbildungen nach Medikamenteinnahme wahrend der Graviditat? *Dtsch. Med. Wochenschr.* **86**, 2555-2556.

Lenz, W. and Knapp, K. (1962) Die thalidomide-embryopatie, *Dtch. Med. Wochenschr.* **87**, 1232-1242.

Leridon, H. (1977) *Human fertility: The basic components,* Chicago University Press, Chicago, USA.

Leridon, H. (1987) Spontaneous fetal mortality. Role of maternal age, parity and previous abortion, *J. Gynecol. Obstet. Biol. Reprod.* **16**, 425-431.

Levy, F., Barr, C. and Sunohara, G. (1998) Directions of aetiologic research on attention deficit hyperactivity disorder, *Aust. N.Z.J. Psy.* **32**, 97-103.

Linnet, K.M., Dalsgaard, S., Obel, C., Wisborg, K., Henriksen, T.B., Rodriguez, A., Kotimaa, A., Moilanen, I., Thomsen, P.H., Olsen, J. and Jarvelin, M.R. (2003) Maternal lifestyle factors in pregnancy risk of attention deficit hyperactivity disorder and associated behaviors: review of the current evidence, *Am. J. Psych.* **160**, 1028-1040.

Lohstroh, P.N., Overstreet, J.W., Stewart, D.R., Nakajima, S.T., Cragun, J.R., Boyers, S.P. and Lasley, B.L. (2005) Secretion and excretion of human chorionic gonadotropin during early pregnancy, *Fertil. Steril.* **83**, 1000-1011.

Lukacs, G.L., Szakall, S., Kozma, I., Gyory, F. and Balazs, G. (1997) Changes in the epidemiological parameters of radiation-induced illnesses in East Hungary 10 years after Chernobyl, *Langenbecks Arch. Chir. Suppl. Kongressbd.* **114**, 375-377.

Mastroiacovo, P., Spagnolo, A., Marni, E., Meazza, L., Bertollini, R., Segni, G. and Borgna-Pignatti, C. (1988) Birth defects in the Seveso area after TCDD contamination, *JAMA* **259**, 1668-1672.

McBride, W.G. (1961) Thalidomide and congenital abnormalities, *Lancet* **2**, 1358.

McMillen, M.M. (1979) Differential mortality by sex in fetal and neonatal deaths, *Science* **204**, 89-91.

McElreavey, K., Vilain, E., Abbas, N., Herskowitz, I. and Fellous, M. (1993) A regulatory cascade hypothesis for mammalian sex determination: SRY represses a negative regulator of male development, *PNAS* **90**, 3368-3372.

Michnovicz, J.J., Naganuma, H., Hershcopf, R.J., Bradlow, H.L. and Fishman, J. (1988) Increased urinary catechol estrogen excretion in female smokers, *Steroids* **52**, 69-83.

Miller, J.F., Williamson, E., Glue, J., Gordon, Y.B, Grudzinskas, J.G. and Sykes, A. (1980) Fetal loss after implantation. A prospective study, *Lancet* **2**, 554-556.

Mizuno, R. (2000) The male/female ratio of fetal deaths and births in Japan, *Lancet* **356**, 738-739.

Mocarelli, P., Brambilla, P., Gerthoux, P.M., Patterson, D.G. Jr. and Needham, L.L. (1996) Change in sex ratio with exposure to dioxin, *Lancet* **348**, 409.

Mocarelli, P., Gerthoux, P.M., Ferrari, E., Patterson, D.G. Jr., Kieszak, S.M., Brambilla, P., Vincoli, N., Signorini, S., Tramacere, P., Carreri, V., Sampson, E.J., Turner, W.E. and Needham, L.L. (2000). Paternal concentration of dioxin and sex ratio of offspring, *Lancet* **355**, 1858-1863.

Moore, K.L. and Persaud, T.V.N. (1993) *The developing human.* WB Saunders Company, Philadelphia, USA.

Nicolich, M.J., Huebner, W.W. and Schnatter, A.R. (2000) Influence of parental and biological factors on the male birth fraction in the United States: an analysis of birth certificate data from 1964 through 1988, *Fertil. Steril.* **73**, 487-492.

Nikiforov, Y. and Gnepp, D.R. (1994) Pediatric thyroid cancer after the Chernobyl disaster. Pathomorphologic study of 84 cases (1991-1992) from the Republic of Belarus, *Cancer* **74**, 748-766.

Nishimura, H., Uwabe, C. and Shiota, K. (1987) Study of human post-implantation conceptus, normal and abnormal, *Okajimas. Folia. Anat. Jpn.* **63**, 337-357.

Odlind, V., Haglund, B., Pakkanen, M. and Otterblad Olausson, P. (2003) Deliveries, mothers and newborn infants in Sweden, 1973-2000. Trends in obstetrics as reported to the Swedish Medical Birth Register, *Acta Obstet. Gynecol. Scand.* **82**, 516-528.

Ogris, E. (1997) Exposure to radioactive iodine in pregnancy: significance for mother and child, *Acte. Med. Austriaca.* **24**, 150-153.

Peterka, M., Havránek, T. and Jelínek, R. (1986) Dose-response relationships in chick embryos exposed to embryotoxic agents, *Folia Morphol.* **34**, 69-77.

Peterka, M., Jelínek, R. and Pavlík, A. (1992) Embryotoxicity of 25 psychotropic drugs: A study using CHEST, *Reprod. Toxicol.* **6**, 367-374.

Peterka, M., Peterková, R. and Likovský, Z. (1996) Teratogenic and lethal effects of long-term hyperthermia and hypothermia in the chick embryo, *Reprod. Toxicol.* **10**, 327-332.

Peterka, M., Peterková, R. and Likovský, Z. (1997) Different embryotoxic effect of vitamin A and B-carotene detected in the chick embryo, *Acta. Chir. Plast.* **39**, 91-96.

Peterka, M. and Klepáček, I. (2001) Light irradiation increases embryotoxicity of photodynamic sensitizers (5-aminolevulinic acid and Protoporphyrin IX) in chick embryos, *Reprod. Toxicol.* **15**, 111-116.

Peterka, M., Peterková, R. and Likovský, Z. (2004) Chernobyl: prenatal loss of four hundred male fetuses in the Czech Republic, *Reprod. Toxicol.* **18**, 75-79.

Poppe, K. and Glinoer, D. (2003) Thyroid autoimmunity and hypothyroidism before and during pregnancy, *Hum. Reprod. Update.* **9**, 149-161.

Pratt, A.E. (1966) The Pierre Robin syndrome, *Brit. J. Radiol.* **39**, 390-392.

Racowsky, C. (2002) High rates of embryonic loss, yet high incidence of multiple births in human art: is this paradoxical? *Theriogenology* **57**, 87-96.

Romanova, L.K, Pokrovskaia, M.S., Mladkovskaia, T.B., Kulikova, G.V., Volkova, E.V., Safronova, L.A., Zhorova, E.S., Beliaev, I.K., Gerasiuto, G.I. and Sivakoba, I.S. (1998) Morphological characteristics of lung in embryos and fetuses in women living in region contaminated with radionuclides after at the Chernobyl power plant, *Arkh. Patol.* **60**, 32-36.

Rosa, F.W. (1983) Teratogenicity of isotretinoin, *Lancet* **2**, 513.

Ryan, J.J., Amirova, Z. and Carrier, G. (2002) Sex ratios of children of Russian pesticide producers exposed to dioxin, *Environ. Health Perspect.* **110**, A699-A701.

Scherb, H., Weigelt, E. and Bruske-Hohlfeld, I. (2000) Regression analysis of time trends in perinatal mortality in Germany 1980-1993, *Environ. Health Perspect.* **108**, 159-165.

Shiota, K., Uwabe, C. and Nishimura, H. (1987) High prevalence of defective human embryo at the early postimplantation period, *Teratology* **35**, 309-316.

Shizuma, K., Iwatani, K., Hasai, H., Nishiyama, F., Kiso, Y., Hoshi, M., Sawada, S., Inoue, H., Suzuki, A., Hoshita, N., Kanamori, H. and Sakamoto, I. (1987) Observation of fallout in Hiroshima caused by the reactor accident at Chernobyl, *Int J Radiat Biol.* **51**, 201-207.

Slípka, J., Jelínek, R. and Peterka, M. (1980) Comparative morphogenesis of the secondary palate related to the skull development, *Folia Morphol.* **28**, 207-211.

Sperling, K., Pelz, J., Wegner, R.D., Dorries, A., Gruters, A. and Mikkelsen, M. (1994) Significant increase in trisomy 21 in Berlin nine months after the Chernobyl reactor accident: temporal correlation or causal relation? *Br. Med. J.* **309**, 158-162.

Stenke, L., Axelsson, B., Ekman, M., Larsson, S. and Reizenstein, P. (1987) Radioactive iodine and cesium in travellers to different parts of Europe after Chernobyl accident, *Acta Oncol.* **26**, 207-210.

St Sauver, J.L., Barbaresi, W.J., Katusic, S.K., Colligan, R.C., Weaver, A.L. and Jacobsen, S.J. (2004) Early life risk factors for attention-deficit/hyperactivity disorder: a population-based cohort study, *Mayo Clin. Proc.* **79**, 1124-1131.

Tatham, L.M., Bove, F.J., Kaye, W.E. and Spengler, R.F. (2002) Population exposure to I-131 releases from Hanford Nuclear Reservation and preterm birth, infant mortality, and fetal death, *Int. J. Hyg. Environ. Health* **205**, 41-48.

Ulstein, M., Skeie Jensen, T., Irgens, L.M., Lie. and R.T. and Sivertsen, E. (1990) Outcome of pregnancy in one Norwegian county 3 years prior to and 3 years subsequent to the Chernobyl accident, *Acta. Obstet. Gynecol. Scand.* **69**, 277-280.

Vartiainen, T., Kartovaara, L. and Tuomisto, J. (1999) Environmental chemicals and changes in sex ratio: analysis over 250 years in Finland, *Environ. Health Perspect.* **107**, 813-815.

Vatten, L.J. and Skjaerven, R. (2004) Offspring sex and pregnancy outcome by length of gestation, *Early Hum. Dev.* **76**, 47-54.

Warkany, J. (1977) History of teratology/chapter 1, in J.G. Wilson, FC. Fraser (eds.), *Handbook of Teratology*, Plenum Press, New York and London, pp. 3-45.

Wells, J.C. (2000) Natural selection and sex differences in morbidity and mortality in early life, *J. Theor. Biol.* **202**, 65-76.

Wilson, J.G. (1977) Current status of teratology/chapter 2, in J.G. Wilson, FC. Fraser (eds.), *Handbook of Teratology*, Plenum Press, New York and London, pp. 47-74.

Yamazaki, J.N. and Schull, W.J. (1990) Perinatal loss and neurological abnormalities among children of the atomic bomb. Nagasaki and Hiroshima revisited, 1949 to 1989, *JAMA* **264,** 605-609.

Zoeller, T.R., Dowling, A.L., Herzig, C.T., Iannacone, E.A., Gauger, K.J. and Bansai, R. (2002) Thyroid hormone, brain development, and the environment, *Environ. Health Pespect.* **110,** 350-356.

SECTION 4:

COUNTRY REPORTS

CONGENITAL ABNORMALITIES IN GREECE: FUNCTIONAL EVALUATION OF STATISTICAL DATA 1981 – 1995

E. BRILAKIS[1*], E. FOUSTERIS[2] AND J. PAPADOPULOS[3]
[1] *Paediatric Surgery Department*
"TZANEIO" General Hospital of Piraeus
Aristofanous street 18
Piraeus
GREECE
[2] *Department of Internal Medicine*
General Hospital of Livadeia
Agiou Eleutheriou Street 143
Piraeus
GREECE
[3] *Laboratory of Experimental Pharmacology*
School of Medicine
National and Kapodistrian
University of Athens
75 Mikras Asias Street 75
Athens
GREECE
(author for correspondence, Email: brilakis@mycosmos.gr)*

Summary

This study's objective is to record the regional distribution of mortality and hospitalisation caused by congenital abnormalities in Greece and to compare the mortality and the infant mortality for congenital abnormalities between Greece and other countries.

The data were obtained from the official publications of the National Statistical Service of Greece (NSSG). The specific mortality and infant mortality ratios for congenital abnormalities and the number of discharged patients with congenital

P. Nicolopoulou-Stamati et al. (eds.), Congenital Diseases and the Environment, 323–338.

abnormalities per 100,000 of population are evaluated for Greece as a whole and by geographic region, for the period between 1981 and 1995. Greece was compared to other, randomly selected countries (Bulgaria, France, Italy, Japan, Portugal, Sweden, The Netherlands, and USA). The information data for these countries are derived from the Annuals of the World Health Organization (WHO).

The specific mortality and infant mortality ratios by congenital abnormalities are considerably higher in Greece than in the other countries studied. Among the 10 regions of Greece, Athens and Thrace have the highest mortality ratios, while Athens and Crete appear to have the highest proportion of hospitalised patients due to congenital abnormalities according to their population. In Thrace, a disproportional high number of deceased in comparison to discharged patients was noted.

In conclusion, Greece appears to have more deaths by congenital abnormalities compared to the other randomly selected countries. A significant difference in the distribution of the deaths and discharged patients among the geographic regions of Greece is observed. This study is unable to detect the reasons for this distribution.

1. Introduction

Congenital abnormalities are a significant public health concern that affects 3 to 4 per cent of all live births and causes many elective and spontaneous abortions (Centres for disease control and prevention, 1998). In epidemiological studies they emerge along with genetic diseases as one of the leading causes of infant mortality and result in substantial mortality and morbidity throughout childhood (Cunniff *et al.,* 1995; Yoon *et al.,* 1997). For example, every year in the United States approximately 150,000 babies are born with birth defects (Centres for disease control and prevention, 1998). Although birth defects, account for 15 per cent to 30 per cent of all paediatric hospitalisations, they are responsible for a proportionally higher health care cost than other hospitalisations (Yoon *et al.,* 1997). In addition, they put a significant burden on families and society as huge amounts need to be spent yearly for the medical and rehabilitative care of the affected children (Centres for disease control and prevention, 1992; Pew environmental health commission, 1999).

Congenital abnormalities can be structural, functional or metabolic. They might be ascribed to genetic factors (chromosomal abnormalities, mutations, genes inherited by family members), to environmental factors (medical drugs, infectious disease agents, pollutants) or to culturally inherited behavioural and lifestyle factors (Collins *et al.,* 2001; Moore *et al.,* 1997). Most of them are thought to have a multifactorial aetiology resulting from the interactions between genes, environmental and lifestyle factors. However, some 50 to 60 per cent cannot be attributed in known causes (Khoury *et al.,* 1998; Moore *et al.,* 1997).

Major congenital abnormalities are an important problem with many dimensions for both the family and society. To address the problem, the first approach is to register the incidence of birth defects per country and per geographic area. This registration may lead to a better identification of the local factors that would eventually found to be involved in the generation of this problem. In Greece, an official registration of birth defects does not exist and their incidence cannot be evaluated. Therefore, in this paper, the issue is approached in an indirect way, by evaluating the mortality resulting from congenital abnormalities in Greece and 8 other randomly selected countries (Bulgaria, France, Italy, Japan, Portugal, Sweden, The Netherlands, and USA). In addition, the differences in the number of deaths and hospitalised people by cause birth defects were recorded for the main geographic regions in Greece, to evaluate possible local factors. This study may be a first step up to prospective studies that can reveal the exact factors underlying the high prevalence of congenital abnormalities in Greece.

2. Method

Initially, comparative data on the specific mortality ratio for congenital abnormalities per 100,000 of population and the specific infant mortality ratio (deaths within the 1st year of life) by congenital abnormalities per 1,000 live births between Greece and other countries were looked for. Data concerning Greece were obtained from:

- the monthly Statistical Bulletin by National Statistical Service of Greece (NSSG),
- statistics of Population's Physical Movement by NSSG,
- the annual Statistical Yearbook of Greece by NSSG.

Data concerning other countries were collected from:

- annuals of World Health Organization (WHO).

The selection of European countries (Bulgaria, France, Italy, The Netherlands,

Portugal, Sweden) was made on the basis of the most available data from the annuals of WHO for the study period (1981 – 1995), while the USA and Japan were chosen as international controls (comparative sample). In addition, based on the same sources, we collected data on the distribution of deaths and discharged patients with congenital abnormalities per geographic region of permanent residence in Greece, for the period 1981 – 1995. 'Discharged patients' is an official term that refers to 'cases of hospitalisation' and not to individuals (if one patient was hospitalized 3 times in the same year, he was considered 3 times as 3 different patients). This part of the study is useful for medical care cost estimations.

The permanent residence for infants is defined as the place of permanent residence of their mother. This information is used for conclusions on the possible environmental impact on the fetus. Instead, when the 'general population' is discussed, in which adults are included too, the place of permanent residence is not always identical to that of the mother who was subjected to the environmental factors during pregnancy. This disadvantage is due to the limited power of the primary available data.

3. Results

The specific mortality ratio for congenital abnormalities per 100,000 of population between 1981 and 1995 shows a progressive decline in all the selected countries (table 18 and figure 19).

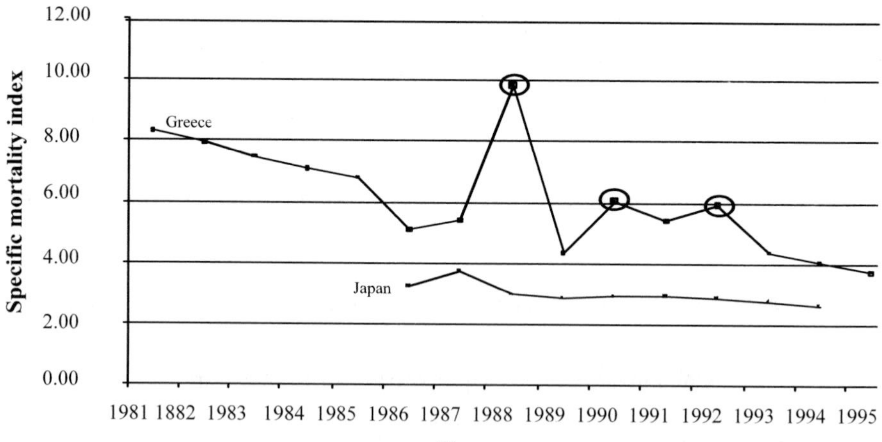

Figure 19. Specific mortality ratio for congenital abnormalities per 100,000 of population in Greece per year from 1981 to 1995. (NSSG – Statistics of Population's Physical Movement (1981 – 1995))

Table 18. Specific mortality ratio for congenital abnormalities per 100,000 of population in Greece and in 8 selected countries. The absolute and percentage difference of the specific ratio of each country between the years 1987 and 1994 has been marked in the bottom lines. The highest ratio among the comparative countries for each year is marked with bold (N/A: Data are not available) (World Health Organization - Annuals (1981-1995)

Year	Greece	Bulgaria	France	Italy	Netherlands	Portugal	Sweden	USA	Japan
1981	8.32	N/A	N/A	N/A	N/A	N/A	N/A	N/A	N/A
1982	7.94	N/A	N/A	N/A	N/A	N/A	N/A	N/A	N/A
1983	**7.48**	N/A	N/A	4.83	N/A	N/A	N/A	N/A	N/A
1984	**7.08**	N/A	N/A	4.74	N/A	N/A	N/A	5.51	N/A
1985	**6.79**	6.47	4.08	4.10	5.18	N/A	N/A	5.33	N/A
1986	5.09	**5.68**	3.69	3.63	5.35	5.29	N/A	5.49	3.22
1987	5.39	**6.02**	3.59	3.90	5.35	5.50	4.32	5.07	3.71
1988	**9.82**	5.93	3.52	3.65	4.87	4.86	4.30	5.19	2.99
1989	4.33	**5.78**	3.33	3.82	4.63	4.83	4.56	5.20	2.85
1990	**6.06**	N/A	3.21	3.33	5.02	4.49	4.80	5.26	2.91
1991	**5.41**	N/A	3.07	3.35	4.74	4.49	4.91	5.00	2.91
1992	**5.91**	N/A	2.65	3.36	4.78	4.11	4.45	4.9	2.86
1993	4.38	N/A	2.9	3.43	4.59	4.45	4.03	**4.83**	2.76
1994	4.03	N/A	2.79	3.26	**4.67**	3.76	3.68	4.62	2.65
1995	3.72	N/A	N/A	N/A	**4.11**	3.28	N/A	N/A	N/A
Diff.	-1.36		-0.8	-0.64	-0.67	-1.75	-0.64	-0.5	-1.06
%	-25%		-22%	-16%	-13%	-32%	-15%	-9%	-29%

The smallest reduction is observed in USA (9 per cent), while the most pronounced one is in Portugal (32 per cent). Greece shows an overall reduction of the ratio by 25 per cent and in general a progressive decrease, which is interrupted by 3 peaks, observed in the years 1988 (possibly a mistake registration from the NSSG happened in this year), 1990 and 1992 (figure 19). Greece presents the highest ratio in 7 out of the 15 years studied. The specific infant mortality ratio for congenital abnormalities per 1,000 live births shows an overall decrease in all the selected countries (Greece included), where the percentages of decrease are fluctuating from 10.7 per cent (Japan) up to 25.7 per cent (Italy) (table 19)

The lowest percentages of decrease are observed in Greece (12.9 per cent) and in Japan (10.7 per cent) but in Japan, the ratio is very low. Greece presents the highest specific infant mortality ratio for congenital abnormalities in comparison to the other countries studied, in 10 out of the 15 years the study covers.

Table 19. Specific infant mortality ratio by congenital abnormalities per 1,000 live births in Greece and in 8 selected countries. The absolute and percentage difference of the specific ratio of each country between the years 1987 and 1994 has been marked in the bottom lines. The highest ratio among the comparative countries for each year is marked with bold (N/A: Data are not available) (World Health Organization – Annuals (1981-1995)

Year	Greece	Bulgaria	France	Italy	Netherlands	Portugal	Sweden	USA	Japan
1981	4.84	N/A	N/A	N/A	N/A	N/A	N/A	N/A	N/A
1982	4.77	N/A	N/A	N/A	N/A	N/A	N/A	N/A	N/A
1983	**4.65**	N/A	N/A	3.09	N/A	N/A	N/A	N/A	N/A
1984	**4.44**	N/A	N/A	3.04	N/A	N/A	N/A	2.33	N/A
1985	**4.57**	3.8	1.79	2.71	2.67	N/A	N/A	2.28	N/A
1986	3.37	**3.5**	1.67	2.45	2.52	3.08	N/A	2.19	1.68
1987	**3.66**	3.6	1.73	2.61	2.74	3.35	2.23	2.07	1.65
1988	3.49	**3.6**	1.54	2.46	2.34	2.89	1.95	2.08	1.68
1989	3.26	**3.73**	1.53	2.36	2.15	3.10	2.16	2.01	1.65
1990	**4.83**	N/A	1.48	2.04	2.40	2.83	2.04	1.98	1.66
1991	**4.46**	N/A	1.36	1.97	2.26	2.71	2.16	1.87	1.60
1992	**4.81**	N/A	1.25	2.00	2.40	2.60	1.82	1.83	1.69
1993	**3.40**	N/A	1.42	2.17	2.30	2.58	1.66	1.78	1.54
1994	**3.19**	N/A	1.32	1.94	2.11	2.55	1.67	1.73	1.47
1995	**3.03**	N/A	N/A	N/A	1.85	2.22	N/A	N/A	N/A
Diff.	-0.47		-0.42	-0.67	-0.63	-0.8	-0.56	-0.34	-0.18
%	-12.90%		-24.10%	-25.70%	-22.90%	-23.80%	-25.10%	-16.20%	-10.70%

Also noticeable is the increase of the ratio in Greece during the years 1990 – 1992. This phenomenon is not observed in any other country under study (figure 20).

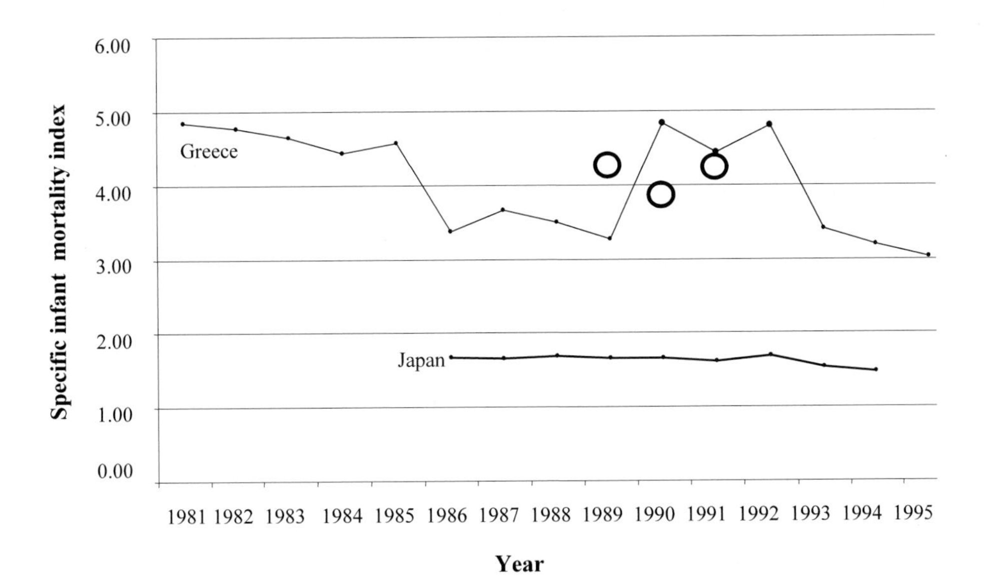

Figure 20. Specific infant mortality ratio by congenital abnormalities per 1,000 live births in Greece from 1981 to 1995 (NSSG – Statistics of Population's Physical Movement (1981-1995))

The quotient of discharged patients with congenital abnormalities in Greece per 100,000 of population is shown in figure 21. This figure remains rather constant throughout the period 1981 – 1995

Discharged patients

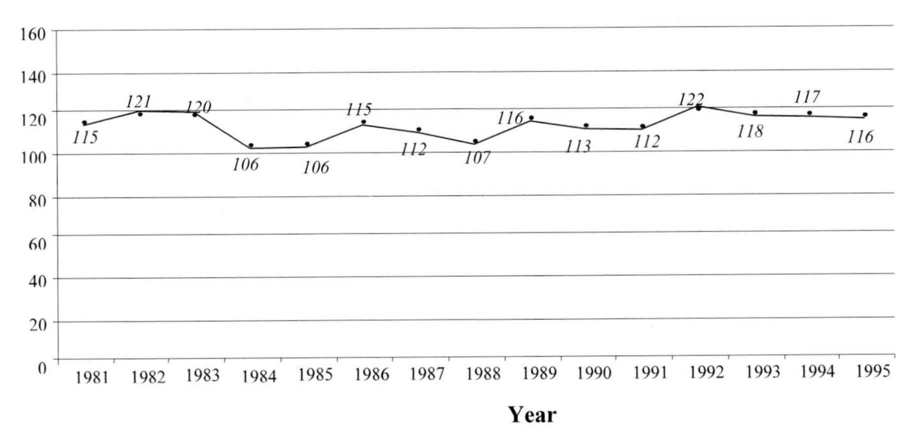

Figure 21. Discharged patients with congenital abnormalities per 100,000 of population in Greece per year from 1981 to 1995 (NSSG – Monthly Statistic 315 (1981-1995))

Table 20. Number of babies born with congenital abnormalities in Greece per category of congenital abnormality from 1981 to 1995 (NSSG – Annual Statistic Yearbook (1981-1995)

Category of congenital abnormality	Number of babies born during the period 1981-1995	Percentage
Spina bifida and hydrocephalus	4.591	2
Other deformities of C.N.S.	2.971	25
Congenital anomalies of heart and circulatory system	46.075	2
Cleft palate and cleft lip	4.358	7
Other deformities of digestive system	13.747	5
Congenital dislocation of hip	9.2	12
Other congenital anomalies of musculoskeletal system	22.033	44
Other congenital anomalies	80.342	3
TOTAL	183.317	100

Table 20 shows the number of babies born with congenital abnormalities between 1981 and 1995 in Greece, per category of congenital abnormality (Annual Statistical Yearbook of Greece by NSSG) and the percentage of each category of the total of births of individuals with congenital abnormalities. The most frequent defects are the congenital anomalies of the heart and the circulatory system. These account for 25 per cent of all birth defects in Greece. Malformations of the musculoskeletal system (17 per cent) are the second most frequent category, while malformations of the digestive system account for 7 per cent.

Figure 22 shows the differences between the geographic departments of Greece. The specific mortality ratios for congenital abnormalities per 100,000 of population in the 10 Greek departments show an overall decrease during the study period, except for Crete (table 21). The reduction fluctuates from 25 per cent in Thrace up to 65 per cent in greater Athens, while Crete shows an increase of 11 per cent. The reduction noted in the 9 departments is not progressive and has been interrupted by severe peaks that do not necessarily coincide with the peaks of the ratio of the entire country (table 18). In 1981, greater Athens and Thrace show the highest ratios. 10.65 and 9.97 respectively, while the lowest ratio is found in Crete (4.38) and in the Peloponnesus (6.31). In 1995, the last year of the study period, the highest ratios are observed in Thrace (7.43) and on the islands of the Aegean Sea (5.32). In the same year 1995, the lowest figures are recorded in central Greece and Evia (2.90) and Macedonia (3.12). Greater Athens shows a higher ratio as compared to Greece as a whole (table 18) for every year studied.

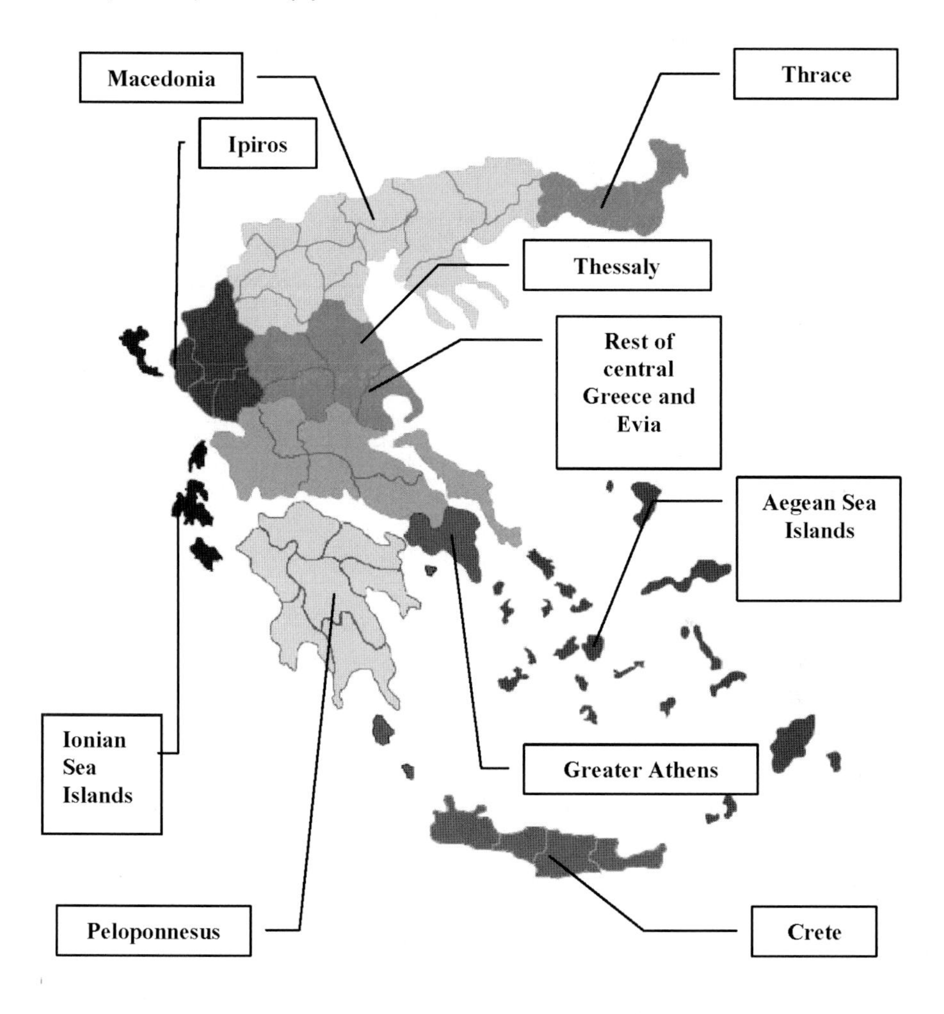

Figure 22. Geographic departments in Greece (the National Statistical Service of Greece)

The specific mortality ratio in Crete shows a peculiar pattern. The increases and reductions do not correspond at all with the ratio of the country as a whole. The number of discharged patients with congenital abnormalities per 100,000 of population in each geographic department is shown in table 21. A reduction of the number of hospitalised persons for birth defects is seen in 7 departments out of the

10: greater Athens (10 per cent), rest of central Greece and Evia (31 per cent). Peloponnesus (22 per cent), Ionian Sea islands (5 per cent), Thessaly (11 per cent). Aegean Sea islands (12 per cent) and Crete (4 per cent). An increase is recorded for Ipiros (3 per cent). Macedonia (20 per cent) and Thrace (21 per cent). Both in the beginning and by the end of the study, the highest number of hospitalisations for congenital malformations is noticed in greater Athens (142 and 128 respectively) and in Crete (116 and 112 respectively).

Table 21. Deaths due to congenital abnormalities per 100,000 of population (specific mortality ratio) per deceased's permanent residence from 1981 to 1995. All higher ratios than the respective ratios of entire Greece in the same year are marked with bold letters. (See table 18) (NSSG – Statistics of Population's Physical Movement (1981-1995))

Year	Greater Athens	Rest of central Greece & Evia	Pelopon-nesus	Ionian sea islands	Ipiros	Thessaly	Macedonia	Thrace	Aegean sea islands	Crete
1981	**10.65**	7.45	6.31	6.58	6.48	7.62	7.7	**9.97**	7.48	4.38
1982	**11.11**	5.48	6.74	6.01	5.54	**8.39**	5.97	**10.96**	4.42	6.72
1983	**8.7**	5.95	6.58	4.36	**9.82**	5.07	6.79	**9.88**	**7.66**	**8.84**
1984	**8.08**	6.06	5.85	3.8	6.43	6.98	6.72	6.66	5.09	**10.54**
1985	**8.58**	5.05	5.71	6.5	5.8	6.78	6.32	**8.88**	**7.63**	2.72
1986	**6.52**	4.41	4.36	3.79	4.27	2.48	4.26	**7.69**	**5.54**	4.06
1987	**7.47**	3.43	4.92	3.24	4.57	3.98	3.88	**7.06**	**5.52**	**5.96**
1988	**10.68**	7.94	**10.3**	**12.95**	7.9	6.31	9.56	**10.09**	**11.93**	**10.88**
1989	**5.1**	2.54	4.16	2.68	3.65	**4.66**	4.02	**7.32**	2.73	**5.29**
1990	**7.03**	4.68	4.58	3.19	3.3	6.02	**6.42**	**10.07**	5.2	5.42
1991	**6.96**	4.07	4.21	4.19	**6.15**	3.67	4.96	**6.73**	4.49	**6.1**
1992	**7.23**	3.98	5.12	4.14	5.73	5.01	5.75	**9.52**	5.14	5.5
1993	**4.96**	2.88	4.06	2.56	**5.05**	4.05	**4.42**	**6.79**	2.9	**5.83**
1994	**4.37**	3.43	3.84	2.54	3.6	2.56	**4.09**	**9.27**	3.78	3.45
1995	**3.75**	2.9	3.19	3.54	**4.39**	**3.91**	3.12	**7.43**	**5.32**	**4.86**
Diff.	-65%	-61%	-49%	-46%	-32%	-49%	-59%	-25%	-29%	**11%**

Table 22. Discharged patients with congenital abnormalities per 100,000 of population per place of permanent residence from 1981 to 1995. All higher ratios than the respective ratios of entire Greece (see table 20) in the same year are marked with bold letters (NSSG – Monthly Statistical Bulletin (1981-1995))

Year	Greater Athens	Rest of central Greece & Evia	Pelopon-nesus	Ionian sea islands	Ipiros	Thess-aly	Macedonia	Thrace	Aegean sea islands	Crete
1981	**142**	98	101	89	75	93	92	77	97	**116**
1982	**151**	91	111	83	96	102	95	83	87	**125**
1983	**133**	108	104	**142**	103	104	99	91	**130**	**129**
1984	**122**	81	91	81	93	87	94	95	84	102
1985	**117**	86	96	82	86	85	98	93	93	99
1986	**139**	88	95	87	108	95	96	92	95	**121**
1987	**132**	82	94	101	108	74	95	94	84	**139**
1988	**119**	85	86	86	**115**	91	93	98	88	**124**
1989	**117**	92	105	96	109	93	112	114	93	**144**
1990	**115**	96	100	102	109	96	106	110	99	113
1991	**116**	79	95	101	103	89	**114**	98	91	**118**
1992	**124**	86	94	90	85	94	118	102	100	**123**
1993	**131**	82	92	112	70	91	115	117	100	**129**
1994	**127**	85	95	111	98	92	115	117	102	**138**
1995	**128**	68	79	84	77	82	110	93	85	112
Diff.	-10%	-31%	-22%	-5%	**3%**	-11%	**20%**	**21%**	-12%	-4%

The number of hospitalizations with congenital anomalies is higher in greater Athens than in the rest of Greece during the whole study period. A similar pattern exists in Crete, for 12 out of the 15 years of study. The ratio for Thrace increases by 21 per cent (77 – 93) between 1981 and 1995. The number of discharged patients in Thrace never exceeds the corresponding figure for Greece as a whole. This contrasts with the number of deaths as a result of congenital abnormalities (table 23).

Table 24 shows the results of a 'peculiar' percentage: the total number of deaths by congenital abnormalities in each geographic region was divided by the total number of discharged patients with congenital abnormalities in the same geographic region for the 15 year period. This quotient is the 'percentage of deaths/discharged patients per geographic region'. In 9 out of the 10 departments, it is close to 5 per cent. Thrace ha, however, an extremely high percentage of deaths/discharged patients (8.74 per cent) (considerably more deaths per discharged patient than the other regions in Greece).

Table 23. Average annual specific mortality ratio for congenital abnormalities per 100,000 of population and average annual ratio of discharged patients per 100,000 of population of each geographic region of Greece for the time period 1981-1995. (NSSG – Statistics of Population's Physical Movement/Monthly Statistical Bulletin (1981-1995))

Geographic region	Average annual specific mortality ratio for congenital abnormalities per 100.000 of population (1981-1995)	Average annual specific number of discharged patients with congenital abnormalities per 100.000 of population (1981-1995)
Greater Athens	**7.3**	**125**
Rest of central Greece and Evia	4.8	91
Peloponnesus	5.37	97
Ionian sea islands	4.71	98
Ipiros	5.61	97
Thessaly	5.12	91
Macedonia	5.63	105
Thrace	**8.61**	99
Aegean sea islands	5.67	96
Crete	6.06	**123**
Entire Greece	**6.09**	**108**

4. Discussion

4.1. Specific mortality and infant mortality ratios

The specific mortality and infant mortality ratios for congenital abnormalities present a gradual decline, both in Greece and in the selected countries (tables 18 and table 19). This is attributed to the continuous improvement in the diagnostic and therapeutic means. Greece is characterised both by higher prevalence figures in comparison to the other countries in this study, but also by a decline in the same prevalence figures over the years. This has been achieved by pre-birth control and medical care. However, the peaks in the figures of the prevalence should be investigated in more detail to identify their causes. In particular, the question arises. whether they are related to environmental incidents.

Table 24. "Percentage of deaths/discharged patient": Percentage ratio of the number of deaths by congenital abnormalities to the number of discharged patients with congenital abnormalities, for each geographic region of Greece from 1981 to 1995 (NSSG – Statistics of Population's Physical Movement/Monthly Statistical Bulletin (1981-1995))

Geographic region	Percentage of deaths/discharged patients
Greater Athens	5.83
Rest of central Greece & Evia	5.30
Peloponnesus	5.53
Ionian sea islands	4.81
Ipiros	5.76
Thessaly	5.63
Macedonia	5.37
Thrace	**8.74**
Aegean sea islands	5.92
Crete	4.92
Entire Greece	**5.37**

4.2. Variations of the ratios within Greece

The department of greater Athens and the department of Thrace show a high mortality as compared to the country as a whole (tables 21 and table 23). However, in Athens a 65 per cent reduction of the deaths by congenital abnormalities has been achieved during the study period, while in Thrace only a 25 per cent reduction was noticed. The population of Athens is about 1/3 of Greece's total population, its density is very high and people are exposed to several environmental stresses (air pollution. toxic waste. endocrine disrupters. etc). On the other hand, Thrace is an area in which part of the population is Muslims that behave as a closed social group. These differences in population characteristics might point to factors that cause the high incidence of deaths by congenital abnormalities in Thrace. Such factors entail:

weddings among relatives, lack of systematic pre-birth control, lack of easy access to the hospitals, and lack of social concern. This requires further investigation to confirm or reject these hypotheses. An advantageous position with regard to deaths by congenital abnormalities is occupied by central Greece and Evia (2.90 in 1995), the Peloponnesus and the Ionian Sea islands. The reduction in the mortality ratios in the different geographic departments varies between 25 and 65 per cent. Crete is the only region with an overall increase (11 per cent) and a lot of variation during the study period. The net increase observed from 1981 to 1984 and from 1985 to 1988 needs further investigation (table 21).

The most frequent causes of death in babies with birth defects (table 20) are the congenital anomalies of the heart and the circulatory system and those of the musculoskeletal system. This corresponds also to the international data. The deaths caused by congenital abnormalities have not been classified by the NSSG by geographic area. This is unfortunate, as this information is helpful to identify possible causal factors.

4.3. Hospitalisation due to birth defects within Greece

The rather constant tendency of hospitalisation of patients with birth defects (figure 21) can be attributed to the reduction of the number of deaths by congenital abnormalities. Another possible explanation is the increase of social concern on this problem. The regions with higher hospitalisation incidences for congenital abnormalities are greater Athens and Crete (tables 22 and table 23). This might refer to an easier access to the hospitals, but it can also reflect an increase in the number of individuals born with congenital abnormalities. Another important issue is the gradual reduction of both hospitalisations and death of babies with congenital abnormalities in the rest of central Greece and Evia after 1990 and in the Peloponnesus after 1989. This may be the result of better antenatal follow up or other prevention actions in these regions.

4.4. Deaths/hospitalisation ratio due to births defects in Greece

When the percentage in table 24 increases, either more deaths occur in the particular region or the number of hospitalised patients decreased. Taking into consideration this factor, we underscore the exceptionally high percentage of 8.74 per cent that Thrace presents. In this region it is obvious that specific ‘factors’ have affected the observed 15 year time period, which caused the elevation of the number of deaths to high levels. Possible explanations of this fact could be the impedimental access to the hospitals, the occurrence of major birth defects, human intervention etc.

5. Conclusions

This study describes the problem of congenital abnormalities (deaths and hospitalization) in Greece. It provides a picture of the distribution of congenital abnormalities in Greece by department during the period 1981–1995. The results were compared to data from other European countries with a lower, similar or higher socioeconomic level than Greece and to results from the USA and Japan. These latter served as international reference points. Although many studies in Greece have been published analyzing specific congenital abnormalities, none of them provides a general overview of the problem. Moreover, this study is unique in comparing the mortality ratios of congenital defects between regions in Greece and between Greece and other countries. The results provide information on the different factors that influence the incidence of congenital abnormalities in the regions, i.e. medical care, environmental, social, behavioural, and lifestyle factors. In Greece and in other countries with a high incidence of congenital abnormalities, there is a need for more information on the prevention methods used by other countries (i.e. Japan. USA. etc.). There is equally a need to establish centres of birth defects registration, to evaluate the genetic and environmental factors associated with congenital abnormalities. The USA project (Centre for Birth Defects Research and Prevention) can be used as a model. If this analysis contributes towards making authorities more sensitive to the problem, this might mean a significant step forwards.

References

Centers for disease control and prevention. (1992) The economic costs of birth defects and cerebral palsy – United States. *MMWR. Morb. Mortal. Wkly. Rep.* **44**, 694 – 699.

Centers for disease control and prevention. (1998) Trends in infant mortality attributable to birth defects – United States 1980 – 1995. *MMWR. Morb. Mortal. Wkly. Rep.* **47**, 773 – 778.

Collins, F.S. and McKusick, V.A. (2001) Implications of the human genome project for medical science. *JAMA* **285**, 540 – 544.

Cunniff, C.M., Carmack, J.L., Kirby, R.S. and Fiser, D.H. (1995) Contribution of heritable disorders to mortality in the pediatric intensive care unit. *Pediatrics* **95**, 678 – 681.

Khoury, M.J.and Yang, Q. (1998) The future of genetic studies of complex human diseases: an epidemiologic perspective. *Epidemiology* **9**, 350 – 354.

Moore, K.L., Persaud, T.V.N. and Shiota, K. (1997) *Color Atlas of Clinical Embryology*. Pashalides Medical Publications. Athens, Greece, pp. 105.

National Statistical Service of Greece (1981-1995) *Annual Statistical Yearbook of Greece*. National Statistical Service of Greece. Athens, Greece.

National Statistical Service of Greece (1981-1995) *Monthly Statistical Bulletin*. National Statistical Service of Greece. Athens, Greece.

National Statistical Service of Greece (1981-1995) *Statistics of Population's Physical Movement*. National Statistical Service of Greece. Athens, Greece.

Pew environmental health commission (1999) *Why America needs a better system to track and understand birth defects and the environment: Companion report of healthy from the start*. Johns Hopkins School of Public Health. Baltimore.

World Health Organization (1981-1995). *Annuals*. World Health Organization, Geneva, Switzerland.

Yoon, P.W., Rasmussen, S.A. and Lynberg, M.C., (2001) The National Birth Defects Prevention Study, *Pub. Health Rep.* **116**(supp 1), 32 – 40.

Yoon, P.W., Olney, R.S., Khoury, M.J. and Sappenfield, W.M., Chavez, G.F., Taylor, D. (1997) Contribution of birth defects and genetic diseases to pediatric hospitalisations. *Arch. Pediatr. Adolesc. Med.* **151**, 1096 – 1103.

CONGENITAL ANOMALIES IN BULGARIA

E. TERLEMESIAN AND S. STOYANOV*
University of Chemical Technology and Metallurgy
Ecology Centre
blvd. "Kl. Ohridski" 8
Sofia
BULGARIA
(author for correspondence, Email: stoyan@uctm.edu)*

Summary

The paper presents data and analysis concerning congenital anomalies in Bulgaria. Assessment is based on statistical data reported in the National Statistical Institute Yearbook: Health Protection and the Sofia registry of congenital anomalies in the period 1996 – 1999. Forty subgroups of isolated congenital anomalies and congenital diseases, detectable at birth during in the first year of life, have been detected out of 34,124 pregnancies, registered during the period. The rates of live births, stillbirths and induced abortions due to anomalies, per 10,000 pregnancies, are selected as indicators. Results are compared with EUROCAT rates of cases per 10,000 births of 85 subgroups with congenital anomalies in the EUROCAT full member registers in the period 1996 – 2001. The analysis which is made by Simeonov and Dimitrov shows that only 2 per cent of the total incidence of congenital abnormalities has a purely environmental origin. Most of the cases have been attributed to multifactorial etiology. Comparison is made with the experience of the TIS in Jerusalem and the data reported in the frame of OTIS.

Trends of reduction in the total infant mortality rates and the mortality rates due to congenital anomalies and certain other conditions originating in the prenatal period 1990 – 2003 descised. It is shown that, in 2003, these two reasons could be ranked at the first two places among the causes of the infant's deaths with rates 260/100,000

P. Nicolopoulou-Stamati et al. (eds.), Congenital Diseases and the Environment, 339–357.

live births and 386/100,000 live births respectively. Territorial distribution of infant mortality in 2003, caused by congenital anomalies and due to reasons originating in the prenatal period, is connected with the territorial distribution of some known teratogenic agents e.g. pesticides, dioxins, heavy metals, etc.

Increases in age of Bulgarian mothers at the birth of their age of the mother at the last delivery are associated with increasing rates of chromosomal anomalies.

1. Introduction

Over recent years, public awareness has been growing for the developmental risks of the embryo due to exposures in the uteros. Reproductive hazards of chemical, physical or infectious agents on fetuses are related with the rates of spontaneous abortions and stillbirths, gestational age, birth weight and the rate of congenital anomalies. Prenatal complications and postnatal growth, neurological, psychomotoric and behavioural development are equally important aspects in assessmening of the effects of environmental agents on the outcome in human pregnancies.

Congenital anomalies are structural errors, or disturbed chemical function due to a metabolic deficiency. They constitute a considerable burden for the affected person, his family and the society. Over all women have a 3-4% risk of having a child with a significant birth defect. About 10 percent of, the birth defects are related to drugs, chemicals, and infections, known maternal illnesses or other exposures.

It is estimated that 1:15 children born today has an inborn defect. Physical anomalies can be overt, or occult. Common locations for congenital anomalies today are the skull, spine, sacrum, hip, hands and feet (Beers and Berkow, 2005).

Birth defects incompatible with survival as a rule result in a spontaneous abortion. Inborn metabolic abnormalities involve utilization of carbohydrates, lipids (fat), pigments, and minerals. Abnormal metabolites may be stored, excreted, or absent. Some deviant metabolic processes are gene transmitted and some are sporadic occurrences.

Development is the process of growth and differentiation. Developmental anomalies can be pre- or post-natal. Those appearing after birth can be inborn or orginate during the *in utero* development.

Teratology is the study of environmentally induced congenital anomalies. A teratogen is an environmental agent, such as a drug, a virus or a pollutant, which can cause a structural or functional anomaly by acting on the developing embryo or fetus. As a part of the development of prenatal care programs, Teratogen Information Services (TIS) emerged in the late 1970's in many parts of the world. Sixteen national TISes constitute the European Network of Teratogen Information Services (ENTIS). In the North America and Canada, TISes are organized to form the Organization of Teratogen Information Services (OTIS).

In this paper, statistical data are presented on the incidence rates, types and trends in the congenital diseases and infant mortality rates in Bulgaria during recent years and the causal factors are discussed.

2. Congenital diseases in Bulgaria: The Sofia register database

Sofia Registry of Congenital Anomalies (SORCA) emerged in 1996 in response to a growing need for up to date scientific information about the effects of drugs and other environmental agents on the developing human embryo and fetus. The registry was organized by the Bulgarian Society of Human Genetics and Sofia Municipality and was supported by the State and private sponsors until 2001. The registry covers all mothers who are residents of the region of Sofia and covers approximately 10,000 births annually. In 1996, it became a member of EUROCAT. The registry is not currently funded. In 2000, Simeonov and Dimitrov presented to the Euroworkshop on reproductive and developmental toxicity of pesticides, helded in Sofia, the four years experience of the Sofia register of congenital anomalies, based on EUROCAT criteria. According to their report, 40 isolated subgroups of congenital anomalies (CA) and congenital diseases, detectable at birth or to the end of the first year of life have been detected, out of 341,24 pregnancies, registered in the period 1996 – 1999. Almost one third of the registered families has been contacted personally and has provided diagnostic information by competent specialists. The overall incidence rates of CA are presented in table 25.

Table 25. Sofia register database, 1996 – 1999 (after Simeonov and Dimitrov, 2001)

Type of registrations	Total number	Percentage	Percentage of CA
Pregnancies	39,124	100	-
Total congenital anomalies, including:	778	1.9	100
- Live births	668	-	89.3
- Fetal deaths	38	-	5.1
- Induced abortions	42	-	5.6

Details about the number and incidence rates of the groups of anomalies of the registered cases in Sofia are shown in table 26.

Table 26. Registered number of congenital anomalies by type, 1996 – 1999 (after Simeonov and Dimitrov, 2001; EUROCAT, 1996 – 2001)

Type of congenital anomalies	Live births	Fetal deaths	Induced abortions	Total	Incidence rate per 10000 pregnancies	
					SORCA	EUROCAT
Total number	668	38	42	748	191.49	**215.47***
Nervous system	83	35	31	148	37.89	21.75
Neural tube defects	35	18	23	76	**19.46**	10.03
Spina bifida	34	5	8	47	**12.03**	5.11
Heart diseases	245	3	2	250	**64.00**	60.84
Cleft lip with or without cleft palate	32	2	1	35	8.96	9.11
Cleft palate	9	0	0	9	2.30	5.84
Digestive system anomalies, excluding pyloric stenosis	65	2	2	69	**17.66**	16.06
Internal urogenital system	27	0	2	29	7.42	**27.86**
External genital organs	39	0	0	39	9.98	**13.11**
Limb anomalies	54	5	2	61	15.62	**36.62**
Musculoskeletal and connective tissue anomalies	62	4	4	70	17.72	**22.17**
Down's syndrome	44	0	3	47	12.03	**18.85**
Other chromosomal anomalies, incl.	16	1	8	19	4.86	**13.29**
Trisomy 13	2	1	0	3	0.77	**1.53**
Trisomy 18	1	0	1	2	0.51	**3.61**
Eye anomalies (total)	40	1	1	44	**11.26**	4.44
Ear anomalies (total)	59	2	3	64	16.38	3.89

The values in bold show the highest prevalence rates in the Sofia registry or the EUROCAT register.

For comparison, the number of cases among live births, stillbirths and terminations of pregnancy per 10,000 births, of 85 congenital anomaly subgroups in the EUROCAT full member registers in the period 1996 – 2001 are presented in the same table (EUROCAT report, 1996-2001). Unfortunately, the SORCA data are not enough to make comparison with equal significance with the EUROCAT results. EUROCAT has 41 members in 20 countries. More than one million births per year are surveyed, one quarter of the births in EU member states and more than half of births in seven non-EU countries. The central database holds a total of more than 350,000 cases of congenital anomalies since 1980, including live births, stillbirths and terminations of pregnancy following prenatal diagnosis.

In general, the overall and individual incidence rates from SORCA do not differ significantly from the average of EUROCAT (1996 – 2001). According to the incidence rates, at the first places are ranked the heart diseases (64.00), nervous system abnormalities (37.89), neural tube defects (19.46), musculoskeletal and connective tissue anomalies (17.72), digestive system anomalies (17.66), limb anomalies (15.62), ear and eye anomalies (16.38 and 11.26 respectively), spina bifida (12.03). In spite of the smaller number of anomalies per 10,000 pregnancies registered in Sofia than EUROCAT's value, there are several groups with incidence rates higher than the European register data. The table shows that in Bulgaria one baby with neural tube defects is born per 1084 newborns and one with spina bifida per 1,120 live births. The worldwide incidence rate of neural tube defects is 2.6 per 1,000 (Wyszynski, 2005).

Spina bifida occurs in the first month of pregnancy, often before the woman knows that she is pregnant. It is known that women taking certain medication for epilepsy and women with insulin dependant diabetes have a higher chance of having a child with a neural tube defect. In the United States, spina bifida occurs in about one in every 1,000 – 2,000 births. More children in the U.S. have spina bifida than have muscular dystrophy, multiple sclerosis, and cystic fibrosis combined (NINDS, 2005).

In Western Australia, until 1996, about 2 children in every 1,000 have been born with a neural tube defect. Since 1996, as a result of the folic acid campaign, the value has dropped to 1.3 children per 1,000 births.

The Sofia register database indicates congenital facial clefs in 1: 925 live births – a ratio comparable to ratios in the United States. At the same time, according to William Johnson the birth certificate rate for the Native Americans was 1:512 (Johnson, 1982; Rosano, 1999). As the data in the last report do not come from clinical examination, but from birth certificate data, they frequently are considered of questionable validity. The best demographic information about inborn anomalies in Missouri River Basin Native Americans relates to the craniofacial area. In 1963, cleft lip and palate were reported as 1: 276 Montana Native American live births. The general population ratio was 1: 583. The incidence rate of lip and palate clefts in the Sofia register are less than the values shown in the EUROCAT database (11.16 incidence rate vs. 14.95 incidence rate). The prevalence of non-chromosomal congenital anomalies has been positively associated with deprivation of the area of residence. Summarized data, mentioned above for comparison are shown in table 27.

The total chromosomal anomalies registered in Sofia per 10,000 pregnancies including Down's syndrome, trisomy 13 and 18 are almost twice less than those registered by EUROCAT diseases, as shown in the table. No case of trisomy 21 is registered in SORCA in the period studied. The prevalence of chromosomal anomalies in some regions was negatively associated with deprivation, because of higher average mother age in more affluent areas (Smith, 1988).

Table 27. Comparison of some nervous system diseases and facial clefts from SORCA and worldwide

Type of congenital anomalies	SORCA	Worldwide
Spina bifida	0.89 babies per 1000 live births	About one in every 1,000 - 2,000 births (USA)
Neural tube defects	0.92 babies per 1000 newborns	2.6 babies per 1,000 newborns (worldwide) About 2 children per 1,000 newborns (W. Australia) – until 1996 1.3 children per 1000 births (W. Australia) after 1996
Congenital facial clefts	1.12 per 1000 live births	1.71 per 1,000 live births (US) 1.49 per 1,000 live births(EUROCAT)

The age-specific fertility of Bulgarian women in 1998 – 2002 is shown in table 28. It could be noticed that the number of births fell from 1998 to 2002 and that the average age of mothers has increased.

Table 28. Live births in Bulgaria according to the age of the mother (NSI, Population, 1999 – 2003)

Year	Total number of live births	Mother age groups (percent)					
		Under 20	20 – 24	25 – 29	30 – 34	35 – 39	40 – 44
1998	45,082	9.74	42.43	32.28	11.60	3.27	0.64
1999	47,256	8.80	39.36	34.40	13.16	3.59	0.67
2000	45,724	7.21	37.24	35.84	15.01	3.90	0.74
2001	39,760	6.42	34.96	36.99	16.50	4.30	0.79
2002	38,255	5.54	32.68	38.60	18.00	4.36	0.77

A reduction of the total number of births and the age delivery of the first child is noticed in mothers aged up to 24 years old recently in Bulgaria, on the contrary to the increased women in their mid twenties and thirties. The same trends are shown worldwide (Health, US, 2004).

3. **Mortality rates**

Infant mortality rate per 1,000 live births, compared with total mortality rate for all ages (per 1,000 of population) are listed in table 29.

During in the period studied a reduction is observed in the infant mortality rates, both in the cities and in the countryside. Nevertheless, the infant mortality rate in Bulgaria is 2 to over 4 times higher than the corresponding of the EU countries (5.9/1,000 in Greece and 2.8/1,000 – in Sweden) (Health, US, 1996 - 1997). Infant mortality rate is stabilized at about 14 deaths per 1,000 live births till 2001. In 2002, reduction to 13 3 per 1,000 is registered, which goes on decreasing to 12.3 in 2003. Mortality is higher in the countryside than in the towns. At the same time, mortality rate per 1,000 adult rises, mainly on behalf of the countryside.

During the last years, an increased trend of delivery at home is registered in Bulgaria. This is a reason for the high maternal mortality. In the rural areas, mother mortality is 25.5 per 100,000 births (UN Population Fund Report, 2004). The high percentage of delivery at home is typical for specific ethnic groups with low living standards. The main reasons are socio-economical, lack of health insurance, etc.

Table 29. Mortality rates in Bulgaria during the period 1980 – 2003 (NSI Health Protection, 1981 – 2004

Year	All ages, per 1000 people			Infant mortality rate per 1,000 live births		
	Total	Urban	Countryside	Total	Urban	Countryside
1980	11.1	8.2	15.7	20.2	18.0	24.9
1990	12.5	9.4	18.6	14.8	13.8	17.1
1995	13.6	10.7	19.9	14.8	14.0	16.7
2000	14.1	11.5	19.6	13.3	12.4	15.5
2001	14.2	11.5	20.4	14.4	12.9	18.2
2002	14.3	11.7	20.3	13.3	12.0	16.9
2003	14.3	11.8	20.1	12.3	10.7	16.5

In 2003 certain conditions, originating in the prenatal period (386/100,000 live births) and congenital anomalies (259.8/100,000 live births) were at the two main causes of infant death (figure 23).

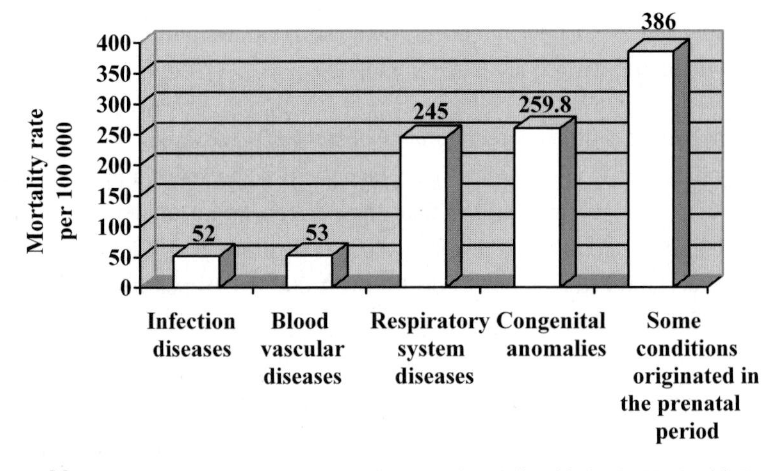

Figure 23. Infant mortality rates in Bulgaria per 100,000 live births (NSI, Health Protection, 2004)

For comparison, the three leading causes of death to infants in the US are congenital anomalies and disorders relating to a short gestation period, low birth weight and sudden infant death syndrome (SIDS) (Health, US, 1996 - 1997). It is known that major birth defects cause 20 per cent of infant mortality and are responsible for a substantial number of childhood hospitalizations (Ornoy, 2001).

Trends in the mortality rates per 100,000 people from all ages due to the most common reasons for the infant mortality in the period 1990 – 2003 are shown in table 30.

Table 30. Trends in the mortality rate in Bulgaria during the period 1990 – 2003 (NSI, Population, 2002, 2003)

Mortality cause	Mortality rate (per 100,000)			
	1990	2000	2002	2003
Total congenital anomalies	5.8	3.4	3.0	2.7
Conditions, originating during the prenatal period	4.7	4.0	3.4	3.3

Trends of reduction are noticed for both reasons for death. These decreases in the mortality rates could be explained by the increase of deaths, because of the higher mean age of the constantly aging population (e.g. cardiovascular diseases or cancer).

Reduction of the infant mortality rates as a result of congenital malformations and prenatal period conditions are shown in table 31. The data show that a general decreasing trend, except from 1995.

A possible interpretation is that the 1990's are a decade of dramatic political and economic changes in Bulgaria, characterized by on industrial and agricultural crisis, lack of environmental control, enhanced diversification and poverty, impaired health care system, concerning a big proportion of the population. Improvement from the beginning of the new millennium is due to the stabilization of the Bulgarian economy and involvement of principles of prevention and control in the new environmental legislation.

A comparison of infant mortality rates attributable to congenital anomalies by five year periods for five countries is shown in table 32.

In the countries studied during the period, the infant mortality rates attributable to congenital anomalies have increased by a factor of 1.7 (Bulgaria), or stabilized at high levels (Rumania, Greece), while in the USA and Belgium they have decreased by factors of 1.8 and 2.2 respectively. During the studied period, the proportional impact of congenital anomalies on infant mortality varied significantly from country to country but everywhere it increased.

Table 31. Infant mortality rates in Bulgaria due to congenital anomalies and certain conditions, originating during prenatal period 2003 (NSI, Health Protection, 2002, 2003)

Reasons for death	Mortality rates per 100,000 live births				
	1990	1995	2000	2002	2003
Congenital anomalies (total) including:	401.2	418.2	320.3	282.7	259.8
Congenital anomalies (total) including:	**401.2**	**418.2**	**320.3**	**282.7**	**259.8**
Spina bifida	20.9	22.2	13.6	16.5	10.4
Other nervous system anomalies	38.0	66.7	43.4	51.1	20.8
Anomalies of bulbus cordis	30.4	26.4	19.0	33.1	38.6
Other congenital heart anomalies	136.9	105.6	103.2	66.2	75.7
Congenital respiratory system anomalies	37.1	59.7	16.3	12.0	13.4
Congenital digestive system anomalies	37.1	37.5	28.5	25.6	32.7
Conditions, originating in the prenatal period, including:	**393.6**	**464.1**	**438.4**	**398.5**	**386.0**
Disorders related to a preterm delivery and low birth weight	130.3	84.8	62.4	120.3	154.4
Difficult breathing syndrome (sudden infant death syndrome)	42.8	62.5	69.2	49.6	44.5

Table 32. Infant mortality rates (IMR) attributable to congenital anomalies per 10,000 live births by five year periods in five countries 2003 (Rosano *et al.*, 2000)

Country	1965-1969		1970-1974		1975-1979		1980-1984		1985-1989		1990-1994	
	IMR	%*	IMR	%	IMR	%	IMR	%	IMR	%	IMR	%
Bulgaria	24.9	8	29.6	11	30.9	14	35.4	20	36.5	25	43.1	27
Rumania	33.6	7	33.5	8	37.3	12	42.9	16	50.4	19	38.9	16
Greece	39.9	12	43.	16	45.6	22	46.5	30	36.7	31	41.4	47
Belgium	47.4	21	42.3	22	35.1	25	29.6	27	24.4	26	21.4	26
USA	33.2	15	29.0	16	25.9	18	24.4	21	21.3	21	18.7	21

* *per cent of total infant mortality*

For instance, in Bulgaria it increased from eight per cent in the 1960s to 27 per cent in the early 1990s. The increase is highest in Greece – almost fourfold and reached 47 percent. It is established (Br. Med. J. Publishing Group, 2000) that the infant mortality rates attributable to congenital anomalies show a strong inverse correlation with per capita gross domestic product (GDP). At the same time, the proportional impact of congenital anomalies on infant mortality correlates directly with per capita gross domestic product. This explains why in Bulgaria later, when GDP per capita was enhanced, IMR dropped to 28.3 in 2002 and 26.0 in 2003, with proportional impact on infant mortality of 21 per cent in both years.

Territorial distribution of infant mortality in 2003 caused by different types of congenital anomalies and due to reasons originating in the prenatal period, is shown in table 33.

Table 33. Geographical distribution of infant mortality in Bulgaria in 2003, caused by congenital anomalies and due to reasons originating during the prenatal period (NSI, Population, 2002)

District	Live births rate per 1,000 inhabitants	Still births rate per 100,000 inhabitants	Infant mortality (per 1,000 live births)		Infant mortality rate (per 1,000 live births) due to	
			Urban	Country-side bordering towns	Congenital anomalies	Reasons from the prenatal period
Average for the country	**8.6**	**7.0**	**10.7**	**16.5**	**2.6**	**3.9**
Vratzsa	8.1	6.4	**12.3**	15.6	**2.8**	**4.0**
Montana	7.7	8.0	**12.3**	**22.4**	1.5	**6.0**
Dobrich	**8.8**	8.6	**12.5**	13.9	3.8	**4.9**
Varna	**9.9**	**6.7**	9.4	**19.2**	3.9	**3.7**
Silistra	**8.7**	**9.4**	**12.3**	**18.1**	5.8	**4.7**
Targovishte	**9.4**	6.4	**14.3**	**23.6**	5.3	**6.6**
Shumen	**9.5**	**20.9**	**13.3**	**23.9**	3.1	**4.7**
Burgas	**9.6**	**8.6**	**12.7**	16.6	1.2	1.2
Sliven	**11.9**	**15.9**	**26.2**	**38.3**	**3.1**	**13.7**
Yambol	8.2	4.6	**12.0**	**44.9**	**4.9**	**4.1**
Kardjali	**9.6**	4.9	**13.2**	**17.9**	1.9	**5.2**
Plovdiv	**8.8**	6.5	**14.3**	**18.4**	**2.9**	**5.9**
Kyustendil	7.2	5.1	**15.5**	**17.9**	**3.5**	0.9
Sofia-capital/ district.*	**8.9/7.5***	4.7/**7.3***	6.9	**20.1***	1.6/**4.0***	**2.0/3.5***

The values in bold show the rates exceeding the averages for the country and the problematic regions. The asterisks mark values concerning Sofia-district

The higher birth rates in some regions could be explained both by the higher standard of living in the big towns (Varna, Sofia-town, Burgas) or by the higher percentage of ethnic groups with faster population growth in them (e.g. Sliven – with the highest percentage of gipsy people and Kardjali, Shumen and Targovishte – with a high proportion of Muslims). Infant mortality rate varies in large range – from 5.9 in the region of Vidin to over 30.6 per 1,000 live births in the region of Sliven. In half of the regions, the values of this index are over the average for the country. In most of the regions with higher infant mortality rates than the average for the country (Silistra, Targovishte, Shumen, Sliven, Yambol, Kardjali, Plovdiv, Montana), the proportion of congenital anomalies and conditions, originating in the prenatal period, as reasons for death exceed the averages for the country. As was noticed in our previous AREHNA Report (2004), these are developed agricultural areas or highly industrialized regions with production of energy, or centres of metallurgy. Therefore, the environment is more polluted, due to increased use of pesticides, emissions of dioxins and heavy metals.

4. Discussion

The primary registration form of SORCA contains environmental information – e.g. drugs, and other exposures. Approximately:

- 60 per cent of the registered fetus/children cases have been identified as isolated, probably multifactorial, congenital anomalies, and
- 40 per cent – with multiple congenital anomalies, distributed as follows:
    - chromosomal aberrations – 8 per cent,
    - single gene disorders – 10 per cent,
    - purely environmental – 2 per cent, and
    - multiple congenital anomalies of unknown etiology – 20 per cent (Simeonov and Dimitrov, 2001).

The experience of the TIS in Jerusalem (Ornoy, 2001) reported 38,000 calls over twelve years. 75% of them were due to drug exposure, 9% to diagnostic exposure, 6% suspected to have intrauterine infection, 5% due to vaccination and only 2% following exposure to environmental pollutants (Ornoy *et al.,* 1999).

In the frame of OTIS, more than 40 fact sheets summarize the effects of environmental agents on the developing embryo and fetus for six different categories of exposures: medications, infections, illicit substances, herbal products, maternal medical conditions and other common exposures. The types of exposures and the reproductive status of those involved in the OTIS are presented in table 34 and table 35.

Table 34. Causes of congenital anomalies (OTIS, 2003-2004)

Cause	Relative importance (%)
Mental illness	4
Environmental agent	8
Radiation	3
Occupational agents	17
Herbal products	4
Drugs abuse	6
Medication	58

Data show that the biggest proportion of anomalies is trigged by medications and drugs exposure (64 per cent). Environmental factors including radiation, occupational agents and herbal products rank in second, with 32 per cent.

This classification is not very convincing. Data about the environmental quality in general may reveal little information about a person's actual contact with environmental agents. Living in an agricultural region does not necessary mean exposure to pesticides or other agents which might have teratogenic activity. When a drug has been taken by mother and a birth defect was investigated, she was asked for the dose, time and duration of the treatment. In the case of chemicals, she could be asked only whether she has been working with pesticides or if there has been an indoor use against insects, or if the nearby field has been sprayed, or if her husband has worked with pesticides. When a chemical is used as a pesticide, paternal exposure is most often considered (Vergieva, 2001; Wakefield, 2000). Environmental monitoring which determines what is in the soil, air, food and water is not equivalent to the individual exposure.

Risk assessment of prenatal exposures requires knowledge about chemical structure, dose and timing of exposure and any contributory factors regarding medical and family history. The human embryo, like all mammalian embryos, is only partially protected from the harmful impacts of the environmental factors. Hence many chemicals, drugs, infections and physical agents may adversely affect the developing embryo and fetus. The degree of susceptibility of these agents depends on the developmental stage at exposure, duration of exposure and dosage. Very little is known about the effects of exposure to low doses of pesticides, solvents and other toxic substances. Anomalies are inherited through gene transmission (genotype), or are started by some stimulus during embryogenesis, when developing structures are vulnerable (phenotype). As the OTIS sheets data show, 74 per cent of the anomalies registered started during the stage of pregnancy (table 35).

Table 35. Origin of congenital anomalies (OTIS, 2003-2004)

Reproductive status	Relative importance (%)
Pregnancy	74
Paternal	3
Preconception	9
Breast feeding	8
Unknown	6

Factors that are suspected to induce developmental anomalies include paternal and maternal exposure to dioxins, PCB and pesticides. Epidemiological studies in Russian cities heavily polluted with dioxins (Chapaevsk) showed higher frequency of newborns with congenital hydrocephalus. In a city with high exposure PCBs (Serpukov), higher incdense rates of with congenital anomalies (4.5 per 100 in newborns were register, compared to 2.2 per 100 newborns an average Russian cities) (Revich *et al.,* 2000)

An increased rate of birth defects has been associated with the use of specific pesticides or with pesticide mixtures. A comprehensive review of the relevant epidemiological studies is produced by Garcia (1998) and Hanker and Houseman (2000). Over all, the analysis indicated that parental employment in agriculture increases the risk of congenital malformations in their offspring. In particularly orofacial cleft, birthmarks, musculoskeletal and nervous system defects occur more frequently than expected. The general public is exposed, however, at levels far below the levels at which these effects have been established.

The total amount of pesticides used in Bulgaria during the period 2002 – 2004 is shown in table 36.

The areas where pesticides are used are limited. As an example, the surface of herbicide-treated vineyards is presented in figure 24 and comparison is shown with the entire area of vineyards in Bulgaria. The compression shows that less than five percent of the area is treated with plant defence products (PDP).

Table 36. Pesticides used in Bulgaria (in tons) during the period 2002 – 2004 (National Office for Plant Defence, 2004)

Type of pesticides	Year		
	2002	2003	2004
Herbicides	2535.30	2422.09	2669.82
Insecticides	577.19	568.52	584.99
Fungicides	2096.13	1510.78	1549.76

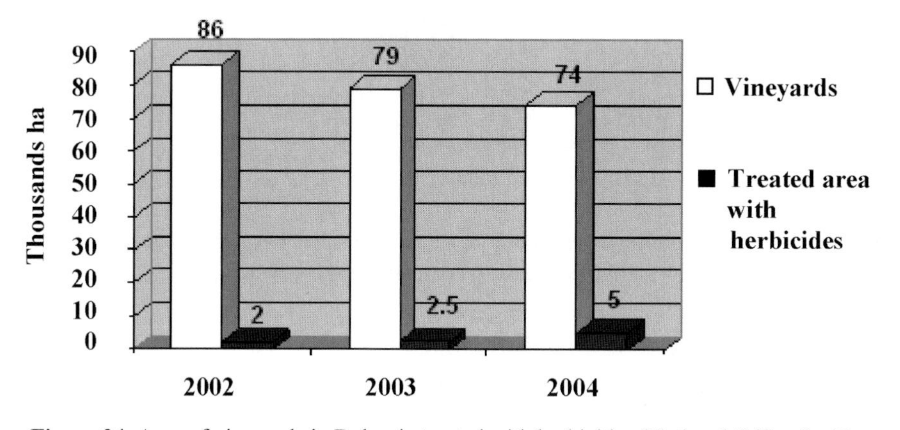

Figure 24. Area of vineyards in Bulgaria treated with herbicides (National Office for Plant Defence, 2004)

This is the result of the Bulgarian likely to reduce the use of pesticides. This strategy includes systematic control of the appearance, dissemination, density and the degree of invasion of the plant diseases, insects and weeds and their natural regulators. An aspect of the strategy includes e.g. the localization of the locusts in the unplanted areas. Areas unwanted by locusts could be used as agricultural land. Also the timing of pesticide use and optioning for more effective chemicals are important. The executive body, carrying out this policy, is the National Office for Plant Defence. Its formal responsibility includes biological testing of the pesticides according to the Best Experimental Practice; requirements for permission according to Directive 91/414/EEC and Annex I of PDP and their active substances and next official registration; control of the import of PDP and control in the market network (Agrarian Report for 2003, 2004). Fifty-four of the active substances included in Annex I ninety-one are components of PDP permitted in Bulgaria (National Office for Plant Defence, 2004).

Outdated pesticides, which are deposed in ruined, unprotected stores as a rule, without control, contribute a significant problem. An elimination policy is in place nowadays. Financial resources, spent for the collection and permanent disposal of these old pesticides, are an indicator of the measures taken to protect the environment against the dangerous substances (figure 25).

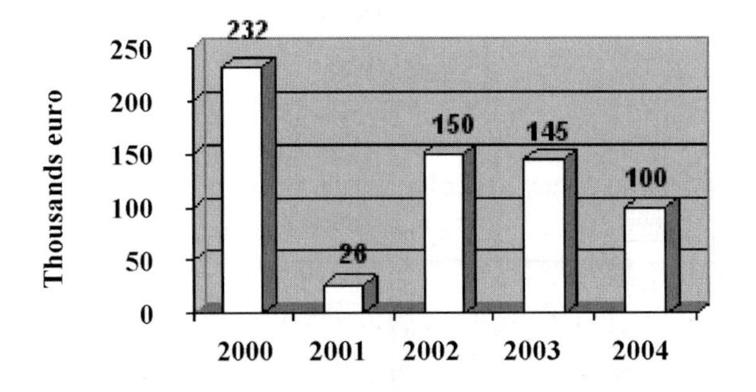

Figure 25. Financial resources, spent in Bulgaria for the collection and permanent disposal of those pesticides forbidden to use (National Office for Plant Defence, 2004)

An effort is in progress to limit the harmful health impacts of currently use pesticides nowadays, by reducing their occurrence in soil, water, plants (tobacco, herbs, forage) and food (tomatoes, lettuces, apples, strawberries). In addition, the content of heavy metals and arsenic, some micro- and macro- elements in plants, soil, water, sludge, composts, organic and mineral fertilizers are analyzed systematically. Although the number of analysed samples is increases as steadily, it is not sufficient to obtain a reliable view on the health situation. Till now the monitoring does not cover all the regions of the country.

5. Conclusions

During recent years, Bulgaria faced serious demographic problems. Negative population growth was accompanied by increased mortality and impaired reproduction. During the most recent period, reduction trends of the total number of births and increase of average mother's age at birth, including the first child delivery, are noticed in Bulgaria. Increase of delivery at home becomes more frequent, typically for specific ethnic groups with low economical level and resulting in high mother mortality. Reduction of infant mortality is observed in both the cities and countryside in Bulgaria. Nevertheless, the rate remains higher than in the other European countries. Congenital anomalies and certain conditions originating in the prenatal period are reported as the main reasons of infant death, with trends of reduction. The favourable trend from the beginning of the new millennium is a most likely consequence of the improvement of the environmental conditions in Bulgaria, due to more effective control on environment polluting activities and involvement in the principles of prevention.

Comparison the data of the Sofia Registry of Congenital Anomalies for the region of Sofia the results with EUROCAT data, shows that the overall and individual incidence rates in Sofia do not differ significantly from the average EUROCAT values. Heart diseases and nervous system abnormalities are the most frequent congenital diseases in Sofia.

Simeonov and Dimitrov (2001), by analyzed the primary information of the SORCA register and concluded that about 60 per cent of the registered fetus/children cases of birth defects have a multifunctional etiology two per cent are considered as purely environmental. The territorial distribution of the infant mortality rates, including congenital anomalies and certain conditions originating in the prenatal period and stillbirths, confirms the harmful influence on infant anomalies of environmental pollution with known teratogen agents as pesticides, dioxins and heavy metals. Regions with intensive agriculture or metallurgy and energy production centres shown the highest incidence rates.

The new strategy for the use of pesticides is discussed in connection with the increase of the rate of birth defects associated with pesticides. It includes biological testing of the pesticides, official registration after permission, control of their import and in the market network. Elimination of prohibited pesticides is also one of the measures taken reduced the health risk according to the Stockholm Convention.

References

Beers, M. and Berkow, R. (2005) The merck manual of diagnosis and therapy, seventeenth ed., Section 19 pediatrics, Chapter 261 Congenital Anomalies.

EUROCAT - European Surveillance of Congenital Anomalies (1996-2001), *EUROCAT Report 1996-2001*.

Garcia, A. M. (1998) Occupational exposure to pesticides and congenital malformations: a review of mechanisms, methods and results, *Am. J. Ind. Med.* **33**, 232-240.

Hanke, W. and Hausman, K. (2001) *Study on reproductive health of workers exposed to pesticides: an overview of epidemiological evidenc*e, Euro workshop on reproductive and developmental toxicity of pesticides, 2000, Sofia.

Johnson, W. (1982) *Vital statistics division*, South Dakota Health Department, in OTIS.

Ministry of Agriculture and Forestry (2004) *Agrarian Report for 2003*, Sofia.

National Office for Plant Defence (2004) *Annual Report*, Sofia.

NINDS, National Institute of Neurological Disorders and Stroke (2005), US National Institutes of Health, Spina Bifida Fact Sheet.

NSI - National Statistical Institute (1981) *Annual Handbook, Health Protection*, Sofia.

NSI - National Statistical Institute (1991) *Annual Handbook, Health Protection*, Sofia.

NSI - National Statistical Institute (1996) *Annual Handbook*, Health Protection, Sofia.

NSI - National Statistical Institute (1997), *Annual Handbook*, Health Protection, Sofia.

NSI - National Statistical Institute (1998) *Annual Handbook*, Health Protection, Sofia.

NSI - National Statistical Institute (1999) *Annual Handbook*, Health Protection, Sofia.

NSI - National Statistical Institute (1999) *Statistical Yearbook*, Population and demographic processes, Sofia.

NSI - National Statistical Institute (2000) *Annual Handbook*, Health Protection, Sofia.

NSI - National Statistical Institute (2000) *Statistical Yearbook*, Population and demographic processes, Sofia.

NSI - National Statistical Institute (2001) *Annual Handbook*, Health Protection, Sofia.

NSI - National Statistical Institute (2001) *Statistical Yearbook*, Population and demographic processes, Sofia.

NSI - National Statistical Institute (2002) *Annual Handbook*, Health Protection, Sofia.

NSI - National Statistical Institute (2002) *Statistical Yearbook,* Population and demographic processes, Sofia.

NSI - National Statistical Institute (2003) *Annual Handbook*, Health Protection, Sofia.

NSI - National Statistical Institute (2003) *Statistical Yearbook*, Population and demographic processes, Sofia.

NSI - National Statistical Institute (2004) *Annual Handbook*, Health Protection, Sofia.

Ornoy A. (2001) *Teratogen information services and embryo cultures as tools to study the effects of pesticides and toxicants on the human embryo and fetus*, Euroworkshop on reproductive and developmental toxicity of pesticides, 2000, Sofia.

Ornoy, A., Shechtman, S. and Arnon, J. (1999) The Israeli Teratogen Information Service: Our experience in the last 10 years activity regarding 20631 cases, *Harefuah* **136**, 1-9.

OTIS, Organization of Teratology Information Services, *2003-2004 Annual Report,* USA.

Revich, B,. Sergeev, O., Zeilert, V., Ivanova, I., Zhuchenko, N., Korrich, S., Altshul, L. and Hauser, R. (2001) *Dioxins, PCB and Reproductive health – two case studies in Russia*, Euroworkshop on reproductive and developmental toxicity of pesticides, 2000, Sofia.

Rosano, A., Botto, L., Botting, B. and Mastroiacovo P. (2000) Infant mortality and congenital anomalies from 1950 to 1994: an international perspective, *J. Epidemiol. Com. Health* **54**, 660 - 666.

Rosano, A., Smithells, D., Cacciani, L., Bottig, B., Castilla, E., Cornel, M., Erikson, D., Goujard, J., Irgens, L., Merlob, P., Robert, E., Siffel, C., Stoll, C. and Sumiyoshi, Y. (1999) Time trends in neural tube defects prevalence in relation to preventive strategies: an international study, *J. Epidemiol. Com. Health* **53**, 630-635.

Simeonov, E. and Dimitrov, B. (2001) *Registration of congenital anomalies in genetic/environmental interactions studies*, Euroworkshop on reproductive and developmental toxicity of pesticides, 2000, Sofia.

Smith's Recognizable Patterns of Human Malformation (1988) *Saunders.*

Stoyanov, S. and Terlemesian, E. (2006) *Environment and reproductive health in Bulgaria*, in Reproductive health and the environment P. Nicolopoulou-Stamati, Luc Hens and C.V Howards (eds). Springer, Derdrecht, the Netherlands, 342-357.

UN Population Fund (2004) *Report "Goals for the Millennium*", New York, Washington DC, USA.

US Department of Health and Human Services Center for Disease Control and Prevention, National Center for Health Statistics (2004) *Health, Chart book on trends in the health of Americans*, US Government Printing Office, Washington DC, USA.

US Department of Health and Human Services, Center for Disease Control and Prevention, National Center for Health Statistics (1996) *Health, US Chart book on trends in the health of Americans*, US Government Printing Office, Washington DC, USA.

US Department of Health and Human Services, Center for Disease Control and Prevention, National Center for Health Statistics (1997) *Health, US Chart book on trends in the health of Americans*, US Government Printing Office, Washington DC, USA.

Vergieva, T. (2001) *Overview of the experimental and epidemiological data on developmental toxicity of pesticides*, Euroworkshop on reproductive and developmental toxicity of pesticides, 2000, Sofia, Bulagira.

Wakefield, J. (2000) Human exposure assessment: finding out what is getting in. *Environ. Health Perspect.* **108**, A24-A26.

Wyszynski, D. (2005) *Neural tube defects: From origin to treatment*, Oxford Univ, UK. In press.

CONGENITAL ANOMALIES IN THE BRITISH ISLES

J. RANKIN
School of Population & Health Sciences
University of Newcastle upon Tyne
UNITED KINGDOM
Email: J.M.Rankin@newcastle.ac.uk

Summary

The first congenital anomaly register in the British Isles was established in 1949, with a national system for England and Wales introduced in 1969 in the wake of the thalidomide epidemic. There are now 14 regional congenital anomaly registers and three disease-specific registers. These registers involve an extensive local network of notifiers, use multiple sources of case ascertainment, consistent coding, and include cases resulting in termination of pregnancy for fetal anomaly following prenatal diagnosis. They have optimised the coverage, completeness and ascertainment of congenital anomalies within their population, and are therefore able to provide better quality data than the national system. There are notable variations in the prevalence of congenital anomaly subtypes within these regions that cannot be accounted for in terms of differences in case ascertainment or registration practices. This underlying variation in prevalence should be recognised and taken into consideration in the design of epidemiological studies that are investigating the contribution of environmental influences to congenital anomaly risk, to ensure correct interpretation of the findings. Local congenital anomaly register data has been used to investigate congenital anomaly risk in populations within the British Isles living close to landfill sites and incinerators, and to possible contaminants in drinking water. Whilst the inclusion of high quality data from established congenital anomaly registers enhances the quality of outcome data, such studies are currently limited by the lack of detailed information on exposure. Causal pathways will only

P. Nicolopoulou-Stamati et al. (eds.), Congenital Diseases and the Environment, 359–377.

be determined if future studies combine high quality congenital anomaly data with increased information on exposure assessment.

1. Introduction

Congenital anomalies are a significant cause of stillbirth and infant mortality, accounting for approximately 15% of stillbirths and one third of all infant deaths per year in England (Office for National Statistics, 2001). In 2003, 1.1% of all births in England and Wales were found to have a congenital anomaly (Office for National Statistics, 2004). Congenital anomalies are also important contributors to morbidity in the first year of life and beyond. The congenital anomaly rate for England and Wales has decreased from 168.1 per 10,000 live births in 1964 to 114 in 2001 (Office for National Statistics, 2004). Although the birth prevalence of certain types of congenital anomalies has been declining in recent years (Murphy *et al.,* 1996; Rankin *et al.,* 2000; Dastgiri *et al.,* 2002), they continue to be a major cause of childhood mortality and morbidity. It is therefore essential to accurately monitor their occurrence to identify possible clusters and trends, to address concerns about putative environmental influences and to investigate possible causes.

1.1. Recording of congenital anomalies in the British Isles

The importance of registering the type and number of congenital anomalies has long been recognised within the British Isles. The first formal register of children with congenital anomalies was established in Birmingham in 1949. Further registers were set up, including one in Liverpool (Smithells, 1962; Owens *et al.,* 1988), South Wales (Richards and Lowe, 1971), and Glasgow (Stone, 1986). These early registers recorded information on congenital anomalies among those resident in these cities. The vulnerability of the developing fetus to chemical exposures was highlighted by the thalidomide tragedy in the 1960s (McBride, 1961; Lenz, 1962), in which women who took the drug between days 20-35 post-conception gave birth to children with severe congenital limb defects (Lenz, 1962). This episode led to the establishment of malformation registers in many parts of the world including Canada, Australia, New Zealand, and Japan. Within the British Isles, the national collection and notification of information on congenital anomalies present at birth was proposed by the Chief Medical Officer in 1963 to facilitate the detection of hazards (Office for National Statistics, 2001). This led to the establishment of a national surveillance system, the National Congenital Anomaly System (NCAS), in 1964. The NCAS was set up with the aim of detecting visible congenital anomalies at birth, to identify changes in the frequency of reporting particular congenital anomalies or groups of anomalies (Office for National Statistics, 2001). The NCAS contributes data to the

International Clearinghouse for Birth Defects Monitoring Systems, a World Health Organisation recognised body representing 39 malformation monitoring programs worldwide (International Clearinghouse for Birth Defects Monitoring Systems, 2002).

1.1.1. The national congenital anomaly system

NCAS is a voluntary system, involving the passive collection of notifications, run by the Office for National Statistics (ONS). It collects information on congenital anomalies arising in live and stillbirths in England and Wales. As legal abortion was not introduced in England and Wales until 1968, four years after NCAS was established, NCAS does not include information on congenital anomalies associated with termination of pregnancy following prenatal diagnosis, or spontaneous abortion. Information is supplied to NCAS from acute hospital trusts, child health systems, and public health departments within primary care trusts, or midwifery units that are required to complete and return a birth notification form (Misra *et al.,* 2005). Minor anomalies are not included on the system. The ONS uses the same exclusion list for minor anomalies developed by the European Surveillance of Congenital Anomalies (EUROCAT). However, minor anomalies are reported if they are found in conjunction with other anomalies. The information collected includes partial surname and forename of the child, place of birth, date of birth, sex, live/stillbirth, multiplicity, last menstrual period, gestation in weeks, birth weight, home address, parent's occupation, mother's date of birth, number and outcome of previous pregnancies, and the congenital anomalies reported. The original system registered cases notified within seven days of birth, but this age cut-off limit for registration has now been withdrawn. All congenital anomalies are coded to the International Statistical Classification of Diseases and Related Health Problems version 10 (ICD-10) (World Health Organisation, 1994). These codes are grouped together into ONS monitoring groups for surveillance purposes.

The national notification system has long been known to be incomplete. The completeness of certain data items is known to be particularly poor, for example, gestational age is present in only 76% of notifications and less than half of all notifications contain information on parents' occupation (Misra *et al.,* 2005). Underascertainment of cases is also known to be high, especially where the diagnosis is confirmed only some time after birth (Knox *et al.,* 1984; Payne, 1992; Office for Population, Census and Surveys, 1995; Dolk *et al.,* 1998a; Boyd *et al.,* 2005; Misra *et al.,* 2005). This under ascertainment varies by congenital anomaly type. A comparison of the number of notifications of Down's syndrome made to NCAS with those on the National Down's syndrome Cytogenetic Register, revealed a fifty to sixty per cent underascertainment in the NCAS (Botting, 2000). A recent assessment of the completeness of NCAS data compared to four local congenital anomaly registers, found that, overall, NCAS only identified forty per cent of congenital anomalies arising in live and stillbirths during the study period 1991-99 (Boyd *et al.,* 2005). Compared to local register data, the lowest ascertainment was for neural tube defects; NCAS recorded sixty-eight per cent of all neural tube defects (when terminations of pregnancy were excluded) recorded by the local registers and thirteen per cent of cardiac anomalies (Boyd *et al.,* 2005).

Despite the known shortcomings, a number of which are currently under review, the NCAS system provides the most extensive coverage of data on congenital anomalies in England and Wales at present.

1.1.2. Regional congenital anomaly registers

Local congenital anomaly registers have been established in a number of regions within the British Isles, to meet local needs and to carry out a variety of audit and research projects including audits of prenatal diagnosis and research into the causes of specific anomalies within their locality. To complement the passive system of notification used by NCAS, data from local registers began to be fed into the national system by electronic data exchange in 1998. This dramatically increased the notification rate. For example, the notification rate for Wales was approximately 150 per 10,000 live births in 1997, which increased to 310 per 10,000 live births in 1998 once the Congenital Anomaly Register and Information System for Wales (CARIS) began reporting to NCAS (CARIS, 2001).

A congenital anomaly register was established in Glasgow in 1972, and this was extended to include Greater Glasgow in 1974. In 1985, a register was set up in the Northern region of England (the Northern Congenital Abnormality Survey, NorCAS) and since then seven further regional registers have been established in England (West Midlands, Merseyside and Cheshire, North West Thames, Wessex, East Midlands and South Yorkshire (formerly Trent), Oxfordshire, Berkshire and Buckinghamshire, the South West region), one in Scotland, three in Ireland (Dublin, South Eastern Health Board, South East), and the South Wales register has been extended to cover all of Wales. The geographical area covered by each register and the total number of annual births is summarised in table 37.

Table 37. Regional congenital anomaly registers in the British Isles

Register	Year started	Geographical area covered	No. of annual births covered[1]	Age limit for data collection (years)
		Regional registers		
Glasgow EUROCAT Register	1972	Greater Glasgow Health Board	9,680	No limit
Dublin EUROCAT Registry	1980	Counties Dublin, Wicklow & Kildare	23,270	5
Galway EUROCAT Register	1981	County of Galway	3,000	5
Northern Congenital Abnormality Survey (NorCAS)	1985	Northumberland, Tyne & Wear, Durham, Tees, North Cumbria	30,330	12
North West Thames Congenital Malformation register	1990	North West London, Hertfordshire, Bedfordshire	48,470	1
Congenital Anomaly Register for Oxfordshire, Berkshire and Buckinghamshire (CAROBB) (previously the Oxford CAR which covered Oxford births only from 1991)	2003	Oxfordshire, Berkshire & Buckinghamshire	27,360	1
West Midlands Congenital Anomaly Register	1995	West Midlands	66,280	2
Wessex Clinical Genetic Service (WANDA)	1994	Hampshire, Dorset & part of Wiltshire	26,310	No limit
Merseyside and Cheshire Congenital Anomaly Survey	1995	Merseyside & Cheshire	25,000	1
Cork and Kerry Congenital Anomaly Register	1996	Counties of Cork and Kerry, Ireland	7,500	No limit

Table 37. (Continued)

Register	Year started	Geographical area covered	No. of annual births covered[1]	Age limit for data collection (years)
East Midlands and South Yorkshire Congenital Anomaly Register (formerly Trent)	1997	East Midlands, South Yorkshire	68,800	No limit
South Eastern Health Board (Ireland)	1997	Counties of Kilkenny, Carlow, Wexford, Waterford, South Tipperary	6,320	No limit
Wales Congenital Anomaly Register & Information Service (CARIS)	1998	Wales	31,330	1
South West Congenital Anomaly Register	2002	Cornwall, Devon, Avon Gloucestershire, Wiltshire (not South), Somerset	44,820	18
Scottish Congenital Anomaly Register	1994	Scotland	52,730	1
		Disease-specific registers		
National Down's syndrome Cytogenetic Register	1989	England & Wales	621,470[2]	No limit
Audit of Screening for Chromosomal Anomalies in London (ASCAL)	2001	London	105,000	Diagnosed after April 2001
CRANE	2000	England & Wales	621,470[2]	All live births

[1] *Live and stillbirths only, figures are rounded*
[2] *Livebirths only*

Most of the regional congenital anomaly registers are population-based, with registration dependent on the mother's place of residence within the boundaries of each region, even if they were delivered outside the region. Two registers (North West Thames and Wessex) are hospital based, that is, they cover all deliveries within all hospitals within a geographically defined region. These local registers share a standardised minimum dataset (table 38), methodology, multiple source notification system and confidentiality and information technology security protocols. They are proactive in their method of data collection and regularly check the completeness of the notification of groups of anomalies, facilitated by their close links with collaborating clinicians.

Table 38. The core dataset for registers belonging to the British Isles Network of Congenital Anomaly Registers (BINOCAR)

Core items
Mother's National Health Service (NHS) number
Mother's date of birth/age
Mother's ethnic origin group
Postcode delivery/mother
Baby unique identifier (ID)
Anomaly description
Anomaly ICD10 code
Anomaly status (suspected)
Antenatal detection
Date of diagnosis
Diagnosis method
Birthweight
Date of delivery
Infant's sex
Date of death
Estimated date of delivery
Gestation
No. of fetuses
Pregnancy outcome
Identifiers
Baby hospital NHS number
Baby forename
Baby surname
Mother's forename
Mother's hospital ID
Mother's surname
Administrative fields
Address delivery/mother
Notifier name/details
Place of delivery text
Baby hospital ID 1

As well as collecting information on cases of congenital anomalies in live and stillbirths, the local registers also record cases occurring in late miscarriages (> 20 weeks gestation) and in terminations of pregnancy following prenatal diagnosis for fetal anomaly. All such cases should be included, when calculating a total prevalence rate for a particular congenital anomaly. When comparing prevalence rates between registers, it is essential to know whether or not all cases occurring in all pregnancy outcomes have been included, so that accurate comparisons can be made.

The integration of the regional registers with the national notification system has gone a long way to improve the completeness of NCAS. However, regional congenital anomaly registers only cover approximately half of the population of England and Wales, and about sixty per cent of the population of Ireland. There are, therefore, large populations in the British Isles where this improved system of congenital anomaly notification is not currently available. The under-reporting of all congenital anomalies in areas not covered by a local register is of major public health concern.

There is currently no central source by which the regional registers are funded. This means that an enormous amount of time and effort is spent by individual registers to secure funding to maintain their existence. Inevitably, this is time deflected from central activities such as data collection, validation, analyses and monitoring. The feasibility of providing national funding for congenital anomaly registers in England and Wales is currently being investigated.

1.1.3. Disease-specific registers

In addition to local congenital anomaly registers, three disease-specific registers have been established. The National Down's syndrome Cytogenetic Register, set up in 1989, collects information on all cases found to have a karyotype characteristic of Down's syndrome, from all cytogenetic laboratories in England and Wales (Mutton *et al.,* 1991). In 2001, an Audit of Screening for Chromosomal Anomalies in London (ASCAL) was established and includes all cases diagnosed after April 2001. The CRANE database, managed thorough the Royal College of Surgeons of England, collects information from each of the nine regional cleft teams in England on all live born children with a facial cleft. This register introduced a web-based system for notifying cases in 2000.

1.1.4. *The British Isles network of congenital anomaly registers*

The regional and disease-specific registers joined together in 1998 to form the British Isles Network of Congenital Anomaly Registers (BINOCAR). The aim of BINOCAR is to provide continuous epidemiological monitoring of the frequency, nature, cause and outcomes of congenital anomalies for the population of the British Isles, by means of national, regional and disease-specific registers of congenital anomalies. BINOCAR was set up for:

(i) the surveillance and analysis of congenital anomalies,

(ii) the monitoring and audit of health provision, detection and outcomes for congenital anomalies,

(iii) the planning and administration of the provision made for health and social care for pregnancies and infants affected by congenital anomalies,

(iv) medical research into the causes and consequences of congenital anomalies approved by research ethics committees,

(v) the provision of information to clinicians to support their clinical practice.

BINOCAR is a collaborative network which aims to collectively address issues specific to the functioning of a register, including ensuring data quality, common approaches to coding and finding sources of funding, as well as combining data to monitor regional and secular trends in the prevalence of congenital anomalies, and investigations into possible causes. For example, using the BINOCAR network, Abramsky *et al.* (1999) demonstrated that the decline in neural tube defect rates noted in the 1980s, did not continue in the 1990s (Abramsky *et al.,* 1999) despite the UK Medical Research Council's recommendation that increasing periconceptional folate supplementation could reduce the occurrence of neural tube defects (Medical Research Council Vitamin Study Research Group, 1991). The prevalence of gastroschisis, a congenital anomaly of the anterior abdominal wall, has been increasing globally over the past 30 years, especially among young mothers (Torfs *et al.,* 1994; Tan *et al.,* 1996; Nichols *et al.,* 1997; Rankin *et al.,* 1999). Recent data from BINOCAR has shown that the prevalence of gastroschisis from the UK congenital anomaly registers (4.4 per 10,000 total births in 2003) is higher than that reported by NCAS (2.1 per 10,000 births in 2003) (Chief Medical Officer, 2005). The reported rate for BINOCAR reflects the higher case ascertainment by the local registers, facilitated by the close collaboration with, for example, paediatric surgeons and the inclusion of cases occurring in terminations of pregnancy and late miscarriages. A significant increase in the total prevalence rate of gastroschisis has been reported by the BINOCAR network (2.5 per 10,000 births in 1994 to 4.4 in 2003) as shown in figure 26.

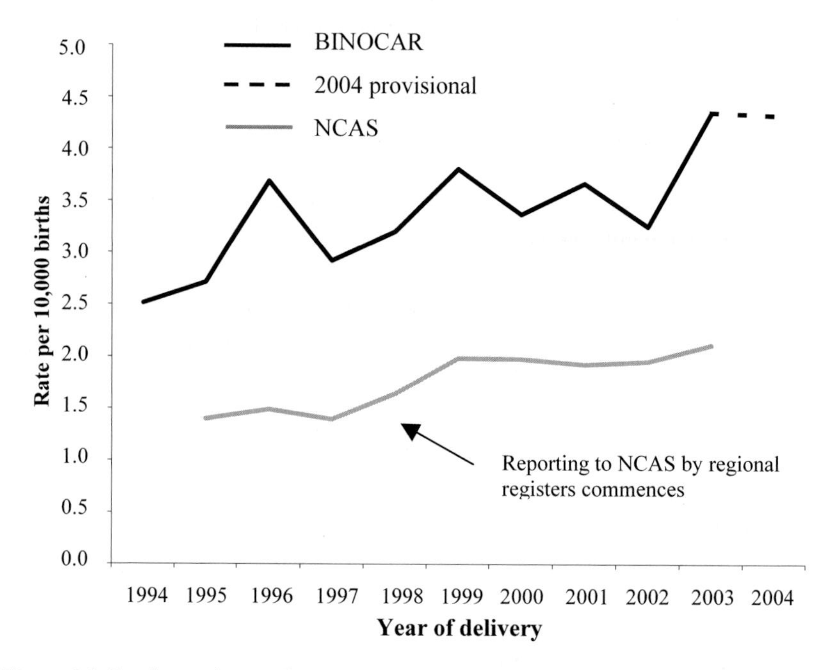

Figure 26. Total prevalence of gastroschisis for BINOCAR registers and England & Wales from NCAS per 10,000 births, 1994 – 2004 (BINOCAR and NCAS data for England and Wales, 1994–2004.) Annual Report of the Chief Medical Officer, 2004. Crown Copyright

2. Variations in the prevalence of congenital anomalies in the British Isles

The pattern of prevalence and of trends in prevalence of congenital anomaly groups and subtypes can vary over time and across regions and countries.

2.1. Secular trends in congenital anomaly prevalence

Figure 27 shows the total prevalence rate of selected congenital anomalies reported to EUROCAT by seven registers (Dublin, CARIS, NorCAS, North West Thames, Oxford, Trent and Wessex) for the years 1999-2003. There is obvious year on year fluctuation, but a decreasing trend in the prevalence of Down's syndrome is suggested (21.4 per 10,000 births in 1999 to 19.9 per 10,000 in 2003). The total prevalence rate for neural tube defects has remained stable at approximately 12 per 10,000 births for these registers during this five-year period. It has previously been suggested that the prevalence of neural tube defects may have reached a plateau in the UK (Rankin *et al.,* 2000). Conversely, there was a trend towards an increase in the total prevalence of malformations of cardiac valves notified to these selected registers, an increase from 7 per 10,000 births in 1999 to 7.8 per 10,000 in 2003.

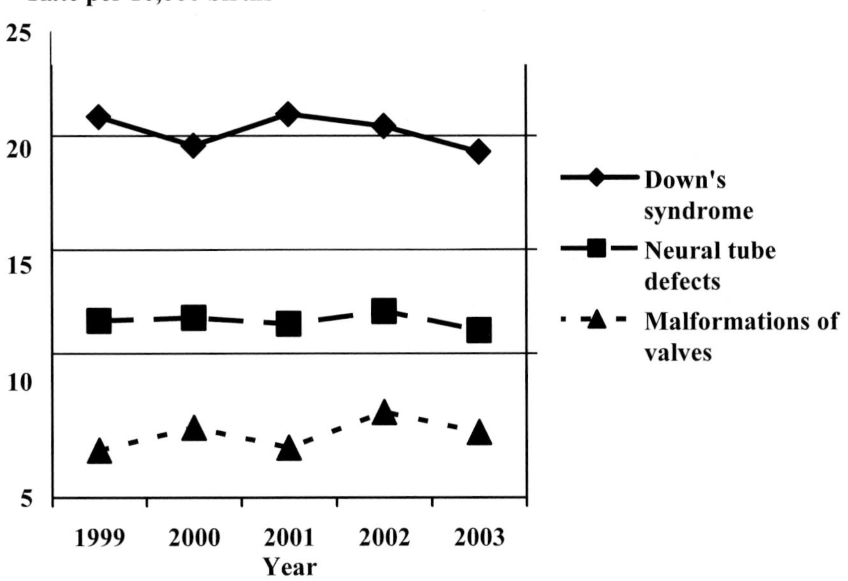

Figure 27. Total prevalence rate of selected congenital anomalies per 10,000 births, 1999-2003

2.2. *Regional variations in congenital anomaly prevalence*

There are also regional differences in the prevalence of congenital anomalies within the British Isles. Historically, Glasgow had a relatively high prevalence of neural tube defects compared to other parts of Europe (EUROCAT, 1991), although this is no longer the case. A recent study which described the prevalence of selected congenital anomalies within five well-defined geographical areas of Britain (Glasgow, Northern, Oxford, North West Thames, Wessex) between 1991-99, showed clear differences between registers in the total prevalence of certain congenital anomaly subtypes (Rankin *et al.,* 2005). The congenital anomalies considered did not include all congenital anomalies that occurred within the resident population but were well-defined, major anomalies consistently notified and coded across the five registers. A higher proportion of non-chromosomal anomalies were notified to the Glasgow and Northern registers. The Glasgow register had higher rates for malformations of cardiac septa, transposition of the great vessels, digestive system anomalies, ano-rectal atresia and stenosis, limb reduction defects and diaphragmatic hernia, but lower rates of chromosomal anomalies. The Northern region had the highest rate for malformations of valves (18.3 per 10,000 births). Conversely, rates for isolated cleft palate were lowest in Wessex and Oxford.

There are a number of underlying factors that may explain variations in geographical and secular trends in prevalence: case ascertainment methods including data collection methods, sources of information, and demographic (e.g. maternal age) and environmental factors (e.g. preconceptional vitamin supplementation). In this study, the registers showed similar rates in cardiac defects which can be diagnosed antenatally, e.g. transposition of the great arteries, hypoplastic left heart and atrioventricular septal defect, but greater variation was seen across the registers for late postnatally diagnosed cardiac defects e.g. Tetralogy of Fallot, malformations of valves (Rankin *et al.,* 2005). The higher rates of some cardiac defects, reported by the Glasgow and Northern registers, reflect the very close relationship these registers have with Paediatric Cardiology Departments, which facilitates the ascertainment of postneonatal diagnoses.

Another source of ascertainment variation is the ascertainment of less severe forms of congenital anomalies. For example, in this study differences were also noted in the prevalence of limb reduction defects, with a higher rate in Glasgow (Rankin *et al.,* 2005). This may reflect differential ascertainment of (or registration of) less severe cases (e.g. missing parts of fingers or toes).

Other examples of regional variation in prevalence cannot be readily explained as ascertainment variation or differences in registration practice and may reflect true differences in prevalence. Glasgow has a particularly high prevalence rate of cleft palate (6.9 per 10,000 births) (Womersley and Stone, 1987). The Glasgow and Northern registers experienced a higher rate of neural tube defects, gastroschisis and diaphragmatic hernia during 1991-99 than the other registers (Rankin *et al.,* 2005). Indeed, geographically, a north-south divide in the total prevalence rate of gastroschisis in the British Isles has been reported. Table 39 shows the average three-year prevalence of gastroschisis (2002-2004) for the ten congenital anomaly registers in England and Wales. Registers covering Wales, northern England and Glasgow had higher total prevalence rates than those covering southern England (Chief Medical Officer, 2005). The reasons for this geographical gradient are currently not known.

It is important to appreciate that regional differences in prevalence exist, as variation can easily occur due to ascertainment differences or variation in registration practices. The underlying prevalence of congenital anomalies in a particular locality must be understood and taken into account in the design of epidemiological studies that aim to assess the contribution of environmental exposures to congenital anomaly risk, otherwise incorrect interpretation of findings may result.

Table 39. Average three-year (2002-04) prevalence of gastroschisis for ten BINOCAR registers

Region	Prevalence rate (per 10,000 births)	95% confidence limits
Wales	6.2	4.6 –7.8
NorCAS	5.2	3.7 – 6.7
Glasgow	4.9	2.3 – 7.4
Merseyside & Cheshire	4.7	3.2 – 6.3
East Midlands & South Yorkshire (formerly Trent)	4.7	3.7 – 5.7
West Midlands	4.2	3.2 – 5.1
Wessex	3.3	2.0 – 4.5
South West	3.3	2.4 – 4.2
Oxford	2.4	0.3 – 4.5
North West Thames	1.6	1.0 – 2.2
All BINOCAR registers	**4.0**	**3.6 – 4.4**

(Source: BINOCAR registers, 2002-04.) Annual Report of the Chief Medical Officer, 2004. Crown Copyright.

3. Use of data from congenital anomaly registers in environment and health research in the British Isles

The effect of maternal exposure to environmental influences and congenital anomaly risk is a rapidly growing area of research in the British Isles. Within the last ten years, data from regional congenital anomaly registers and from NCAS have contributed to a number of studies. These include studies of congenital anomaly risk among populations living in close proximity to landfill sites (Dolk *et al.,* 1998b; Vrijheid *et al.,* 2000a, 2002b; Elliott *et al.,* 2001; Morris *et al.,* 2003; Boyle *et al.* 2004; Palmer *et al.,* 2005), and to incinerators (Cresswell *et al.,* 2003; Dummer *et al.,* 2003), as well as to possible contaminants present in water such as lead (Macdonell *et al.,* 2000) and fluoride (Lowry *et al.,* 2003), and in soil (Elizaguirre-Garcia *et al.,* 2000). The findings from such studies have ranged from no excess risk to small, elevated risks in particular congenital anomaly subtypes. In addition to these published studies, there are ongoing studies of the association between maternal exposure to ambient air pollutants (particulate matter, sulphur dioxide) and to chlorination by-products, and congenital anomaly risk.

The use of data from congenital anomaly registers has a number of advantages: good case ascertainment through active data collection methods within a geographically defined area, inclusion of cases of miscarriages and terminations of pregnancy following prenatal diagnosis, consistent registration procedures, and availability of local expertise. However, the limitation to the use of this data is the lack of detailed information on the exposure. Registers do not routinely collect this information, as undertaking analytic investigations of environmental exposures was not the reason the registers were originally set up. If further progress is to be made in the study of the effect of maternal environmental exposures on congenital anomaly risk, the collection of high quality exposure data is essential, to reduce exposure misclassification.

3.1. Cluster investigations

Congenital anomaly registers are valuable sources of information when clusters of particular types of anomalies are suspected. For example, reports of possible clusters of limb reduction defects in populations living near the sea, initiated studies using data from congenital anomaly registers to investigate whether there was any difference in the reported rates of these anomalies in babies born in coastal areas compared with those living inland. No differences in rates were found (Botting, 1994; Wright *et al.,* 1995). Media reports of alleged clusters of anophthalmia and microphthalmia in England and a possible link to exposure of mothers to a fungicide, led to the establishment of a register of all children born with these conditions in England during 1988-94. An investigation into the geographical variation and clustering of anophthalmia and microphthalmia found little evidence of localised environmental exposures causing clusters of these conditions (Dolk *et al.,* 1998a).

3.1.1. Geographical variation in congenital anomaly rates

The suggestion of a small increased risk of selected congenital anomalies in pregnancies of mothers living near environmental exposures such as landfill sites (Dolk *et al.,* 1998b) has highlighted questions around spatial variation in congenital anomaly rates. Using data from five congenital anomaly registers (Glasgow, Northern, Oxford, Wessex, and North West Thames), geographical variation and clustering were analysed at four geographical levels: register, hospital catchment area, ward, and census enumeration district (Dolk *et al.,* 2003). The analyses concerned cases of selected congenital anomaly among live births, fetal deaths from 20 weeks gestation, and terminations of pregnancy following prenatal diagnosis, identified by the five registers from 1991-1999. Only congenital anomalies that were well defined and recorded, and could plausibly be related to environmental exposures, were included. For non-chromosomal anomalies, there was evidence of clustering at hospital catchment area level but not below this. For chromosomal anomalies, there was clear geographical heterogeneity in rates between registers but not below this. The lack of evidence of generalised clustering of congenital anomalies below hospital catchment level in Britain suggests that the interpretation of results from studies of proximity to landfill sites is substantiated (Dolk *et al.*, 2003).

4. Conclusion

The importance of recording and monitoring the occurrence of congenital anomalies in the British Isles has long been recognised. Regional congenital anomaly registers have been established to monitor the prevalence of congenital anomalies and to carry out investigations into increasing trends and possible clusters. Within the British Isles, a large proportion of the population is not covered by a congenital anomaly register, resulting in underreporting of congenital anomalies within these areas, which is of major public health concern. The precise aetiology of most congenital anomalies is not fully understood, but is likely to be multifactorial involving both genetic and environmental factors. The incomplete understanding of the causes of many important anomalies means that the scope for primary prevention is limited at present. As the causes of many congenital anomalies are still unknown, continued monitoring of, and investigation into, their occurrence is a priority for public health. High quality congenital anomaly registers play an essential role in providing the means to investigate how environmental exposures and individual lifestyle factors influence the occurrence of congenital anomalies. Networks such as BINOCAR and EUROCAT have been established to provide a framework for such investigations and to enable collaborative work to facilitate aetiological research.

The link between environmental exposures and congenital anomaly risk is an emerging area of research within the British Isles. The effects of living in close proximity to waste disposal sites (landfill sites and incinerators), of exposure to ambient air pollutants, and to possible contaminants in drinking water, are increasingly attracting public concern. There are a number of potential limitations to current studies of the effect of environmental exposures on congenital anomaly risk, including misclassification of exposure, ascertainment bias, migration bias and the influence of occupational and industrial exposures. Further investigations of the risk from such exposures are needed. Such studies within the British Isles will have the advantage of using high quality, population-based data on congenital anomalies from local registers within geographically defined areas that are maintained by highly motivated and dedicated staff. However, causal pathways will only be determined if good quality data on exposure is ascertained, and this is the direction that future research should take.

Without regional registers that can identify all cases of congenital anomaly, including those occurring in termination of pregnancy following prenatal diagnosis, evaluations of developments in preventive strategies and prenatal diagnosis will not be amenable to formal assessment; planning health care provision for children with long-term disabilities will not be accurate and assessing the impact of environmental exposures will not be possible. The current insecure funding of the regional congenital anomaly registers within the British Isles is of grave concern.

Acknowledgements

My thanks to Dr. Elizabeth Draper and Dr. Martine Vrijheid for their helpful comments and to Nigel Physick from the Office for National Statistics. I would like to acknowledge the hard work and commitment of all the co-ordinators, staff and contributing clinicians who support the regional congenital anomaly registers. Dr. Rankin is Secretary to the BINOCAR Management Committee, and is funded by a UK Department of Health National Career Scientist Award.

References

Abramsky, L., Botting, B., Chapple, J. and Stone, D. (1999) Has advice on periconceptual folate supplementation reduced neural tube defects? *Lancet* **354**, 998-999.

BINOCAR. http://www.binocar.org.uk

Botting, B. (2000) Improving the completeness of Down syndrome notification, *Health Stat. Quarterly Summer*, 14-17.

Botting, B.J. (1994) Limb reduction defects and coastal areas, *Lancet* **343**, 1034.

Boyd, P.A., Armstrong, B., Dolk, H., Botting, B., Pattenden, S., Abramsky, L., Rankin, J., Vrijheid, M. and Wellesley, D. (2005) Congenital anomaly surveillance in England-ascertainment deficiencies in the national system, *Br. Med. J.* **330**, 27-31.

Boyle, E., Johnson, H., Kelly, A. and McDonnell, R. (2004) Congenital anomalies and proximity to landfill sites, *Ir. Med. J.* **97**, 16-18.

Chief Medical Officer (2005) *Gastroschisis: a growing concern.* Annual Report of the Chief Medical Officer 2004: On the State of Public Health. Department of Health.

Congenital anomaly register and information service (CARIS), Wales. (2001) *Annual Report.*

Cresswell, P.A., Scott, J.E.S., Pattenden, S. and Vrijheid, M. (2003) Risk of congenital anomalies near the Byker waste combustion plant, *J. Pub. Health Med.* **25**, 237-242.

Dastgiri, S., Stone, D.H., Le-Ha, C. and Gilmour, W.H. (2002) Prevalence and secular trend of congenital anomalies in Glasgow, UK, *Arch. Dis. Child. Fetal Neonat. Ed.* **86**, 257-263.

Dolk, H., Abramsky, L., Armstrong, B., Botting, B., Boyd, P., Dunn, C., Grundy, C., Jordan, H., Pattenden, S., Physick, N., Rankin, J., Stone, D., Vrijheid, M. and Wellesley, D. (2003) *A study of the geographical variation in overall rates of congenital abnormalities and the rates of specific abnormalities.* Report to the UK Department of Health online: www.eurocat.ulster.ac.uk/pubdata; last asscessed 21/04/2006.

Dolk, H., Busby, A., Armstrong, B. and Walls, P. (1998a) Geographical variation in anophthalmia and microphthalmia in England, 1988-94, *Br. Med. J.* **317**, 905-910.

Dolk, H., Vrijheid, M., Armstrong, B., Abramsky, L., Bianchi, F., Garne, E., Nelen, V., Robert, E., Scott, J.E.S., Stone, D. and Tenconi, R. (1998b) Risk of congenital anomalies near hazardous-waste landfill sites in Europe; the EUROHAZCON study, *Lancet* **352**, 423-427.

Dummer, T.J.B., Dickinson, H.O. and Parker, L. (2003) Adverse pregnancy outcomes around incinerators and crematoriums in Cumbria, north west England, 1956-93, *J. Epidemiol.Comm. Health* **57**, 456-461.

Elizaguirre-Garcia, D., Rodriguez-Andres, C. and Watt, G.C. (2000) Congenital anomalies in Glasgow between 1982-89 and chromium waste, *J. Public Health Med.* **22**, 54-58.

Elliott, P., Briggs, D., Morris, S., de Hoogh, C., Hurt, C., Jensen, T.K., Maitland, I., Richardson, S., Wakefield, J. and Jarup, L. (2001) Risk of adverse birth outcomes in populations living near landfill sites, *Br. Med. J.* **829**, 363-368.

EUROCAT, Onlien at [http://www.eurocat.ulst.ac.uk last], [last assessed 21/04/2006].

EUROCAT working group. (1991) *Eurocat Report 4,* Department of Epidemiology, Catholic University of Louvain, Brussels, Belgium.

International Clearing House for Birth Defects Monitoring Systems. ICDBM *Annual Report 2002,* Online at [http://www.icbd.org/programme.htm], [last assessed 21/04/2006].

Knox, E.G., Armstrong, E,H. and Lancashire, R. (1984) The quality of notification of congenital malformations, *J. Epidemiol. Comm. Health* **38**, 296-305.

Lenz, W. (1962) Thalidomide and congenital abnormalities, *Lancet* **6,** 45.

Lowry, R., Steen, N. and Rankin, J. (2003) Water fluoridation, stillbirths, and congenital abnormalities, *J. Epidemiol. Comm. Health.* **57**, 499-500.

Macdonnell, J.E., Campbell, H. and Stone, D.H. (2000) Lead levels in domestic water supplies and neural tube defects in Glasgow, *Arch. Dis. Child. Fetal. Neonat. Ed..* **82**, 50-53.

McBride, W.G. (1961) Thalidomide and congenital abnormalities, *Lancet* **2**, 1358.

Medical Research Council Vitamin Study Research Group. (1991) Prevention of neural tube defects: results of the Medical Research Council Vitamin Study, *Lancet* **338**, 131-137.

Misra, T., Dattani, N. and Majeed, A. (2005) Evaluation of the National Congenital Anomaly System in England and Wales, *Arch, Dis. Child. Fetal. Neonat. Ed.*. **90**, F368-F373.

Morris, S.E., Thomson, A.O., Jarup, L., de Hoogh, C., Briggs, D.J. and Elliott, P. (2003) No excess risk of adverse birth outcomes in populations living near special waste landfill sites in Scotland, *Scot. Med. J.* **48**, 105-107.

Murphy, M., Seagrott, V., Hey, K., O'Donnell, M., Godden, M., Jones, N. and Botting, B. (1996) Neural tube defects 1974-94 - down but not out, *Arch. Dis. Child.. Fetal. Neonat. Ed.* **75**, F133-F134.

Mutton, D.E., Alberman, E., Ide, R. and Bobrow, M. (1991) Results of first year (1989) of a national register of Down's syndrome in England and Wales, *Bri. Med. J.* **303**, 1295-1297.

Nichols, C.R., Dickinson, J.E. and Pemberton, P.J. (1997) Rising incidence of gastroschisis in teenage pregnancies, *J. Mater. Fetal Invest.* **6**, 225-229.

Office for National Statistics. (2001) *The National Congenital Anomaly System: a guide for data users and suppliers*. London: ONS.

Office for National Statistics. (2004) *The health of children and young people*. Chapter 11 Congenital Anomalies, pp.1-29.

Office for Population, Census and Surveys. (1995) *The OPCS Monitoring Scheme for Congenital Malformations*. A review by a Working Group of the Registrar General's Medical Advisory Committee. London: OPCS Occasional Paper 43.

Owens, J.R., Simpkin, J.M., McGuinness, L. and Harris, F. (1988) The Liverpool Congenital Malformations Registry, *Paedia. Perinat. Epidemiol.* **2**, 240-252.

Palmer, S.R., Dunstan, F.D.J., Fielder, H., Fone, D.L., Higgs, G. and Senior, M.L. (2005) Risk of congenital anomalies after opening of landfill sites, *Environ. Health Perspect.* **113**, 1362-1365.

Payne, J.N. (1992) Limitations of the OPCS congenital malformation notification system illustrated by examination of congenital malformations of the cardiovascular systems in districts within the Trent region, *Pub. Health* **106**, 437-448.

Rankin, J., Glinianaia, S.V., Brown, R. and Renwick, M. (2000) The changing prevalence of neural tube defects in the North of England, 1986-96, *Paedia. Perinat. Epidemiol.* **14**, 104-110.

Rankin, J., Pattenden, S., Abramsky, L., Boyd, P., Jordan, H., Stone, D., Vrijheid, M., Wellesley, D. and Dolk, H. (2005) Prevalence of congenital anomalies in five British regions, 1991-99, *Arch. Dis. Child. Fetal Neonat. Ed.* **90**, 374-379.

Rankin. J., Dillon, E. and Wright, C. (1999) Congenital abdominal wall defects in the northern region, 1986-95: occurrence and outcome, *Prenat. Diagno.* **19**, 662-668.

Richards I.D.G. and Lowe C.R. (1971) Incidence of congenital defects in South Wales, 1964-6, *Bri. J. Prevent. Soc. Med.* **25**, 59-64.

Smithells, R.W. (1962) The Liverpool Congenital Abnormalities Registry, *Dev. Med. Child Neurol.* **4**, 320-324.

Stone, D.H. (1986) A method for the validation of data in a register, *Pub. Health* **100**, 336-342.

Tan, K.H., Kilby, M.D., Whittle, M.J., Beattie, B.R., Booth, I.W. and Botting, B.J. (1996) Congenital anterior abdominal wall defects in England and Wales 1987-93: retrospective analysis of OPCS data, *Br. Med. J.* **313**, 903-906.

Torfs, C.P., Velie, E.M., Oechsil, F.W., Bateson, T.F. and Curry, C.J. (1994) A population-based study of gastroschisis: demographic, pregnancy and lifestyle factors, *Teratology* **50**, 44-53.

Vrijheid, M., Dolk, H., Armstrong, B., Abramsky, L., Bianchi, F., Fazarinc, I., Garne, E., Ide, R., Nelen, V., Robert, E., Scott, JES., Stone, D. and Tenconi, R. (2002a) Chromosomal congenital anomalies and residence near hazardous waste landfill sites, *Lancet* **359**, 320-322.

Vrijheid, M., Dolk, H., Armstrong, B., Boschi, G., Busby, A., Jorgensen, T., Pointer, P. and the EUROHAZCON collaborating group. (2002b) Hazard potential ranking of hazardous waste landfill sites and risk of congenital anomalies, *Occup. Environ. Med.* **59,** 768-776.

Womersley, J. and Stone, D.H. (1987) Epidemiology of facial clefts, *Arch. Dis. Child. Fetal. Neonat.* **62**, 717-720.

World Health Organisation. (1994) *International Statistical Classification of Diseases and Related Health Problems*. Tenth Revision, Geneva, Switzerland.

Wright, M., Newell, J.N., Charlton, M.E, Hey, E.N., Donaldson, L.J. and Burn, J. (1995) Limb reduction defects in the northern region of England, 1985-92, *J. Epidemiol. Comm. Health.* **49**, 305-308.

EUROPEAN UNION-FUNDED RESEARCH ON ENDOCRINE DISRUPTERS AND UNDERLYING POLICY

T. KARJALAINEN
European Commission
Research Directorate General –Unit E2 (Food quality)
Brussels
BELGIUM
Email: Tuomo.KARJALAINEN@cec.eu.int

Summary

As a result of public concerns about hormone-disrupting chemicals – such as publication of 'Silent Spring' – and calls from the European Parliament, the Commission took early, decisive actions to tackle this problem. A strategy was adopted in 1999, recognising that the final goal should be the protection of human health and the environment. A call for more research in the Strategy resulted via the publication of calls for proposals in the Fifth Framework of Research (1998-2002), in multi-centre, pan-European projects. They addressed a wide range of issues related to endocrine disruption ranging from declining sperm counts in humans to long-term effects of exposure to low doses of chemicals in laboratory animals. In addition, methods for detection and testing have been developed. This concentrated effort has produced results useful for regulatory purposes. Taking into consideration projects funded by the ongoing Sixth Framework Programme (2002-2006), the total expenditure on Commission-sponsored endocrine disrupter projects now exceeds €120 million. The purpose of this chapter is to provide to an overview of Commission-sponsored research projects dealing with endocrine disrupters and underlying policy decisions, sponsored mostly by the Key Action 'Environment and Health'.

P. Nicolopoulou-Stamati et al. (eds.), Congenital Diseases and the Environment, 379–406.

1. Introduction

1.1. Path to adoption of community strategy for endocrine disrupters

As a result of public concerns about hormone-disrupting chemicals – such as publication of 'Silent Spring' – and calls from the European Parliament in the 1990's, the Commission took various actions to tackle this problem. One of the first ones took place in 1996, in the form of a workshop in Weybridge, UK, bringing together 70 scientists and policy-makers from the EU, USA and Japan as well as from organisations such as OECD, WHO, European Science Foundation (ESF) and The European Chemical Industry Council (CEFIC) and non-governmental organisations. The workshop concluded, among other things, that certain phenomena such as the apparent decline in sperm counts in some countries is likely to be genuine and that exposure to endocrine disrupters should be dealt with by measures in line with the "precautionary principle".

The European Parliament adopted in 1998 a report and a resolution on the topic of endocrine disruption, calling upon the European Commission to take specific actions, in particular with a view to improving the legislative framework, to reinforcing research efforts and to making information available to the public.

In 1999, the European Commission's Scientific Committee for Toxicity, Ecotoxicity and the Environment (SCTEE) presented its opinion "Human and Wildlife Health Effects of Endocrine Disrupting Chemicals (EDCs), with Emphasis on Wildlife and on Ecotoxicology Test Methods", which concluded that "impaired reproduction and development causally linked to endocrine disrupting chemicals are well-documented in a number of wildlife species and have caused local and population changes". Public policy-makers were urgently requested to address this issue. It was proposed that the European Commission adopt a strategy with short-, medium- and long-term actions (see below), in order to respond quickly and effectively to the problem.

As a consequence, the European Commission adopted, the same year, a Communication to the Council and European Parliament on a Community Strategy for Endocrine Disrupters (European Commission, Environment Directorate-General, 1999). Recommendations were made for short-, medium- and long-term actions (see below).

In 2000, the Environment Council adopted conclusions on the Commission's Communication, in which it stressed the need to develop quick and effective risk management strategies and the need for consistency with the overall chemicals policy. The Council invited the Commission to report back on the progress of the work at regular intervals: thus the first implementation report was published in 2001 (European Commission, Environment Directorate-General, 2001) and, the second one in 2004 (European Commission, Environment Directorate-General, 2004).

In the same year, the European Parliament adopted a resolution on endocrine disrupters, emphasizing the application of the precautionary principle and calling on the Commission to identify substances for immediate action.

Since 2000, the recommendations outlined in the Communications, by the Environment Council, and the European Parliament, are being implemented through the short-, medium-, and long term strategy of the European Commission, managed by the Environment Directorate-General, with research efforts being supported by the Directorate-General for Research.

The views expressed in this publication may not necessarily reflect the official position of the European Commission. The author is solely responsible for its contents

1.2. Community strategy for endocrine disrupters

1.2.1. Short-term strategy

One of the first key short-term actions identified in the European Commission's Communication (European Commission, 1999) was the establishment of a priority list of substances for further evaluation of their role in endocrine disruption. The priority list has been established in two steps, first an independent review of evidence of endocrine-disrupting effects and human/wildlife exposure, and second a priority-setting exercise in consultation with stakeholders and the Commission's Scientific Committees. The establishment of this list is managed by the Environment Directorate General of the European Commission.

Short-term action also encompassed the need for communication to the public and international co-operation. Further information can be retrieved from "Information Exchange and International Co-ordination on Endocrine Disrupters" (Institute for Environment and Health, 2003). As to the communication to the public, the increased activities in the field of research, sponsored by the European Commission's Research DG, have considerably increased the visibility of the phenomenon of endocrine disruption, as project results have been made available to the public through leaflets, press releases, websites, and workshops. The European

Commission has also created two websites: one policy related (European Commission, Environment Directorate-General, 2003) and one focused on research activities.

As far as international co-operation is concerned, the European Commission and the WHO have co-operated in the past, through the International Programme for Chemical Safety, on the maintenance of a global research inventory, which is housed at the Commission's Joint Research Centre, and on the compilation of a "Global state-of-the-science of endocrine disrupters" report, published in 2002 (The International Programme on Chemical Safety, 2002). In December 2003, the Ministry of Environment of Japan and the WHO/UNEP/ILO International Programme on Chemical Safety hosted in Tokyo a joint workshop on "Endocrine Disrupters: Research Needs and Future Directions" (United Nations Environment Programme *et al.,* 2004). It followed an International Symposium on Environmental Endocrine Disrupters held in Sendai, Japan. A workshop to improve international collaboration was also held in Brussels in 2005 (European Commission, Research Directorate-General, 2005).

Both the European Commission and the WHO are supporting the efforts of the OECD (Organisation for Economic Co-operation and Development, 2005) and ECVAM (European Commission, Joint Research Centre Directorate-General, 2005) to develop agreed test methods for endocrine disrupters (EDs). A workshop to transfer research results to these organisations was held in 2005.

1.2.2. Medium-term strategy

Medium-term action includes:

(i) Identification and assessment of EDCs. The availability of agreed test strategies/methods to identify and assess endocrine-disrupting chemicals is a basic requirement for comprehensive legislative action aimed at protecting people and the environment from the potential dangers posed by these chemicals,

(ii) Research and development. A need for further research has been identified on test methods and testing strategy; effects on humans/wildlife; ED mechanism of action, effects at key stages of life cycle; models to estimate exposure; development of monitoring tools). As a result, the Research DG of the European Commission has substantially increased spending on research projects linked to endocrine disruption.

1.2.3. Long-term strategy

The Community Strategy for Endocrine Disrupters includes as a long-term action the review and adaptation of relevant legislation. A recent assessment showed that a considerable number of endocrine disrupters were already subject to bans or restrictions or were being addressed under existing Community legislation.

Major steps forward have been made in integration of the endpoint 'endocrine disruption' in existing or upcoming legislation over the last few years. In the proposal for a new chemicals policy (REACH) (European Commission, Enterprise and Industry Directorate-General, 2003), for instance, endocrine disrupters are covered by the authorisation procedure for substances of very high concern. Substances of very high concern are defined as: substances that are category 1 and 2 carcinogens or mutagens; substances that are toxic to the reproductive system of category 1 and 2; substances that are persistent, bioaccumulative and toxic or very persistent and very bioaccumulative; and substances such as endocrine disrupters, which are demonstrated to be of equivalent concern. However, many of the endocrine disrupter candidate substances, such as pesticides, by-products, and drug ingredients, are not within the scope of REACH, but are covered by other legislative approaches (e.g. for drug ingredients via the European Agency for the Evaluation of Medicinal Products (EMEA)-registration based approach).

The current chemicals policy addresses the issue of endocrine disruption under risk assessment strategy for dangerous substances and preparations and includes various exposure scenarios, including those to endocrine disrupters. The water framework directive, the dangerous preparation and the limitations directive, the directive on general product safety and the directive prohibiting the use of substances having a hormonal action for growth promotion in farm animals, also address endocrine disrupters.

Although there is lack of agreed testing methods, endocrine disruption is already being considered in the discussions on data requirements and principles for risk assessment of plant protection products and in the evaluation scheme for biocidal products in the framework of the OECD test guidelines programme on endocrine disrupters (Organisation for Economic Co-operation and Development, 2005). Further efforts are currently being carried out on test method validation. This task is complicated as non necessary animal testing should be avoided and testing has to be based on firm science. The European Centre for Alternative Methods (ECVAM), as part of JRC (European Commission, Joint Research Centre Directorate-General,

2005), is strongly involved in establishing this scientific basis and making available validated tests and test strategies.

The main aim of the Environment and Health Strategy (European Commission, Environment Directorate-General, 2003), launched by the European Commission in June 2003, is to achieve a better understanding of the complex relationship between environment and health and to identify and reduce diseases caused by environmental factors. This so-called SCALE initiative focuses on five elements: Science, children, awareness, legislation, and evaluation. The strategy will be implemented in cycles. One focus of the first cycle, 2004 - 2010, will be on endocrine disrupting effects of chemicals. Extensive consultations with experts and stakeholders from the environment, health and research fields in all parts of the enlarged Europe have taken place and an Action Plan has been published, in which a number of priority research actions are described. It is published on the DG Environment website dedicated to this strategy (European Commission, Environment Directorate-General, 2003).

2. European Commission-sponsored research on endocrine disrupters

2.1. Research activities in the fourth research framework programme (1994-1998)

Even prior to increased policy action focused on endocrine disruption, the European Commission had launched actions in the Fourth Research Framework Programme to tackle the problem of endocrine disrupters, since the topic of endocrine disruption had emerged as a research priority in response to rising public concerns. Around 12 million euros were spent on projects dealing with endocrine disruption. Since most effects of endocrine-disrupting chemicals had at that time been observed in the environment and, in particular, in the aquatic world, many of the projects focused on fish populations, to understand the mechanisms involved, to develop test methods and to identify potential endocrine disrupters (European Commission, Research Directorate-General, 2003).

2.2. Research activities in the fifth framework programme of research (FP5: 1998-2002)

As a direct response to the Community Strategy on Endocrine Disrupters described above, research activities in this field were greatly enhanced. These activities coincided with the introduction of FP5. Projects dealing with endocrine disruption were covered mainly in the following thematic programmes:

(i) The Quality of life and management of living resources thematic programme, which spent over €40 million. The human health issues associated with ED are mainly funded through Key action 4: Environment and Health 15,

(ii) The Energy, environment and sustainable development programme (EESD), which has financed 8 projects with a total budget of around €16 million touching on ED via the key actions 'Sustainable management and quality of water' and 'Sustainable marine ecosystems' (European Commission, Research Directorate-General, 1999).

2.2.1. Key-action 4: environment and health

Close to 30% of the Quality of Life Programme budget was devoted to projects dealing with endocrine disruption. Other areas covered included indoor and outdoor air pollution, electromagnetic fields, noise, UV light, allergy and asthma, among other things.

There has been an increase in funding and number of projects since the first call, published in 1999. The importance of the topic is highlighted by the publication of a special call solely dedicated to endocrine disrupter research in 2001. The evaluation of incoming proposals resulted in the creation of the CREDO cluster (cluster of research into endocrine disruption in Europe) (The cluster for Research into Endocrine Disruption in Europe, 2003). This cluster includes four research projects with over 60 laboratories across Europe participating until 2006, or even beyond. The 18 projects funded by the Quality of Life programme are presented on a dedicated website (European Commission, Research Directorate-General, 2003).

A large majority of the 143 institutions involved in Commission-sponsored endocrine disrupter projects come from universities or research institutes, although there are some industrial participants. Countries that are most active in endocrine disrupter research sponsored by the Quality of Life Programme are Germany, Sweden, the Netherlands, the United Kingdom, Italy, and Finland. The average budget per project is €2.2M with close to 8 laboratories involved. Around 10% of participants are from outside the EU, be it from associated countries to the framework programme or overseas participants.

Table 40. Selected issues dealt with in Key action 4-sponsored projects dealing with endocrine disrupters

Total number of Key Action 4 research projects dealing with endocrine disruption	18
Of these, number of projects	
▪ Dealing with human health effects	14 (78%)
▪ Dealing with non-human effects	14 (78%)
▪ Dealing with reproductive effects	13 (72%)
▪ Dealing with neuroendocrine effects	10 (56%)
▪ Dealing with multi-organ effects	10 (56%)
▪ Looking at exposures at low doses	14 (78%)
▪ Looking at exposures to chemical mixtures	7 (39%)
▪ Looking at long term effects of exposures (low doses; sensitive life stages)	13 (72%)
▪ Using birth cohorts or other epidemiological approaches	10 (56%)
▪ Developing new animal models	7 (39%)
▪ Developing new or improved *in vitro* assays or biosensors	14 (78%)

Table 40 describes the scope of issues dealt with in FP5 on projects dealing with endocrine disrupters. The multidisciplinary nature of the projects is evident from the table. Most projects deal both with human health effects and/or study human relevant animal models. New animal models being developed are usually transgenic animals. In addition, many of the projects are developing new or improved validated test methods for endocrine-disrupting compounds, ranging from cell assays to biosensors.

'Hot' topics being investigated increasingly in the 5th FP projects, as compared to those in the 4th FP, are health outcomes of exposures to low doses and chemical mixtures over longer periods and at sensitive life stages. Over half of the projects have access to sometimes very large cohorts of human populations representing various exposure scenarios (high or low) and these have in some cases been followed over many years. An overview of topics dealt with is given in the table 41, for some of the projects.

In general, a multitude of issues related to endocrine disruption and risk assessment are covered by the FP5 projects. Human health effects are being investigated through epidemiological studies or in human-relevant animal models including transgenic animals, and some projects are undertaking cross-species comparisons to find out whether observations in one species can be relevant in another. An ever increasing list of tissues or organ systems influenced by endocrine disruption under scrutiny ranging from the popular reproductive tissues to the less studied effects, e.g. those on the immune system. A large array of novel biomarkers is under development, usable in epidemiological investigations of population cohorts. Increasingly sophisticated technologies employing approaches using 'omics' (transcriptomics, proteomics. etc.), biosensors and cellular bioassays are being used to investigate the exact cellular structures and targets being affected, and the role of individual genetic susceptibility has not been neglected.

Table 41. Endocrine disrupter research projects: Compounds studied in selected ongoing or finished FP5 projects

Project[a]	Chemicals/chemical contaminants studied
Anemone	PCB congeners, hydroxylated PCBs, polybrominated diphenyl ethers, hexabromocyclododecane, methylmercury
Bonetox	2,3,7,8-tetrachlorodibenzo-p-dioxin, dioxin-like PCBs (PCB126), non-dioxin-like PCB congeners, polybrominated diphenylethers
Compare	PCB mixture Aroclor 1254, PCB metabolites: 4-OH-CB107, 4-OH-CB187, polybrominated phenols 2,4,6-tribromophenol, tetrabromobisphenol A, polybrominated diphenyl ethers BDE47, 6-OH-BDE47
Easyring	17-beta-ethinylestradiol, tamoxifen, methyldihydrotestosterone, flutamide, oestrone (E1), oestriol (E3), nonylphenols (NPs), teroctylphenol (OP), bisphenol A (BPA).
Eden	Phthalates (dibutyl [DBP]-, diethyl hexyl – [DEHP], benzylbutyl – [BBP], monoester -); flutamide, dihydrotestosterone (DHT), finasteride; organochlorine pesticides: o,p-DDT, p,p'-DDT, o,p-DDD, p,p'-DDE, metoxychlor, mirex, lindane, aldrin, endrin, dieldrin, endosulfan I, endosulfan II, endosulfan ether, endosulfan lactone, endosulfan diol, endosulfan sulphate, hexaclorobenzene (HCB), gamma-hexachlorocyclohexane (gamma-HCH), inclozolin, depone, procymidon, fenthrothion, cemetidine, linuron, prochloraz; steroidal oestrogens (E2, EE2, DES, oestrone); phytoestrogens; bisphenol A and its halogenated derivatives; alkylphenols 2',5'-dichloro-4-hydroxybiphenyl 2',4',6'-trichloro-4-hydroxybiphenyl; parabenes (n-propyl, n-butyl, methyl-); benzophenone; polychlorinated dioxins and furans (PCDD/F); polychlorinated biphenyls (PCB) and their hydroxylated metabolites ; polybrominated biphenyls(PBB), polybrominated diphenyl ethers (PBDE); cadmium, lead

Table 41. (Continued)

Project[a]	Chemicals/chemical contaminants studied
Dioxin risk assessment	2,3,7,8-tetrachlorodibenzo-p-dioxin
Edera	17 beta-oestradiol, bisphenol A, genistein, cadmium, DDT
Endisrupt	Antiandrogens (flutamide, finasteride)
Endomet	Dibutyl, dioctyl, diisononyl, diisodecyl, butylbenzyl phthalates, bis(2-ethylhexyl) phthalate, bis(2-ethylhexyl)adipate, resorcinol, 2,4-dichlorophenol, 4-chloro-3-methylphenol, bisphenol-A, bisphenol-A dimethacrylate, 2-phenylphenol, 4-tert-octyl-, 4-n-octyl-, 4-n-nonyl-phenols
Estrogens and disease	Oestradiol dipropionate, oestradiol and diethylstilboestrol (DES), bisphenol-A, phytoestrogens: isoflavone, genistein
Expored	Halogenated hydrocarbons, polybrominated biphenyls, selected pesticides (Aldrin, cis-, trans-, oxy-chlordane, DDT and metabolites, HCH, alpha-, beta-endosulfan, heptachlor-endo, exo-epoxide, methoxychlor, heptachlor, heptachloroepoxide, mirex), phthalate monoesters (MEHP, MEP, MBP, MNIP, MMP, MBzP), bisphenol A, PCDD/Fs, PCBs, and PBDEs
Fire	Organohalogen compounds (OHCs): 19 polybrominated diphenyl ether (PBDE) congeners, tetrabromobisphenol A (TBBPA), 2,4,6-tribromophenol, 3 hexabromocyclododecanes (HBCDs)
Gendisrupt	17-β oestradiol, mono(2-ethylhexyl) phthalate or MEHP, zeralenone; bisphenol A, γ-hexachlorocyclohexane (γ-HCH or lindane)
Inuendo	2,2’,4,4’,5,5’ hexachlorobiphenyl (CB-153) and p,p' dichlorodiphenyldichloroethylene (p,p’-DDE)
Mendos	Atrazin, alachlor, styrene, resorcinol, bisphenol A, linuron, vinclozolin, maneb, metam, zineb, thiram, acetochlor, DBP, BBP, DEHP, butyl benzyl phthalate, fentin acetate, benzo[a]pyrene, nonylphenol (mixed isomers), 3,4-dichloranilin, 17ß-estradiol
Comprendo	Organotin compounds (mono-, di-, tributyltin, triphenyltin), linuron, diuron, p,p’-DDE, fenarimol, vinclozolin; Positive controls: methyltestosterone, mibolerone, letrozole, CPA, flutamide, emate, clomiphene, finasteride, prochloraz
Pbde-ntox	PBDE 99, Aroclor 1254, PBDE 153, PCB 52

Table 41. (Continued)

Project[a]	Chemicals/chemical contaminants studied
Pcbrisk	PCBs: 28, 52, 101, 138, 153, 170, 180; non dioxin-like PCBs: 18, 22, 33, 44, 47, 49, 66, 70, 74, 87, 92, 95, 96, 99, 107, 110, 117, 128, 132, 133, 136, 137, 146, 149, 151, 171, 172, 176, 177, 178, 183, 187, 194, 195, 196, 200, 202, 206, 209 dioxin-like PCBs: 77, 81, 105, 114, 118, 123, 126, 156, 157, 167, 169, 189; PCDD/Fs: 2378-substituted congeners (17); organochlorine pesticides: HCB, alpha-, beta-, gamma-HCHs, pp'DDT, pp'-DDE; trace elements in plasma, blood and urine; PCB congeners, hydroxylated PCBs, DDT metabolites, 17beta-oestradiol; CB-153; OH-PCBs (4-OH-CB107, 4-OH-CB146, 4-OH-CB187); $MeSO_2$-PCB (4'-$MeSO_2$-CB101, 4'-$MeSO_2$-CB87, 4-$MeSO_2$-CB149), 3-$MeSO_2$-DDE
Eurisked	Octyl-methoxycinnamate (OMC), nonylphenol, 4-methylbenzilidene camphor (4-MBC), bisphenol A (BPA), dibutylphtalate (DBP), benzophenone-2 (BP2), procymidon, linuron, oestradiol-benzoate, androstandiol , 8-prenylnaringenin (8PN), resveratrol, genistein

[a]*More information on projects can be found at http://europa.eu.int/comm/research/endocrine/index_en.html.*

A list of compounds under investigation in the projects is given in table 42. As can be seen from the table, the variety of compounds having been investigated or being investigated in Quality of Life-sponsored research projects is vast and covers all groups of industrial chemicals in addition to natural and synthetic hormones. The results being generated on these compounds should have regulatory relevance as our knowledge of these compounds will increase on several fronts, including exposure and effect assessments and mechanisms of disease.

Table 42. Endocrine disrupter research projects: Assays being developed, animals (incl. humans) studied, and endpoints explored in selected ongoing or finished projects

Project[a]	Animals/humans studied	(bio)assays developed/ used for EDCs	Endpoints
Anemone	Humans (Faroese birth cohort of mother-child pairs); Rat; Fulmar	Serum oestrogenicity	Growth Development; Neurobehavioural parameters; Lymphocyte marcher binding; Thrombocyte MAO-B activity

Table 42. (Continued)

Project[a]	Animals/humans studied	(bio)assays developed/ used for EDCs	Endpoints
Bonetox	Humans (cohort of environmentally and occupationally dioxin-exposed population) Knock-out mice: RALDH, Aryl hydrocarbon receptor (ahr), Epidermal growth factor receptor (EGFR), ERKO, BERKO, Cellular retinol-binding protein (CRBP1)-KO, Retinol binding protein (RBP), CRBP1/RBP, Long Evans and Han/Wistar rat	Analytical method of measurement of chlorinated biphenyls in human samples developed	Development (bone, tooth, kidney, lungs, salivary glands)
Anemone	Humans (Faroese birth cohort of mother-child pairs), Rat Fulmar	Serum oestrogenicity	Growth Development Neurobehavioural parameters Lymphocyte machr binding Thrombocyte MAO-B activity
Edera	ERE-luciferase mouse	In the ERE-Luc mouse the luciferase is expressed ubiquitously in various tissues in the presence of estrogen receptors activated by endocrine disrupters. The luciferase expression is being studied by: (i) immunocytochemistry in tissue slices, (ii) enzymatic assay in tissue extracts; (iii) *in vivo* imaging using a CCD camera	Analysis of the tissue specificity of action of endocrine disrupters

Table 42. (Continued)

Project[a]	Animals/humans studied	(bio)assays developed/ used for EDCs	Endpoints
Eden	Humans (breast cancer patients, cryptorchid boys), Rat, Transgenic zebrafish, Stickleback, Bream	Stickleback assay for oestrogenic and androgenic compounds; transgenic reporter fish (GFP, oestrogenic exposure) MCF-7 E-screen, MCF-7 A-screen, MCF7 DNA array, YES assay	Development Reproduction Neuroendocrine effects
Easyring	Mammalian cells Mouse oocytes and embryos, Fish: carp (Cyprinus carpio as sentinel species), barbel (Barbus plebejus), chub (Leuciscus cephalus), zebrafish (Danio rerio), other fish in the environment: Aspius aspius, Chondrostoma genei, Chondrostoma soetta, Alburnus alburnus alborella, Gobio gobio, Rutilus erythrophthalmus), Silurus glanis, Stizostedion lucioperca. Amphibians: Rana esculenta and Xenopus laevis Mammals: NMRI and FVB mice	Three *in vitro* consolidated tests (amphibian/fish hepatocyte cultures, recombinant yeast assay and breast cancer cells [MVLN]) for evaluation of ED potentiality of water and sediments Analytical methods in GC-MS, GC-MS-MS, LC-MS and LC-MS-MS for the detection and quantification of chemicals and their conjugates in waters and biota To be developed and/or validated: ELISA and dip-stick for the non-invasive detection of known and new biomarkers in the skin of aquatic species. Improvement: sensitive and reliable assays to measure the effects of endocrine disrupters on female gametes, cell cycle control, proliferation and cytodifferentiation in mammals	Mammalian, amphibian and fish development Fish reproduction (sex ratio, morphology of gonads and steroidogenesis) Liver function (histology and enzymatic activity)

Table 42. (Continued)

Project[a]	Animals/humans studied	(bio)assays developed/ used for EDCs	Endpoints
Envir.reprod. health	Human Rat Mouse	Reporter cell assays (AR, ER) Protocol for separate quantification of endogenous and exogenous oestrogenicity in human tissues	Reproduction incl. congenial malformations, endogenous reproductive hormones, sperm quality
Estrogens and disease	C57Bl/6J mouse CBA mouse erβ-KO –mouse C57 black BERKO mouse AROM+ and BERKO/AROM+ mouse	ERE-luc mouse ERE-luc mouse cell line (INS7) U2OS osteosarcoma cell lines, stably transfected with erα/erβ HEC1A and B endometrial cell lines, stably transfected with erα/erβ	Effects of developmental exposure on postnatal brain, HPA axis, mammary gland, prostate, testis, ovaries
Endisrupt	Rat		Reproduction
Expored	Humans (breast milk, placenta)	Phthalate monoester measurements in breast milk and placenta	Cryptorchidism, hypospadias, endogenous hormone levels
Fire	Humans WU Rat Tern Harbour seal Polar bear Flounder Transgenic zebrafish	To be developed: CALUX assays for progesterone and androgen receptor-mediated induction (AR-CALUX, PR-CALUX) Used: TTR binding, DR-CALUX, ER-CALUX, E2-SULT, T-screen, H295R, carphep, ligand trap assays, transient assays	Development Reproduction Immune system Neurobehaviour

Table 42. (Continued)

Project[a]	Animals/humans studied	(bio)assays developed/ used for EDCs	Endpoints
Gendisrupt	Humans Mouse (Mus musculus) Primordial germ cells in culture Primary cultures of Sertoli cells	Gene expression Single Nucleotide Polymorphisms (snps) of target genes and linkage analysis of qtls (Quantitative Trait Loci) Gene expression deregulation and growth in Primordial Germ Cells (pgcs) Expression of ER-β and ERR receptors in gonadal somatic cells; Histo-pathological analysis in testis development of animals exposed to endocrine disrupting compounds. Development and validation of specific DNA microarrays	Mammalian testis development Reproduction
Mendos	N.A	N.A	N.A
Pbde-ntox	Strains of rats and mice	Bioassays *in vitro* and *ex vivo*: several assays, including measurement in primary cultures of neurons as well as neural and glial cell lines to study signal transduction processes, apoptosis and necrotic cell death; cholinergic receptors in different brain areas; steroid receptor expression in different brain areas and reproductive organs; electrophysiological examination of long-term potentiation in different brain areas; determination of circulating steroids. *In vivo*: various neurobehavioral tests, some of which steroid-dependent	Neurobehaviour and other neuroendocrine endpoints Development Reproductive parameters (puberty onset)

Table 42. (Continued)

Project[a]	Animals/humans studied	(bio)assays developed/ used for EDCs	Endpoints
PCB risk	Humans (PCB-exposed cohort in Eastern Slovakia), Lymphocytes	Real time quantification of CYP1A1, CYP1B1expression and polymorphism CYP1B1 Improvement of an HRGC-HRMS method for determination of PCDD/Fs, DL-PCBs and NDL-PCBs Improvement of an SPE method for the isolation of persistent organochlorines from blood serum samples	Human exposure (Neurobehavioural parameters, biomarkers)
Comprendo	Humans, Human cell lines, Rattus norvegicus (strain Wistar). (Mammalia), Gallus gallus domesticus (Aves), Xenopus laevis (Amphibia), Pimephales promelas, Rutilus rutilus (Pisces), Antedon mediterranea, Paracentrotus lividus (Echinodermata), Acartia tonsa, Hyallela azteca (Crustacea), Marisa cornuarietis, Potamopyrgus antipodarum, Nassarius reticulatus (Mollusca)	MCF-7, SK-BR-3, lncap-FGC, PC-3, Jeg-3, HEP-G2, VTG	Mortality, growth, weight, sex ratio, gonad status, sex gland size, size copulatory organs, other secondary sex characteristics, malformation of genital organs, histopathology, virilisation, fecundity (clutch size, number of produced eggs), spermato- and oogenesis, sperm quality, sperm motility, fertilisation success, percentage of sexually active animals, hatching success, larval development, regeneration, apoptosis, cell cycle kinetics, steroid concentrations, phase I + II metabolism, VTG

Table 42. (Continued)

Project[a]	Animals/humans studied	(bio)assays developed/ used for EDCs	Endpoints
Endomet	Human and rat cell lines (especially neuronal, gut, breast, thyroid, skin). Human and porcine primary cell cultures, human platelets. Wistar rats	Human and rat cell lines stably over-expressing oestrogen receptors (ERE-luc), androgen and arylhydrocarbon receptors, NIS transporter in thyroid cells Sulphotransferase (SULT 1A1, 1A2, 2A1, 1E1) assays, cysteine dioxygenase and sulphite oxidase expression (sulphate production), cell signalling assays, basal and hormonally induced production of oestrogen and progesterone, reproductive toxicology of mixtures of plasticizers	Reproduction Steroid synthesis Neurological function Steroid/thyroid function
Inuendo	Humans - pregnant couples from Greenland, Sweden, Poland and Ukraine.	Chemical determination by GC-MS ER, AR and AHR – assays	Time to pregnancy Semen quality
Eurisked	Human (HOSE cells, numerous immortal cancer cell lines) Rats Mice (Knock-out ers, TR, ahr and reporter gene mice) Primary cells from various organs	Reporter cell assay (AR, ER) and mice Transactivation assay, TPO activity assay, Microarray chip (steroid relevant genes), Perfusion of hypothalamic fragments RIA, Bone mineral density (BMD) determination (by CT), Urinary incontinence model	Function of various organs within and outside the reproductive tract (20 organs) Development

[a]*More information on the projects can be found at http://europa.eu.int/comm/research/endocrine/index_en.html.*

As seen from table 42, a number of animal species are being investigated ranging from non vertebrates to humans being investigated by epidemiological studies. Several projects are taking advantage of transgenic animals deficient for various hormonal receptors, to further our knowledge on mechanisms of disease development. Receptor and other assays are used; some have already been developed previously but others constitute improvements of existing assays or even novel assays.

Most endpoints under investigation are focused on reproductive, developmental or neuroendocrine parameters, but some projects have widened their scope to a larger array of endpoints.

2.2.2. Projects funded by the Energy, Environment and Sustainable Development (EESD) thematic programme

The Energy, environment and sustainable development programme has financed 8 projects with a total budget of around 16 million euros touching on endocrine disrupters via the key actions 'Sustainable management and quality of water' and 'Sustainable marine ecosystems'. The focus of these projects is/was on environmental aspects of endocrine disruption. Two projects belonging to the CREDO cluster (COMPRENDO and EURISKED), mentioned in earlier sections of the catalogue, are funded by the EESD programme.

Progress or final reports of the eight projects can be found at http://europa.eu.int/comm/research/endocrine/index_en.html.

2.3. ***Research activities in the sixth framework programme of research (FP6: 2002-2006)***

Endocrine disruption has not been forgotten in FP6, the topic being specifically addressed by **Priority 5** (Food quality and Safety - www.cordis.lu/food) and, to much lesser extent, by **Priority 6** (Sustainable development, global change and ecosystems - www.cordis.lu/fp6/sustdev.htm). **Priority 1** (Life Sciences, genomics, and biotechnology for health - www.cordis.lu/lifescihealth/home.html) has sponsored some toxicology-related projects and projects focused on *in vitro* replacement of animal testing.

2.3.1. Outcome of first, second, and third calls for proposals

As in table 43, some projects (funded especially by Priority 1) are aiming to develop methods to replace animal testing for chemicals, and the ReProTect project in particular is relevant to both regulators and scientists working in the endocrine disrupter field.

Table 43. Projects sponsored under the 6th framework programme

Acronym of project	Title	Aims	More information[a]
Athon 2006-2010	Assessing the toxicity and hazard of non-dioxin-like PCBs present in food	The major objectives: (i) establish experimental *in vivo* and *in vitro* models for studies of ndl-pcbs; (ii) provide toxicokinetic data; (iii) provide toxicity profiles; (iv) provide a new classification strategy forndl-PCB congeners based on effect biomarker information; (v) provide an up-to-date compilation and evaluation of toxicological effect and exposure data on ndl-PCBs and PCB metabolites	N.A.
Biocop (2005-2010)	New technologies to screen multiple chemical contaminants in foods	Focused on chemical contaminant (pesticides, environmental contaminants including heavy metals, natural toxins, therapeutic drugs and endocrine disrupters [phytoestrogens]) monitoring in cereals, meats, seafood and processed foods using a range of new technologies such as transcriptomics, proteomics and biosensors	www.biocop.org
Caesar 2006-2009	Computer assisted evaluation of substances according to regulations	Produce (q)sar models for the prediction of the toxicity of chemical substances. Designed to be used for regulatory purposes, especially for REACH.	N.A.

Table 43. (Continued)

Acronym of project	Title	Aims	More information[a]
Cascade (2004-2009)	Chemicals as contaminants in the food chain: known for research, risk assessment and education	Network of excellence, to structure European research in the field of health risks of exposure to chemicals via food. Focus on chemical residues that act via and/or interfere with cellular regulation at the level of nuclear receptors	www.cascadenet.org
Crescendo (2005-2010)	Consortium for research into nuclear receptors in development and aging	Nuclear receptors: signalling dynamics, target gene responses during development and ageing	N.A.
Devnertox (2003-2006)	Toxic threats to the developing nervous system: *in vivo* and *in vitro* studies on the effects of mixtures of neurotoxic substances potentially contaminating food	Development of experimental models to improve predictive toxicity testing and mechanism-based risk assessment for neurotoxic food contaminants. Effects of mixtures of persistent pollutants (PCBs, methylmercury) on the developing nervous system	www.imm.ki.se/devnertox
Esbio (2005-2007)	Development of a coherent approach to human biomonitoring in Europe	Esbio aims at developing a coordinated approach for human biomonitoring with focus on children in Europe, in line with action 3 of the EU environment and health action plan by providing strong scientific support for the pilot project to be launched end 2006	www.eu-biomonitoring.org

Table 43. (Continued)

Acronym of project	Title	Aims	More information[a]
F&f (2005-2008)	Pharmaceutical products in the environment: development and employment of novel methods for assessing their origin, fate and effects on human fecundity	Develop validated methods for screening and testing of pps (mainly fecundity hormones) and their metabolites, to determine their adverse effects, origin and fate as well as their mechanisms of action; assess risks and propose risk management	http://foodandfecundity.factlink.net/
Newgeneris (2006-2010)	Newborns and genotoxic exposure risks: development and application of biomarkers of dietary exposure to genotoxic and immunotoxic chemicals and of biomarkers of early effects, using mother-child birth cohorts and biobanks	Study whether and to what extent parental exposure to dietary genotoxicants and immunotoxicants induces molecular events in the fetus which may induce cancer and immune disorders in later childhood. Some endocrine related mechanisms are covered.	N.A.
Nhr devtox (2004-2006)	A prospective analysis of the mechanisms of nuclear hormone receptors and their potential as tools for the assessment of developmental toxicity	N.A.	N.A.

Table 43. (Continued)

Acronym of project	Title	Aims	More information[a]
Nomiracle (2004-2009)	Novel methods for integrated risk assessment of cumulative stressors in Europe	Develop new methods for assessing the cumulative risks from combined exposures (including mixtures of chemicals); integrate risk analysis of environmental and human health effects; identify biomarkers of exposure and effects	www.bio.vu.nl/thb/research/proposal/nomiracle.pdf
Norman (2005-2007)	Network of reference laboratories and related organisations for monitoring and bio-monitoring of emerging environmental pollutants	N.A.	www.ineris.fr/index.php?module=cms&action=getcontent&id_heading_object=1046
Phime 2006-2011	Public health impact of long-term, low-level mixed element exposure in susceptible population strata	Assess health impacts of metals, sources, benefits and toxicity. Neuroendocrine parameters will be examined in relation to exposure to methylmercury. POPs will be analysed in cohorts.	N.A.
Pioneer (2005-2008)	Puberty onset – influence of nutritional, environmental and endogenous regulators	Obtain updated data on the age of puberty onset in Europe; identify genetic and nutritional factors involved in the regulation of the onset of puberty; develop novel experimental test models, optimised for the investigation of genetic and nutritional factors regulating onset of puberty	N.A.

Table 43. (Continued)

Acronym of project	**Title**	**Aims**	**More information**[a]
A-cute-tox (2004-2009)	Optimization and pre-validation of an *in vitro* test strategy for predicting human acute toxicity	The aim is to develop a simple and robust *in vitro* testing strategy for prediction of human acute systemic toxicity, which could replace the animal acute toxicity tests used today for regulatory purposes.	N.A.
Reprotect (2004-2009)	Development of a novel approach in hazard and risk assessment for reproductive toxicity by a combination and application of *in vitro*, tissue and sensor technologies	Development of novel *in vitro* and in silico alternatives for reproductive toxicity testing. Validation of a conceptual framework in the area of reproductive toxicity	N.A.
Safefoods (2004-2009)	Promoting food safety through a new integrated risk analysis approach for foods	Improvement of risk assessment methods and risk analysis practices for foods produced by different production practices (high or low input systems) and with different breeding technologies (traditional, molecular, genetic modification). One task: quantitative risk assessment of combined exposure to food contaminants and natural toxins	www.safefoods.nl/default.aspx
Testmetedeco (2006-2008)	Development of test methods for the detection and characterization of endocrine disrupting chemicals in environmental species	Development and validation of test methods intended to be used in Europe and elsewhere for the screening and testing of the hazard assessment of endocrine disrupting chemicals in environmental species	N.A.

[a]*All projects will have their own website in the near future*
N.A.: Not available

In Priority 5, endocrine disruption is covered in particular by the sub-area "Environmental health risks", which has as the main objective to identify the environmental factors that are detrimental to health, understand the mechanisms involved and determine how to prevent or minimise these effects and risks. In addition, the subarea "Methods of analysis, detection and control" is contributing to the development, improvement, validation and harmonisation of reliable and cost-effective sampling and measurement strategies for chemical contaminants and existing or emerging pathogenic micro-organisms in order to control the safety of the food and feed supply and ensure accurate data for risk analysis. From these areas, both large-scale projects (e.g. Cascade Network of Excellence, BioCop Integrated Project) and more targeted research projects (e.g. Pioneer, F&F Specific Targeted Research Projects) are being funded. Various aspects of risk analysis are covered, and new areas include risk/benefit analyses (Beneris, Qualibra projects).

Finally, in Priority 6, endocrine disruption is covered indirectly by the Sub-area "Global change and ecosystems", and, in particular, by so-called "complementary research". The focus of this area is on the development of advanced methodologies for risk assessment of processes, technologies, measures and policies, appraisal of environmental quality, including reliable indicators of population health and environmental conditions and risk evaluation in relation to outdoor and indoor exposure. Relevant pre-normative research on measurements and testing for these purposes will also be necessary.

3. Research activities in the seventh framework programme of research (FP7: 2006-2013)

In view of preparations for the next framework programme, the Commission published in April 2005 its proposal for the FP7 (European Commission, Research Directorate-General, 2005). It will have four programmes:

The 7th Framework Programme will be organised in four specific programmes, corresponding to four major objectives of European research policy:

1) **Cooperation**: Support will be given to the whole range of research activities carried out in trans-national cooperation, from collaborative projects and networks to the coordination of research programmes. International cooperation between the EU and third countries is an integral part of this action.

2) **Ideas:** An autonomous European Research Council will be created to support investigator-driven "frontier research" carried out by individual teams competing at the European level, in all scientific and technological fields, including engineering, socio-economic sciences and the humanities.

3) **People**: The activities supporting training and career development of researchers, referred to as "Marie Curie" actions, will be reinforced with a better focus on the key aspects of skills and career development and strengthened links with national systems.

4) **Capacities:** Key aspects of European research and innovation capacities will be supported: research infrastructures; research for the benefit of SMEs; regional research driven clusters; unlocking the full research potential in the EU's "Convergence" regions; "Science in Society" issues; "horizontal" activities of international co-operation.

The nine themes identified for the "Cooperation" part are:

1. Health
2. Food, agriculture and biotechnology
3. Information and communication technologies
4. Nanosciences, nanotechnologies, materials and new production technologies
5. Energy
6. Environment (including climate change)
7. Transport (including aeronautics)
8. Socio-economic sciences and the humanities
9. Security and space research

Endocrine disrupter research could be funded under the first, second and sixth theme. It is possible that the planned Environment and Health programme in Theme 6 will be the main funder. However, final priorities and budgets will be set in the course of 2006.

4. Conclusion

The ongoing and past research efforts on endocrine disrupters, sponsored by European Commission, have proven that a commitment to elucidate an issue of public concern, backed by adequate financial support, can advance our knowledge in a considerable manner. The results obtained have already elicited interest by regulators and industry as new legislative proposals (such as REACH) will take advantage of some of the work carried out. In this context, in many projects *in vitro* (such as receptor binding assays) and *in vivo* tools (such as transgenic animals) have developed, which have improved our knowledge about detection and mechanisms behind the phenomenon of endocrine disruption. We not only know more about exposures in various populations to a number of industrial chemicals but also about effects, including long term effects and those at environmentally relevant doses. However, these findings are only the first step on a path, which should lead to a more complete picture. New research, relevant to legislators and industry, among others, will be supported in the future to build upon existing knowledge via open calls to the scientific community. The European Commission's DG Research, being the largest funder of research in the world in this field, with its strategic view on the issue and capacity spanning the European Union and beyond, should continue to be a major force and impetus, able via its multidisciplinary, multipartner projects, to mobilise the critical mass of researchers necessary to confront this complex issue.

References

European Commission (1999) Environment Directorate-General. COM(1999)706: Communication from the Commission to the Council and the European Parliament. *Community Strategy for Endocrine Disrupters – a range of substances suspected of interfering with the hormone systems of humans and wildlife*, Brussels, BE, On line: [http://europa.eu.int/eur-lex/en/com/cnc/1999/com1999_0706en01.pdf.] [last accessed 03/03/2006].

European Commission (2001) Environment Directorate-General. (COM(2001)262: *Communication from the Commission to the Council on the implementation of the Community Strategy for Endocrine Disrupters – a range of substances suspected of interfering with the hormone systems of humans and wildlife (COM(1999)706*, Brussels, BE, On line: [http://europa.eu.int/eur-lex/en/com/cnc/2001/com2001_0262en01.pdf.] [last accessed 03/03/2006].

European Commission. Enterprise and Industry Directorate-General (2003) *The new EU chemicals legislation – REACH*, Brussels, BE, On line: [http://europa.eu.int/comm/enterprise/reach/index_en.htm] [last accessed 03/03/2006].

European Commission. Environment Directorate-General (2003) *Endocrine disrupters website: How the European Commission uses the precautionary principle to tackle endocrine disrupters*, Brussels, BE, On line: [http://europa.eu.int/comm/environment/endocrine/index_en.htm] [last accessed 03/03/2006].

European Commission. Environment Directorate-General (2003) *Environment and health action plan,* Brussels, BE, On line: [http://europa.eu.int/comm/environment/health/index_en.htm#1] [last accessed 03/03/2006].

European Commission. Environment Directorate-General (2004) SEC(2004)1372: *Commission staff working document on implementation of the Community Strategy for Endocrine Disrupters – a range of substances suspected of interfering with the hormone systems of humans and wildlife COM(1999)706*, Brussels, BE, On line: [http://europa.eu.int/comm/environment/endocrine/documents/sec_2004_1372_en.pdf] [last accessed 03/03/2006].

European Commission. Joint Research Centre Directorate-General (2005) *European Centre for the validation of alternative methods*, Ispra, IT, On line: [http://ecvam.jrc.it/index.htm] [last accessed 03/03/2006].

European Commission. Research Directorate-General (1999) *Sustainable management and quality of water*, Brussels, BE, On line: [http://www.cordis.lu/eesd/ka1/home.html] [last accessed 03/03/2006].

European Commission. Research Directorate-General (2003) *Endocrine disrupter research in the European Union,* Brussels, BE, On line: [http://europa.eu.int/comm/research/endocrine/index_en.html] [last accessed 03/03/2006].

European Commission. Research Directorate-General (2003) *Final reports of the projects funded by the 4th Framework programme,* Brussels, BE, On line: [http://europa.eu.int/comm/research/endocrine/projects_completed_en.html] [last accessed 03/03/2006].

European Commission. Research Directorate-General (2003) *Key action 4 – Environment and Health,* Brussels, BE, On line: [http://europa.eu.int/comm/research/quality-of-life/ka4/index_en.html] [last accessed 03/03/2006].

European Commission. Research Directorate-General (2005) COM(2005)119: Proposals for a Seventh Framework Programme (FP7) for research, 2007-2013, and for a Seventh Framework Programme of the European Atomic Energy Community (Euratom), 2007 to 2011, Brussels, BE, On line: [http://www.cordis.lu/fp7/home.html] [last accessed 03/03/2006].

European Commission. Research Directorate-General (2005) Workshop report "Enhanced international collaboration in the field of endocrine disrupters: *How to do It in practice?*", Brussels, BE, On line: [http://europa.eu.int/comm/research/endocrine/pdf/ed_workshop_report_may_2nd_2005.pdf] [last accessed 03/03/2006].

Institute for Environment and Health (2003) *Information Exchange and International Co-ordination on Endocrine Disrupters*. Report for the European Commission, Leicester, UK, On line: [http://europa.eu.int/comm/environment/endocrine/documents/mrc_report.pdf#page=] [last accessed 03/03/2006].

Organisation for Economic Co-operation and Development. (2005) *Endocrine disrupter testing and assessment*, Paris, FR, On line: [http://www.oecd.org/document/62/0,2340,en_2649_34377_2348606_1_1_1_1,00.html] [last accessed 03/03/2006].

The cluster of research into endocrine disruption in Europe (CREDO) (2003) *Website* London, UK, On line: [http://www.credocluster.info] [last accessed 03/03/2006].

The International Programme on Chemical Safety (2002) *Global assessment of the state-of-the-science of endocrine disrupters*, Geneva, CH, On line: [http://www.who.int/ipcs/publications/new_issues/endocrine_disrupters/en] [last accessed 03/03/2006].

United Nations Environment Programme, International Labour Organization, World Health Organization (2004) *Report of the joint IPCS-Japan workshop on "endocrine disrupters: research needs and future directions*", Geneva, CH, On line: [http://europa.eu.int/comm/research/endocrine/pdf/japan_workshop_en.pdf] [last accessed 03/03/2006].

SECTION 5:

CONCLUSIONS

ENVIRONMENTAL IMPACTS ON CONGENITAL ANOMALIES - INFORMATION FOR THE NON-EXPERT PROFESSIONAL

L. HENS
Vrije Universiteit Brussel
Human Ecology Department
Laarbeeklaan 103
Brussel
BELGIUM
Email: human.ecology@vub.ac.be

Summary

This paper summarises the main findings on congenital anomalies and environmental pollution. The lines of evidence for environmental causes of congenital anomalies are listed. These are based on clinical, epidemiological, laboratory and wildlife studies. The data show that not only are a (limited) number of pollutants (e.g. lead, methyl mercury) well known teratogens, but also that the number of daily environmental exposures associated with congenital anomalies is increasing. This latter applies, among others, to lead and nitrates in drinking water, living near waste deposit sites and non- occupational exposure to pesticides.

The increasing incidence of hypospadias and cryptorchidism in a number of industrialised countries is noticeable. The "testicular dysgenesis syndrome" (TDS) theory links the epidemiological data with environmental causes and hypothesises one unifying mechanism for which the experimental evidence is significant.

Core concepts related to the mechanisms of teratology are reviewed. They include dose-response relationships, exposure windows, latency periods and multicausality

P. Nicolopoulou-Stamati et al. (eds.), Congenital Diseases and the Environment, 409–450.

in the underlying factors of congenital anomalies. The discussion shows that at different points during recent years, teratology underwent important paradigmatic shifts.

The concluding part of the paper deals with the question: which data are of primary importance for the stakeholders in the discussion on pollution and congenital anomalies: the prenatal health advisors, lawyers, policy makers and the media? Each of these groups has specific agendas and corresponding needs for information. Although main guidelines for handling this information exist, much more research is needed, e.g. on case studies from the past, and on effectiveness and efficiency of information transfer.

1. Introduction

"Congenital anomaly" (CA) is a "container concept". It is used for all types of structural or functional defects with which a baby can be born. Structural abnormalities can be classified as malformations, malformation syndromes, disruptions and deformations. Table 44 shows the definitions and lists classical examples of these groups.

It equally shows that most of these categories have a multifactoral aetiology. Disruptions usually have environmental causes, while malformations and deformations have a partially environmental background.

Functional anomalies concern, for instance, disturbances of the hormonal balance during developmental stages, which might result in health effects many years after birth. Also prenatally impaired functions of the nervous system might result in congenital anomalies.

It is generally accepted that most of the congenital malformations have a genetic basis. Estimations on the share of the environmental causes range between 37% and 2%. Factors such as drugs, radiation, infections and alcohol are believed to be the responsible aetiology in 6-8% of the cases (Seller, 2004). Part of this variation can be explained by the definition of "environment" that is used. While some authors consider environmental causes as meaning any non-genetic factor which increases the risk of a congenital anomaly for the exposed individual (including nutritional excesses and deficiencies, maternal illness, and infection), others will restrict the term environmental to mean chemical and physical contaminants in air, water, soil and food. Moreover, the precise cause of congenital malformation is not known for as many as 50-60% of the total cases. Finally, most CA has a multicausal origin.

Table 44. Categories of congenital anomalies (after Seller, 2004)

Category	Definition	Examples	Causes
Malformation	Localized error of normal development during the morphogenesis of an organ or a tissue	(-) isolated cleft lip and palate (-) spina bifida (-) ventricular septal defect	Multifactoral aetiology
Malformation syndrome	Occurrence, in one individual, of several different malformations arising from the same underlying cause	(-) trisomy 21	Usually simple gene mutation or a chromosome abnormality; occasionally an environmental agent
Disruption	Major destruction or alteration of a body part that had previously formed, or which had the intrinsic potential to form, quite normally	(-) missing limb	Usually environmental (drugs, infections)
Deformation	An alteration of shape or structure imposed on a body part after, or during, its normal formation.	(-) club foot (-) oligohydramnios (-) positional abnormalities of the foot	Genetic, partly genetic, or environmental

The rising attention to environmental exposure as a causal factor for congenital anomalies was the main driver for the workshop on which this book is based. This concluding section deals with four groups of questions:

(1) Why do we think that environmental exposure is related to congenital anomalies? What is the problem?

(2) How do we study the relationship between environmental exposure and congenital anomalies? What is the evidence at this moment?

(3) What makes congenital anomalies different from other issues in environmental health? What are the specific characteristics when it comes to dose-response relationships, exposure windows, latency periods and a variety of causes underlying congenital anomalies?

(4) What is the essential information for the major, non-expert stakeholders in this debate?

2. Problem identification

Registries of congenital anomalies report 2-4% of births with a congenital anomaly, depending on the inclusion criteria and ascertainment methods. These figures relate to neonates having a single major malformation that requires medical treatment, or perhaps can even be lethal. Less than 1% of the neonates have multiple malformations. As shown in table 45 cardiac defects account for over 25% of all cases, limb anomalies for 17%, chromosomal anomalies and urinary system anomalies each for around 15%, central nervous system anomalies including neural tube defects 10% and oral clefts 6% (Dolk and Vrijheid, 2003)

Table 45. Relative occurrence of congenital anomalies – overall figures (after Dolk and Vrijheid, 2003)

1. Cardiac defects	25%
2. Limb anomalies	17%
3. Chromosomal syndromes	12%
4. Urinary system	15%
5. Central nervous system and neural tube	10%
6. Oral clefts	6%

Table 46. Chemicals and exposure conditions associated with congenital anomalies in humans

1. Chemicals
   - Pharmaceuticals (e.g. DES, Thalidomide, Warfarin)
   - Hair dyes
   - Pesticides
   - Non-pesticide endocrine disrupters (bisphenol A, phthalates, TCDD, vinyl chloride)
   - Heavy metals (Pb, Hg, Cd, As, Cr and Ni)
   - Organic solvents (styrene)
2. Exposure conditions associated with congenital anomalies
   - Drinking water (heavy metals, nitrates, chlorinated substances)
   - Residence near (hazardous) waste deposit sites
   - Pesticides in agricultural areas, home gardening
   - Air pollution
   - Food contamination (dioxins, PCBs)
   - Industrial point sources (smelters, incinerators)
   - Disasters (Hiroshima, Minamata)
   - (Working conditions: e.g. dentists, hairdressers, workers exposed to pesticides)

These congenital anomalies are caused by both genetic and environmental factors. In its widest sense, an environmental cause is any non-genetic factor which increases the risk of a congenital anomaly for the exposed individual. Such factors include biological (e.g. infections such as rubella), physical (e.g. medical X-rays or ionizing radiation), and chemical agents (e.g. such as some drugs taken during pregnancy). Table 46 provides an overview of the main chemical groups that are associated with congenital anomalies. These chemical contaminants are not only found in air, food, soil and water (e.g. pesticides or heavy metals) but they are also related to lifestyle (e.g. cosmetics, pharmaceuticals and packaging materials). Moreover, for some products such as organic solvents, it is difficult to make the distinction between the contributions to exposure of the household and the working environment.

Next to these single factors, causes of congenital anomalies are found in more complex environmental exposure conditions (table 46). Situations that are of particular interest include the use of pesticides in agricultural areas and home gardening, endocrine disrupting chemicals e.g. as food contaminants and unspecified releases from landfill sites (Dolk and Vrijheid, 2003). In addition, environmental crisis situations such as Chernobyl, Bhopal, Hiroshima and Minamata have been associated with congenital anomalies.

Overall, international registries do not point to an increase of congenital anomalies over time (Vrijheid, 2006). However, selected congenital anomalies such as hypospadias, cryptorchidism and gastroschisis show an increasing prevalence (the number of existing cases in a population over a time period), in particular in industrialized countries. Figure 28 shows, as an illustration, the evaluation of the prevalence of hypospadias in six industrialised countries during the period 1964 – 1990.

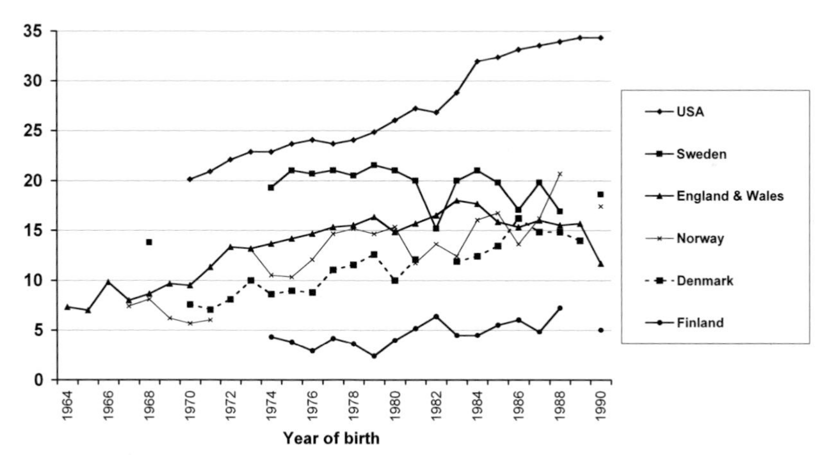

Figure 28. Rate of hypospadias over time in six countries or regions (Toppari, 2006)

The figure shows marked increasing trends in the US and Denmark. In Finland, the prevalence rate is low and remains constant since the second half of the 1970s. In other countries, the rates show a more variable appearance.

Finally, there are noteworthy similarities in the pattern of effects between the various pollutants which cause environmental health-related diseases. Endocrine disrupters such as bisphenol A, phthalates, PCBs and dioxins, pesticides, different solvents, and teratogenic medical drugs are not only causing congenital anomalies, but are equally involved in the aetiology of cancer, reproductive failure and disturbance of the hormonal balance. There is a currently incompletely explained link between contemporary epidemics including hormone related cancers, endometriosis, infertility and pre-term birth. Although part of this convergence might be a coincidence, the finding that the same pollutants are causing different health impairments provides a basis on which to hypothesize common or closely related mechanisms.

In summary, the problem formulation on congenital anomalies and the environment may be summarised in the following way:

(1) Embryonic and fetal life are sensitive periods to environmental exposures.

(2) An increasing number of chemicals and exposure conditions are associated with congenital anomalies.

(3) An increasing prevalence of selected congenital anomalies (hypospadias, cryptorchidism, gastroschisis) has been reported in industrialized countries.

(4) A noticeable similarity exists between pollutants, as regards environmental health outcomes.

This conference overviewed the current information on these issues and their interrelationships.. They are revisited in this concluding chapter.

3. Methodological approaches

3.1. Clinical evidence

The assessment of embryo-fetotoxic risk of therapeutic products is a long process. For new drugs, the evidence is limited to animal and laboratory studies, although it is accepted that predictive values from these tests are limited and insufficient to guarantee human safety. For drugs on the market, epidemiological studies are set up to test for side effects. However, most of the human teratogens have been discovered earlier in man than animals. Alert physicians hypothesized the teratogenicity of many substances from case studies or single observations.

Thalidomide is a historical example. This substance, marketed in Western countries since 1957, was considered a harmless therapeutic drug reputed to have efficacy as a sedative and as an antiemetic for use in early pregnancy. Thalidomide was considered safe, as experimental studies in pregnant mice and rats did not show any side effects at large exposure. Unfortunately, thalidomide only regularly induces the specific pattern of malformations observed in humans, in rabbits and non-human primates. It has been estimated that in the late 1950s and early 1960s, over 6,000 children worldwide were affected with the "thalidomide embryopathy", which is characterized by variable degrees of reduction deformities of the limbs (phocomelia) and many other skeletal and non-skeletal malformations (Clementi *et al.,* 2006).

Diethylstilboestrol (DES) offers a more complex example of the role of the alert clinician in recognising malformations. DES, the first synthetic hormone, when antenatally administered to the mother, causes a variety of congenital abnormalities of the female genital tract in the offspring. In addition, it also causes clear cell carcinoma of the vagina and cervix, in daughters of women treated during their pregnancy ("DES-daughters") (Bernheim, 2001).

Nevertheless, during the period 1955-1975, the use of DES was the subject of an intense medical debate, in particular among clinicians. This was partially related to case studies. What to think about a woman who had five consecutive unexplained spontaneous abortions, who had a healthy boy after DES "protection", then three more spontaneous abortions during unprotected pregnancies, and again a live birth after a pregnancy with DES (Little, 1976). There are a number of anecdotal clinical successes published, that encouraged clinicians to systematically treat threatened pregnancies (Smith and Smith, 1949). This was the current practice in countries such as the United States, France and the Netherlands (Heinonen, 1973; Spira *et al.,* 1983; Hanselaar *et al.,* 1991). It was, however, also clinical observations that resulted in giving up the practice. In 1971, Herbs *et al.* published their observations on eight cases of a rare tumour clear cell carcinoma (CCA) of the vagina or cervix in very young women. The mother of one of the patients had suggested that her daughter's tumour was related to her taking DES during her pregnancy. Seven out of the eight mothers of the CCA cases turned out to have taken DES or another xenoestrogen. This case control study was soon followed by other studies confirming the problem (Greenwald *et al.,* 1971) and the FDA promulgated an absolute contra - indication to DES in pregnancy, followed later on by a similar attitude in other countries. Seller (2004) reviewed the malformations highlighted by clinical studies. A summary of her findings is provided in Box 2.

Box 2. Malformation spectra highlighted by clinical studies (Seller, 2004)

Infectious agents

1. Rubella: before 10 weeks gestation, cataracts and heart defects; 10-16 weeks, hearing loss and retinopathy.
2. Varicella: limb hypoplasia, scarring, microcephaly, chorioretinitis.
3. Cytomegalovirus: hydrocephalus, periventricular calcification, neurological problems.
4. Toxoplasmosis: hydrocephalus, microcephaly, cerebral calcification, neurological problems.

Maternal illnesses

1. Insulin-dependent diabetes macrosomia, caudal regression syndrome, neural tube defects, congenital heart disease especially VSD and transposition of the great vessels. These defects do not occur in children of mothers who develop gestational diabetes.
2. Phenylketonuria (if not controlled by diet): microcephaly, micrognathia, heart defects, mental retardation.
3. Unspecified folate deficiency: neural tube defects and possibly cleft lip and palate.
4. Epilepsy: effects cannot be separated from the teratogenic effect of medication.

Physical agents

1. Radiation: high doses to the fetus in mid to late gestation can produce microcephaly.
2. Hyperthermia: neural tube defects especially anencephaly, microcephaly, microphthalmia, cleft lip and palate; it is difficult to dissociate the effects of viral or other agents producing pyrexia, from the effects of hyperthermia. The data from hot baths and saunas are not convincing. Evidence for these abnormalities arising from pure hyperthermia exists in animals.

Drugs

1. Thalidomide: phocomelia and other limb defects, cardiac defects, gut atresia, renal agenesis.
2. Diethylstilboestrol: females - vaginal adenosis. Males - micropenis, hypospadias, cryptorchidism.
3. Warfarin: hypoplastic nose and bone dysplasia like Conradi disease, choanal atresia, microcephaly, hydrocephaly.

4. Valproic acid: spina bifida, midface hypoplasia, long philtrum, small mouth, cardiac defects.
5. Phenytoin: typical facies of brachycephaly, high forehead, marked eyebrows, long philtrum, depressed nasal bridge, metopic ridging, bowed upper lip; cleft lip and palate; hypoplastic nails and distal phalanges; short neck, hirsutism.
6. Aminopterin/methotrexate: microcephaly, broad nasal bridge, micrognathia, short limbs, talipes, hypodactyly, syndactyly.
7. Retinoic acid (vitamin A congeners): hydrocephalus, microcephaly, cardiac defects especially conotruncal malformations, aortic arch hypoplasia; microtia/anotia, micrognathia, urogenital anomalies, thymic hypoplasia.
8. Alcohol: microcephaly, long face, smooth philtrum, thin vermilion border to upper lip, abnormal palmar creases, short distal phalanges, heart defects, mental retardation.

It should be noticed that these malformations are also related to generalised low birth weight and dysmorphism, and increased risk of fetal loss. The box shows that although different "environmental exposures", in the wider sense of the term, have been shown to cause congenital malformations, clinical studies provide no evidence for malformations resulting from exposure to polluted air, water, soil or food.

3.2. Epidemiological evidence

As clinical studies provide limited evidence for impacts of pollutants on congenital anomalies, we examine the hypothesis that population studies might provide more evidence.

Most data on the prevalence of congenital anomalies are included in congenital malformation registries, where they exist in different countries. As illustrated by the chapters on these registries in Bulgaria (Terlemesian and Stoyanov, 2006) and in Greece (Brilakis *et al.,* 2006) in the case study section of this book, hardly any information on environmental influences on congenital anomalies can be obtained from a general overview of these data. Dolk and Vrijheid (2003) reviewed the reasons why epidemiological studies on congenital anomalies have difficulties in revealing environmental influences, even if they look specifically for chemical exposures associated with environmental pollution.

They list the following main reasons:

- many studies only have the statistical power to detect large increases in risk, and negative results must be assessed in this context,

- many congenital anomalies exist and epidemiologic studies address the problem of variation in diagnosis and reporting of minor anomalies with a varying degree of success,
- many data sets reflect mainly improved diagnosis, medical attitude and treatment aspects. During the period 1995 - 1999, for example, EUROCAT data from 32 European regions showed that 53% of spina bifida and 32% of Down's syndrome cases, that were prenatally diagnosed, also resulted in termination of pregnancy,
- many embryos and fetuses with congenital anomalies are lost as spontaneous abortions and are out of the reach of the registries,
- many congenital anomalies caused by environmental pollutants have a multifactoral background. This means that genetic and other environmental factors interact with exposure and that it might prove difficult to distil out the causative pollutant,
- exposure assessment of environmental pollution is often poor and, consequently, correct dose-effect relationships are difficult to establish,
- environmental epidemiological studies are observational, not experimental and as such, are open to confounding factors. These confounders might mask the real environmental causes of the congenital anomaly under study.

In spite of these major difficulties, epidemiological studies have revealed mainly four lines of evidence that link congenital anomalies with exposure to pollutants:

(1) *Increases in prevalence of selected anomalies.*

The prevalence of male reproductive anomalies, hypospadias and cryptorchidism is increasing in many industrialised countries (Sharpe and Skakkebaek, 1993; Toppari *et al.,* 1996). Hypospadias increases were most marked in the US, in Scandinavia and Japan. As shown in figure 28, Denmark and Norway show rates where hypospadias approximately doubled during the 1970s and 1980s (Paulozzi, 1999). Although the general character of the evidence for these increasing trends is disputed and the fundamental causes are unknown, the possibility that these increases are related to increasing exposures to endocrine disrupters in the environment is an active area of current research (CSTEE, 1999; Nicolopoulou-Stamati *et al.,* 2001).

A number of observations support such an environmental hypothesis:

- lower prevalence rates of hypospadias in remote areas in Finland than in more polluted cities (Olavi Aho et al., 2003),
- an increased risk of hypospadias among sons born on Norwegian farms where pesticides and tractor spraying equipment had been used (Kristensen et al., 1997).

In addition, the prevalence of cryptorchidism (testis maldescent) and testicular cancer is increasing in different industrialised countries. The higher prevalence rates in Denmark than in Finland might be related to higher exposure to pesticides and phthalates in Southern Scandinavia.

As explained by Thonneau *et al.* (2006) in this book, the TDS theory hypothesises that these different phenomena and anomalies are subject to a common mechanism that is mediated by endocrine disrupters during embryonic development. The TDS hypothesis puts this evidence into a noticeable environmental pollution context. The TDS hypothesis refers to the fact that disorders such as cryptorchidism, hypospadias, testicular cancer, and low sperm counts might have a common origin in fetal life. This is explained by hormonal dysfunction of the fetal testis; in particular, reduced production of testosterone. Because testosterone is largely responsible for transforming the fetus into a male, TDS can be viewed as a sexual differentiation (i.e. masculinization) disorder, a congenital disease (Toppari, 2006). Endocrine disrupters such as phthalates are able to reduce the fetal testosterone production and might consequently cause the TDS disorder.

However, not all (recent) increases in the prevalence of congenital anomalies point to environmental interference. The increasing rates in gastroschisis might offer an example. Gastroschisis is a congenital defect of the abdominal wall, characterised by herniation of abdominal viscera outside the abdominal cavity through a defect in the abdominal wall near the umbellicus. Its prevalence at birth has increased in half of the 36 registries for congenital anomalies in Europe, the Americas, Asia, Australia and South Africa (Di Tanna *et al.*, 2002). Although the anomaly is rare (less than 5 cases per 10 000 births), the fast increase is qualified as "epidemic". The aetiology is currently, however, largely unexplained, although first trimester medication, early maternal exposure to X-rays and the use of societal drugs are risk factors (Huges and Flood, 1999).

(2) *Exposure to environmental pollution.*

The teratogenicity of a number of environmental pollutants including lead, organic mercury, TCDD, and PCB's has well been established (see also the section on environmental disasters). Elsewhere in this book, Koppe *et al.* (2006) discuss the teratogenic character of TCDD.

Different exposure conditions have been studied in relation to congenital anomalies. Those for which most information is currently available are summarised in table 47.

Table 47. Environmental exposure conditions associated with congenital anomalies (after Dolk and Vrijheid, 2003)

Type of exposure	Reported issues	Evaluation	References
Drinking water contamination			
1. Nitrates	Nitrates as cause of high incidence of perinatal mortality due to congenital malformations	Dose-response effect	Dorsch *et al.*, 1984
	Potential effect of high nitrate levels on central nervous system defects, anencephaly and cardiac defects		Arbuckle *et al.*,1988 Craen *et al.*, 2001 Cedergren *et al.*, 2002
2. Lead	Positive association between lead pollution from ceramic factories and cardiovascular anomalies, oral clefts and musculoskeletal anomalies. Positive association between lead in drinking water and neural tube defects.	Neurotoxic effects of lead on fetus, neonates and children are beyond any doubt. Nevertheless, there is also a study that failed to demonstrate a link between lead in drinking water and neural tube defects.	Bound *et al.*, 1997 MacDonell *et al.*, 2000 Vincenti *et al.*, 2001
3. Fluoridation	Fluoridation might cause Down's syndrome.	Conflicting results.	Heedleman *et al.*, 1974 Erickson, 1980
4. Chlorination	Chlorinated and aromatic solvents are associated with eye/ear congenital anomalies and nervous system/ oral cleft/ chromosomal anomalies (mostly trisomy 21).		Lagakos *et al.*, 1986

Table 47. (Continued)

Type of exposure	Reported issues	Evaluation	References
Waste disposal, landfill sites	Increase in birth defects among children of parents living near Love Canal, NY.		Goldman *et al.,* 1985
	People living within 1 mile of 590 hazardous waste sites in NY. 12% increase in congenital malformations.	Also studies reporting no effect on congenital anomalies	Geschwind *et al.,* 1992
	Residents living within 3 km of 21 hazardous waste landfill sites in 21 European regions. 33% increase in risk of all non-chromosomal birth defects.		Vrijheid *et al.,* 2002
	Populations living within 2 km of a landfill site in England, Scotland, and Wales. Small (<10%) increase for all congenital anomalies, neural tube defects, hypospadias, abdominal wall defect, low birth weight.		Elliot *et al.,* 2001
	Populations living within 2 km of 24 landfill sites that opened from 1983-1997, show an increase in congenital anomalies by 40%. No increase in data collected from 1998 – 2000		Palmer *et al.,* 2005
Waste incineration	Excess in perinatal and infant mortality due to spina bifida and heart defects near incinerators in Cumbria, UK.		Dummer *et al.,* 2003
	An increase of trisomy 21 was observed in offspring of parents living near a polluting household waste incinerator near Antwerp.	Causal link uncertain	Van Larebeke, 2000
Pesticides at home and in gardens	Many studies pointing to controversy.	Difficult to arrive at conclusions.	Nurminen, 2001
Air pollution	Polydactily has a higher prevalence rate in the highly polluted Brazilian city of Cubatao.		Monteleone Neto and Castilla, 1994
	Increased risk of congenital malformations, in particular, nervous system defects, near PVC polymerisation plants.	No conclusive evidence.	Infante *et al.,* 1976
Food contamination	Organic mercury and persistent organochlorines in fish.	No conclusive evidence.	Clarkson, 2002

The table shows that currently the most conclusive evidence is available for the risks associated with the consumption of high nitrate concentrations in drinking water and for living near (hazardous) waste deposit sites. In particular for the second type of exposure, studies have also been published which report no association between congenital anomalies and living in the vicinity of waste deposit sites (Sosniak *et al.,* 1994; Croen *et al.,* 1997). Nevertheless, the majority of the studies point to effects of which the dose-effect relationships should be established better. An overview of the health impacts of waste deposit sites, including the impacts on congenital malformations, can be found in Hens *et al.* (2000). In the same book, Van Larebeke (2000) describes the case of a waste incinerator in Wilrijk, near Antwerp (Belgium) in which crude data pointed to an increase of trisomy 21 children born from parents living in the nearby neighbourhood. However, after taking into account an extended set of potential confounding factors, no statistically significant increase could be shown. This data contrasts with that from a multi-site study in Cumbria, UK. In this study, birth data from 1956 to 1993 showed excesses in prenatal and infant mortality due to spina bifida and heart defects near incinerators, after controlling for social class (Dummer *et al.,* 2003).

(3) *Congenital anomalies associated with exposure resulting from environmental disasters.*

Next to environmental pollution from normal operations in waste management, agricultural, horticultural, or industrial activities, pollution might equally result from situations that are classified as disasters. These are characterised by accidental or deliberate releases of important amounts or doses of the pollutant(s) involved. In a number of cases, the association with congenital anomalies was studied. The best documented examples are summarised in table 48. The table points to the teratogenic capacities of ionising radiation as extensively evidenced in survivors of the Hiroshima and Nagasaki atomic bombs. The fact that no conclusive evidence on congenital anomalies emerged from the Chernobyl accident, points rather to the special situation after the disaster (induced abortions, clearance of the contaminated area, communication, etc) than to the fact that ionising radiation might not be teratogenic. However, not all environmental exposure to ionising radiation automatically results in an increased observable prevalence of congenital anomalies. In Bosnia - Herzegovina during the 1991 - 1995 war, depleted uranium was used in military activities. A study of 3316 neonates born in 1995 and 2000 did not reveal a significant post-war increase in the prevalence of congenital malformations (Sumanovic-Glamuzina *et al.,* 2003).

The disaster analyses also point to the teratogenic character of dioxin-like products and PCBs. In particular, the Vietnamese data on Agent Orange exposure are most indicative at the case study level, but are in need of a more systematic population analysis of the exposed population never the 17th latitude line in Vietnam.

Table 48. Congenital malformations in disasters involving environmental pollution (after Dolk and Vrijheid, 2003; Kalter, 2003)

Incident	Phenomenology	Congenital anomalies	Remarks	Reference
Hiroshima and Nagasaki (1945), Japan	Atomic bombs Ionising radiation	Microcephaly and mental retardation	Predominant sensitivity: 7-15 weeks of gestation dose-effect dependent	Plummer, 1952 Blot, 1975 Miller and Mulvihill, 1976
Chernobyl (1986) Then USSR, now Ukraine	Nuclear power accident Ionising radiation		No conclusive information on increased prevalence	Anon, 1987
Minamata (1953-1971) Japan	Fish contamination Methyl mercury	Sex ratio changes Central nervous system anomalies		Peterka *et al.*, 2006 Clarkson, 2002 Harada, 1978
Yusho (oil disease) (1968) Fukuoka, Japan	Food contamination Polychlorinated biphenyls (PCB's)	Deep brown pigmentation of skin and nails (cola coloured babies), IUGR, natal teeth, spotty calcification, other irregularities of the skull		Rogan, 1982 Miller, 1985 Rogan *et al.*, 1988
Seveso (1976), Italy	Industrial accident in trichlorophenol production plant Contamination by 2,3,7,8-tetrachloro-dibenzo-p-dioxin (2,3,7,8 TCDD)		No conclusive information on increased prevalence in the most contaminated zone Shift in sex ratio.	Mastroiacovo *et al.*, 1988
Agent Orange (1962-1970), Vietnam	Defoliant 2,4-dichlorophenoxy-acetic acid (2,4-D) 2,4,5-trichlorophenoxy-acetic acid (2,4,5-T) contaminated by 2,3,7,8 TCDD	Association between paternal exposure and congenital anomalies in Vietnamese populations. Increased frequency of spina bifida and cleft lip with or without cleft palate in American children whose fathers were more heavily exposed.	Strength of association varies substantially among studies	Lilienfeld and Gallo, 1989 Skene *et al.*, 1989 Sterling and Arundel, 1986 Constable and Hatch, 1985 Erickson *et al.*, 1984, a,b.

(4) *Occupational exposure.*

Although environmental exposure conditions differ fundamentally from occupational exposure, both types of exposures often provide complementary information. Vrijheid *et al.* (2003) reviewed the information on occupational exposures and congenital anomalies, in relation to environmental exposures to the same pollutants. They showed that occupational exposure studies provide ample evidence for the teratogenicity of organic solvents, both in industrial working conditions, but also at home. Occupational studies also provide information for lead exposure related to effects such as total anomalies, cleft lip and palate and neural tube defects. In contrast, as regards exposure of industry workers and dentists to inorganic mercury, studies showing an effect and others showing no effect both exist. In addition, the possible impacts of arsenic, chromium and cadmium were studied, without indicating significant associations with congenital anomalies. Studies of workers exposed to wood preservatives such as chlorophenol, are in line with home exposure studies. Both point to increased risks of congenital anomalies. Evidence for adverse reproductive effects of dioxins has been extensively reviewed (Lilienfeld and Gallo, 1989; Silbergeld and Mattison, 1987; Skene *et al.,* 1989; Sterling and Arundel, 1986). In this book, Koppe *et al.* (2006) contribute to this discussion. Pesticides are a most heterogeneous group. Both positive and negative relationships with occupational exposure have been reported. However, the number of studies reporting teratogenic effects after both maternal and paternal occupational exposure to pesticides is rising fast.

3.3. ***Evidence from experimental developmental toxicology***

Procedures for testing drugs in animals for potential reproductive toxicity existed prior to the thalidomide event in the late 1950s and the early 1960s. However, in a retrospective evaluation, they were found to be inadequate in detecting possible embryonic or fetal damage (Wilson, 1979).

The thalidomide case and, in the US, the subsequent promulgation in 1963 of new investigational drug regulations, making pre-marketing reproductive and teratological drug testing in animals mandatory (Kelsey, 1982), pushed the establishment of laboratory protocols.

The basic principles can be summarised as follows:

- test animals have to resemble, as closely as possible, the developmental features in humans: in placental anatomy and physiology, reproductive and developmental characteristics, maternal and fetal pharmacology, drug administration route and schedule, etc. The fact that no laboratory species conforms these requirements, does not contradict this principle, that should be matched as closely as possible,

- the procedure further necessitates the use of timed-pregnant female animals of at least two common mammalian laboratory species, to be given several different doses of the test agent, once or on several successive days during the organogenetic and early fetal periods of gestation,

- to find out the amount that causes prenatal death and then, after administering ever smaller doses, to determine the lowest dose causing malformations. This point is called the threshold level, below which no apparent embryotoxicity - growth retardation, malformation or death - occurs. More generally, this approach aims at describing the dose-effect relationship of teratogenic agents.

Since the 1960s, procedures in developmental toxicology have been in continuing evolution. Currently, the protocols developed by OECD are widely used (e.g. OECD guideline 414 to asses birth defects, minor anomalies, fetal growth and variability; OECD guidelines 415 and 416 on one- and two-generation studies; draft TG426 on developmental neurotoxicity studies; TG421 and 422 on reproduction/toxicity screening tests). In this book, Mantovani and Maranghi (2006) discuss the role of developmental toxicology studies in answering questions on long-term exposure to low doses and on non-malformation effects of prenatal exposures, such as disruption of immune development and early predisposition to cancer.

The authors put special emphasis on the challenge that endocrine disrupters (ED) offer to developmental toxicologists. They advocate a combination of two-generational and *in vitro / in vivo* assays, in a tiered approach, to detect the different aspects of functional congenital anomalies that might arise. The complexity of current testing for CA is probably most clear for ED, but should be applied to the testing for new and existing chemicals in general.

New developments in fertility treatment have opened new avenues of research on “preimplantation teratology”. The same applies to the use of stem cells in developmental teratology. New developments should equally address the need to reduce the use of laboratory animals and consequently the cost and time-effectiveness of protocols such as multigenerational studies. New efforts are currently going on in the EU to include validated *in vitro* tests in protocols for evaluating both the embryotoxicity of chemicals and/or their possible influence on the hormonal homeostasis.

3.4. Evidence from wildlife studies

An obvious limitation of laboratory systems is that they hardly provide information on effects resulting from chronic exposure to real world concentrations or mixtures of pollutants. Studies on wildlife aim at overcoming this limitation. They link concentrations, behaviour and bioavailability of pollutants in an environment with biological effects such as congenital anomalies.

A classical example of this approach are studies on bald eagles, herring gulls, night herons, tree swallows, snapping turtles, munk and beluga over the past 30 years in the Great Lakes - Saint Lawrence basin (Canada). These studies revealed a broad spectrum of health effects, including thyroid and other endocrine disorders, metabolic diseases, altered immune function, reproductive impairment, genotoxicity, cancer and developmental toxicity. For an overview, see Fox, 2001. The developmental disorders revealed by these studies are summarized in table 49.

The table shows that in a wide variety of bird species, but also in turtles and salamanders, congenital anomalies were found that are most comparable to malformations in humans, next to species-specific anomalies (such as those of the turtle carapace). As with other health effects, the developmental anomalies were most prevalent and most severe in the most contaminated sites. In particular, association with persistent organic pollutants such as PCBs, PCDDs, PAHs and DDT/DDE was established. In some of the studies, dose-response relationships were documented.

The evidence of the last 10 years that links endocrine disruption with congenital anomalies and related health effects includes: impaired reproduction, deformities of the male reproductive tract, thyroid abnormalities and depressed immune system function.

These wildlife data provide an important complement to the epidemiological and laboratory data. They show a high sensitivity in documenting congenital anomalies in relation to environmental exposure. In particular, the bird data provide additional evidence for the teratogenic character of planar PCBs, TCDD and PAHs.

Table 49. Developmental toxicology anomalies in studies in the Great Lakes - Saint Lawrence basin regions (after Fox, 2001)

Species	Pollutant and phenomenology	Congenital anomalies	Remarks	Reference
Common tern, night heron (Nycticorax nycticorax)	PCB concentration in eggs of the colony (mid 1980s)	Femur length: body mass ratio (measure for growth retardation)	Significant correlation	Hoffman *et al.*, 1993
Nine species of fish-eating birds including terns, gulls, herons, cormorant, bald eagle	Chicks in polluted areas (last 30 years)	Abnormal bills; supernumerary, dwarf or otherwise abnormal appendages; missing or small eyes; club foot and hip dysplasia; other skeletal anomalies	Highest prevalence of bill abnormalities is in cormorants in Green Bay (52/10000 chicks examined) and bald eagles in Michigan (23/10000)	Fox, 1993 Grasman *et al.*, 1998 Gilbertson, 1988 Bowerman *et al.*, 1994
	Embryos in the same areas	Missing jaws, missing skull parts, vertebral anomalies, exencephaly, anencephaly, hydrocephaly and spina bifida, gastroschisis, Siamese twins		Ludwig *et al.*, 1996
Cormorant (Phalacrocorax auritus) Caspian tern (Sterna caspia)	Hatched chicks and live and dead eggs from 37 colonies (1986 - 1991)	Similar malformations to these in chickens exposed to planar PCBs and dioxins		Ludwig *et al.*, 1996
Snapping turtle (Chelidra serpentura)	Eggs with the highest concentrations of PCDDs, PCDFs and PCBs had the highest probability of non hatching and producing deformed turtles	Similar to those in birds plus specific ones including malformations of the tail, carapace and plastron		Bishop *et al.*, 1991 Wren, 1991
Mudpuppies (salamander) (Necturus maculosus)	Highest prevalence in the most contaminated sites of the Saint Lawrence river (early 1990s)	Limb defects		Bishop and Gendron, 1998

Table 49. (Continued)

Species	Pollutant and phenomenology	Congenital anomalies	Remarks	Reference
Mudpuppies (salamander) (Necturus maculosus)	Highest prevalence in the most contaminated sites of the Saint Lawrence river (early 1990s)	Limb defects		Bishop and Gendron, 1998
Western gull (Larus occidentalis)	Channel islands of the coast of Southern California	Sex ratio skewed towards females	Coincident in space and time with high DDE/DDT contamination	Fox, 1992
Herring gull (Larus argentatus)	Lakes Ontario and Michigan (1970s)			Conover, 1984
Caspian tern (Sterna Caspia)	Lake Michigan			Conover, 1984
Herring gull (Larus argentatus)	Embryos and newly hatched chicks in Lake Ontario (1970s)	Feminisation of the reproductive tracts Gonadal histology pathologies		Fox, 1992

Moreover, these data do not only point to associations between exposure to a particular type of pollutant and congenital anomalies, they also offer opportunities to study the mechanisms at the heart of these associations. This is e.g. illustrated by the study of dioxins and co-planar PCBs and the way they cause developmental abnormalities and reproductive failure in Ontario Lake trout.

However, what does the study of these sentinel species mean for human health? There can be no doubt that the mobility, duration of residence and diet of animals residing and feeding in the polluted areas is analogous to that of humans in the same areas. Both wildlife and humans are exposed through water, air, soil and food. Moreover, they share many development mechanisms. Nevertheless, a large amount of work is still required to link the results in the wildlife sentinels with human health effects.

In particular, the biochemical and functional aspects of the biological phenomena need to be better understood, before clear extrapolations can be made. On the other hand, if pollution has marked impacts on congenital anomalies in birds, turtles and salamanders, which show similarity with those found in humans, it might be too optimistic to think that mammals, and humans in particular, are non-sensitive to these exposure conditions.

In this context, the data from the wildlife sentinel species act at least as an important warning sign.

4. Mechanisms causing congenital anomalies

Congenital anomalies are diverse and involve many anatomic and physiological systems, each of them with their specificities in biochemistry and genetics. Congenital anomalies associated with chromosome aberrations are based on different mechanisms than the induction of phocomelia by thalidomide. It was not the aim of this conference to overview the literature describing the different molecular mechanisms of congenital anomalies. Nevertheless, a number of general concepts, with particular specificity, exist on the genesis of congenital anomalies.

(1) *Dose-effect relationships and threshold levels*

The dose-response curve describes the nature of the relationship between the dose of a teratogenic agent and its effects on pregnancy outcome.

For many congenital malformations, sigmoid dose-effect relationships have been described (Kalter, 2003). This includes the notion that it is possible to define not only the dose at which 50% of the test animals show malformations, but more important, the threshold level, below which no apparent embryotoxic effects occur.

On the other hand, there is a long standing relationship between carcinogens, mutagens and teratogens. Most mutagens are held to be carcinogens, many carcinogens to be teratogens and most mutagens might be expected to be teratogens (Ferguson and Ford, 1997). Other aspects that link mutagenesis and teratology relate to the observation that a significant number of (major) malformations are caused by single gene defects or by chromosomal aberrations. Moreover, teratological endpoints that more recently gained attention, such as prenatal predisposition to cancer or prenatal hormone disruption, do not follow a sigmoid dose-effect pattern.

For almost all mutagens and carcinogens, the response is directly and linearly related to the dose of the agent; the line intercepts the effect axis at the point where the background load of cancers and mutations is. The consequence is that there is no dose without an effect. Or phrased otherwise: the risk is linearly related to the increase of the dose, regardless of how small that dose might be. The problem with this point of view is that it is hardly possible to set up experiments that measure long term effects in the low dose area, which is most relevant for pollutants that are carcinogens and/or mutagens.

Nevertheless, the evidence for this straight-line, no threshold, dose-response relation is twofold. The theoretical basis is that mutagens and carcinogens have effects as results of single events in the DNA, and that thus no matter how small the dose of the agent applied, mutated genes or steps leading to cancer occur. The experimental evidence is based on the finding that evidence accumulated over years for selected mutagens, supported the linear dose-effect relationship concept. Ionizing radiation is an example, in relation to mutagenicity and carcinogenicity. In different, subsequent steps during the past 50 years, information has been accumulated on effects at ever lower doses. It has not been possible to demonstrate beyond doubt the existence of a threshold dose. In some experiments, "hormetic" effects were described, whereby low dose irradiation has a protective rather than a deleterious effect (Calabrese and Baldwin, 2002). However, other observations indicated the existence of "Low Dose Hypersensitivity", leading to a relative higher efficiency per dose unit of very low doses ionising radiation, at inducing mutations and cancer (Waldren *et al.,* 1986; Frankenberg *et al.,* 2002). Induction of DNA repair could explain both "Low Dose Hypersensitivity" (the dose is too low to fully induce the relevant DNA repair mechanisms) and adaptation (a first exposure gives some protection against a second larger exposure) and that phenomenon could be important in mutagenesis and carcinogenesis. It seems, however, unlikely that an exposure to a mutagen might, in the absence of subsequent intensive exposures, lower background levels of mutation and cancer. In conclusion, overall, the linear dose effect curve was found at always lower doses of radiation.

However, the main question, from a public health point of view is “do we have evidence that pollutants interfere with congenital malformations, at the concentrations that occur in the environment?” The answer to this question has two main aspects: does the mechanism allow us to expect effects at low levels of exposure; and is experimental evidence available, for this interference? The TDS offers an interesting example in this respect. The TDS and its related congenital anomalies, such as hypospadias and cryptorchidism, are likely to be explained by a reduction in testosterone production during the critical phase of masculinization (Sharpe, 2006). This testosterone production might be impaired by (among other products) phthalates. Swan *et al.* (2006) provide evidence that phthalates, in concentrations to which 25% of the American population is exposed, have similar adverse effects on fetal testosterone production in humans as they do in rats.

In conclusion, while many major malformations might show a sigmoid dose-response curve, an increasing number of congenital anomalies most probably show a linear dose-effect relationship without a threshold value, for exposure conditions as they occur in the environment.

(2) *Exposure windows and latency periods*

In teratology, not only the dose is important. Timing also matters. The sensitive period during which embryonic development is susceptible to the action of teratogens is, as a rule, early in pregnancy. However, it might differ substantially from substance to substance. The sensitive "exposure window" is related to the timing of the organogenesis. On this subject, different excellent reviews exist (see Koren, 1990).

A specific aspect of prenatal exposure is that not only direct effects matter. DES provides an example of how embryonic programming might result in cervical and/or vaginal cancer more than 15 years after the original exposure. The TDS hypothesis links testicular cancer with testosterone deficiency during the prenatal life. These are noticeable examples of the embryonic origin of adult diseases. Moreover, this concept should not be limited to exposure during pregnancy. Preconceptual parental exposure, in particular during the last few months prior to fertilisation, may also be important for transgenerational health, even though no clear epidemiological evidence is currently available on the relevance of such exposure.

(3) *Individual differences in susceptibility*

Congenital anomalies offer clear examples of the interaction between genetic background and environmental influences. Individual, genetically coded differences can be most important to explain variations in the outcome of congenital anomalies. In addition, in related fields, individual variability is important to explain susceptibility. PCBs, for example, cause breast cancer when the CYP 1A1 alleles are present.

(4) *Multicausality*

Environmental impacts on congenital anomalies offer an outstanding illustration of what has been called the complex background of disease. Congenital anomalies arise from a variety of determinants. Genetic condition, individual and familial risk factors, social relationships, living conditions, neighbourhoods and communities, institutions, and health, environmental, social and economic policies - all these influence the prevalence of congenital anomalies.

One should be aware of interactions at each of these levels. Interactions of chemicals and/or physical carcinogens might result in additive, synergistic or antagonistic effects. In addition, the interaction between a chemical teratogen with the body proteins might be more complex than a one cause – one effect type of relationship might be able to forecast.

Table 50. Environmental and health sciences and their main directions of error (EEA, 2005)

Scientific studies	Methodological features	Main[1] directions of error – increases chances
Experimental studies (animal laboratory)	High doses	False positive
	Short (in biological terms) range of doses	False negative
	Low genetic variability	False negative
	Few exposures to mixtures	False negative
	Few fetal-lifetime exposures	False negative
	High fertility strains	False negative (developmental/reproductive endpoints)
Observational studies (wildlife and humans)	Confounders	False positive
	Inappropriate controls	False positive/ more false negative
	Non-differential exposure misclassification	False negative
	Inadequate follow-up	False negative
	Lost cases	False negative
	Simple models that do not reflect complexity	False negative
Both experimental and observational studies	Publication bias towards positives	False positive
	Scientific cultural pressure to avoid false positives	False negative
	Low statistical power (e.g. from small studies)	False negative
	Use of 5% probability level to minimise chances of false positive	False negative

[1] *Some features can go either way (e.g. inappropriate controls) but most of the features mainly err in the direction shown in the table*

This complexity contrasts with the relative simplicity that underlies laboratory, epidemiological and wildlife studies. Table 50 provides a list of the type of errors that might result from contemporary approaches in teratology. The list provides an illustration of how studies designed to accept or reject hypotheses about simple factors have serious limitations for recognising multicausal mechanisms.

It is currently unclear what exactly should be done to bring multicausality on board in the pool of data we have today on the environment-congenital anomalies relationship. A conference organised in 2001 in Stockholm (Sweden) on this issue concluded: "Improved scientific methods to achieve a more ethically acceptable and economically efficient balance between the generation of "false negatives" and "false positives" are needed.

5. Implications for stakeholders

It is of core importance to inform stakeholders in an accessible way on the scientific state of the art on environmental teratogens. In this section, the implications for clinical and public health practitioners, lawyers and legal advisers, policy makers and the media are discussed.

Box 3 summarizes the principles and core facts as they emerged from this seminar and these are relevant for the communication strategy.

Box 3. Core messages on information about teratogens in the environment

PRINCIPLES

1. Precautionary approach: limit exposure to pollutants, also before conception
2. Absence of evidence is not evidence of absence
3. Congenital anomalies resulting from environmental exposure can be avoided

CORE FACTS

1. Environmental pollutants proven to be teratogenic: ionizing radiation, mercury, lead
2. Environmental pollutants that are most likely to be teratogenic: e.g. 2,3,7,8 TCDD, solvents, many pesticides
3. Environmental exposures that are shown to be teratogenic: e.g. nitrates in drinking water, living near landfill sites

Relatively few environmental pollutants and pollution exposures exist for which strong conclusions can be drawn. Even for a known neurotoxic teratogen as lead, studies of lead in drinking water report conflicting results in detecting neural tube defects (Winder, 1993; Bellinger, 1994). Such a negative result is a clear indication that "absence of evidence is not evidence of absence". Moreover, studies of teratogens should be evaluated in a predefined assessment context. "Not proven" does not mean "not teratogenic" or "not harmful" to the conceptus. It rather means that available methodologies and techniques failed to unequivocally prove a cause-effect relationship. On the other hand, many associations between environmental exposure and congenital anomalies have been described. Environmental pollution is usually present at levels predicted to lead to a small but significant excess risk. Because pollution is widespread and large populations might be exposed, even a limited increase in the risk might have a potential public health significance. Another limitation concerns the finding that for most associations that have been described, no mechanism of causation is known. Castilla *et al.* (2001) provide nine criteria for a contextual assessment, but a systematic evaluation of a meanwhile increasing list of suspected teratogens is still lacking.

5.1. Advising in the public health sector

In front of patients, the clinical counsellor faces the border between false reassurance and false alarm. What can be done to prevent teratogenic effects resulting from environmental exposure? A precautionary approach in this case entails the following aspects:

(1) *reduce personal exposure* to environmental pollutants. Be aware of lead and nitrate concentrations in drinking water. Reduce exposure to pesticides at home and in the garden. Realize that this is important not only during (the first weeks of) pregnancy, but also prior to conception,

(2) realize about the *risk - benefit* balance. Limiting personal exposure to pesticides should not result in limiting fresh fruits and vegetables in the diet. However, it should lead to more awareness on the importance of washing and peeling fruits and vegetables prior to consumption. The advice should equally be based on a well understood interpretation of the routes of exposure. For most persistent pesticides (used in the past but still common in the environment), the most common route of exposure is by the consumption of fish and fatty foods (milk, butter, meat). This appears to be even more important than exposure to pesticide residues in fruit and vegetables. In a similar way, the possible problems with lead and nitrates in drinking water should not result in massive consumption of bottled water. In most countries

and areas of the EU, not only is the tap water of excellent drinking quality, but bottled water causes a high environmental cost e.g. in the energy and the emissions to produce, package and transport it,

(3) expand *prenatal services* with advice on environmental issues. Prenatal counsellors and service providers such as general practitioners, midwifes and social nurses should be aware of environmental risks and hazards such as e.g. those related to drinking water. In their practice, they should pay attention to identify whether the drinking water has low lead and nitrates levels.

5.2. Lawyers and legal advisers

Both in the U.S. and the U.K., but much more rarely in other EU countries, court trials have taken place, on cases where parents presented evidence on environmental exposures that might have caused malformations in their child. In particular, cases related to pesticide exposure are well documented.

As evidenced by the data in this book, no methods exist that link a particular congenital anomaly in a newborn with a defined environmental exposure. This means that in each of these court cases, the evidence will be contextual and circumstantial. Nevertheless, it can not be excluded, but might be rare to occur, that the context and the circumstances of exposure are such that they might be the cause of the malformation, beyond any reasonable doubt.

Nevertheless, one might wonder about the effects of a legal DES action in Europe. As explained in section 3.1 of this paper, a most plausible association between exogenous estrogens and the resulting congenital anomalies has been described for DES. Cases on DES have the capacity to provide the much–needed training in legal matters on congenital anomalies.

5.3. Policy makers and advisors

Two main areas of importance for policy makers and advisers have emerged from the discussion on teratogens during the past fifty years: the way one deals with them in environmental policy and health plans; and communication in cases of crisis.

Table 51 lists the 12 out of 38 national environmental health policy plans (NEHAP) from the WHO – Europe area, that mention congenital anomalies. As compared to most issues considered in these plans, CA is of minor importance. Most plans only refer to a possible link between CA and environmental exposure. Only the NEHAPs of Germany and France also announce a policy that targets CA.

Table 51. Congenital anomalies policy in 12 National Environmental Health Action Plans

Country (year of publication or issue)	Prevalence (%)	Context
Belarus (1999)	1.4	Congenital malformations as a main cause of infant mortality
Belgium (2003)		Congenital anomalies linked with pesticide and tobacco use
Bulgaria (2002)		Congenital malformations rank high in infant death causes in areas with industrial pollution by lead and zinc (Svishtov, Rousse, Stara Zagora)
Czech Republic (2002)	3.7	Congenital defects as a cause of neonatal mortality
Croatia (2002)		On the role of registers for congenital anomalies
Estonia (1999)		(-) Congenital malformations as a declining cause of perinatal death during the period 1988-1997 (-) Increased incidence of congenital malformations during the period 1989-1995 (by factor 2.3)
France (2004)		(-) More support for monitoring of congenital anomalies and registries (-) Limiting exposure to reproduction toxins in the occupational environment
Georgia (2001)	0.9	
Germany (1999)		Bans on biocide products, for private use and consumers, which are classified as carcinogenic, mutagenic or teratogenic.
Malta (1997)		Database on congenital anomalies
Poland (2000)		Announces the publication of a "report on the magnitude of population environmental exposure to mutagenic, carcinogenic and teratogenic risk factor".
Romania (1997)		No significant increase in the incidence of congenital malformations in newborns whose mothers were exposed to Chernobyl fall out (1986) in the first three months of pregnancy

The limited attention to CA in NEHAPs is remarkable, as there is no doubt that some pollutants are teratogens. Even if the prevalence of major congenital anomalies is low (approximately 2% of births; 2 in 1,000 births for neural tube defects; 2 in 10,000 for gastroschisis), and even when environmental pollution results in a relatively moderate increase, to have a child affected by a major anomaly (residence near landfill sites has a RR of 1.2 - 1.3), the effect at population level might be important, as environmental pollution generally affects large parts of the population: 20% of the population of Great Britain lives within 2 km of a landfill site (Elliot *et al.,* 2001); 25% of the Americans are exposed to phthalate concentrations that might impair masculinization (Silva *et al.,* 2004)). As with any other pollution-related health effect, congenital malformations should be part of environmental (health) policy plans, on a systematic basis.

Retrospective analysis of major accidents in which teratogenic pollutants were released (Seveso, Chernobyl) showed that authorities were ill prepared to provide correct and effective information on the accident, but also on the preventive measures to be taken under these stressful conditions (Kalter, 2003). The same observation applies to the information that is provided when the press discovers, for instance, small clusters of rare conditions in a limited geographic area. Botting and Stone (2002) report the case of four children who between February 1989 and May 1990 were born with transverse limb reduction defects in the same town on the Isle of Wight (U.K.).

The only common characteristic identified by the mothers was that they had swum in the sea during pregnancy. In January 1994, the Sunday Times, a major British national Sunday newspaper, with a large circulation, broke the news and catalysed a media event. Botting and Stone (2002) show the difficulties that the authorities had to convince the public that swimming in the sea was not the cause of the malformations. Such problems can only be handled in a rational way, when authorities are prepared with credible communication plans to deal with this type of incidents.

5.4. Scientists and media

Media have an essential role in reporting on public health issues in general and consequently on congenital malformations. Media tend to report on these issues when the situation is characterized by "news value". This is the case "*par excellence*" when accidents happen with environmental impacts.

The interaction of the media with scientists, in particular in cases of emergency, is often far from optimal.

A number of objective reasons explain this situation:

- scientists are in general not, or only marginally, trained to deal with media,
- an area such as congenital malformations is characterized by probabilities, hypotheses and scientific uncertainty. This is difficult to communicate, as people often expect "certainties" from science.

Based on the analysis of two separate studies of congenital malformations and the way they were communicated in the British media, the following recommendations can be provided:

- adapt a proactive approach: communicate the significant results of a new study yourself. Time the communication in conjunction with its publication in a scientific journal. Be well prepared in facing the press. Avoid the journalists having the feeling that data are hidden or remain not disclosed,
- adapt a proactive attitude in research. Make sure you have access to good quality monitoring systems. Establish societally significant hypotheses and test them before the media confront you with the situation,
- do not only rely on your own information. Work together with other registries both within the country and internationally,
- adapt a caring, non-directive attitude to a distressed public. Opinion polls show that when it comes to environmental issues, the public trusts scientists (although to a lesser extent than they trust environmental NGO's). This can only be maintained if the public trusts that scientists work in their interests,
- submit your work to the regular scientific quality controls and publish it in international, peer reviewed journals (Botting and Stone, 2002).

In conclusion, dealing with the different professional stakeholders in the societal debate necessitates specific information, with defined targets, tailored on the specific needs of each of these target groups. Making these needs and agendas transparent; and organizing an open debate on what science on environmental teratogens can offer, and what it cannot provide; will improve the societal value of the research data.

6. Discussion and conclusions

Clinical evidence shows that exposure of the embryo to some medical drugs, biological agents such as rubella, or physical pollutants such as X-rays, results in risks of congenital anomalies. This is part of the basis for questioning whether chemical or physical pollutants also contribute to the load of congenital anomalies in human populations.

Next to ionising radiation, chemical pollutants as lead and methyl mercury are also known teratogens. For many pesticides, teratogenic action has been described in laboratory animals systems. Solvents such as styrene and many hair dyes; and endocrine disrupters such as bisphenol A, phthalates, dioxins and vinyl chloride, are substances that are present in air, water, soil and for which (not always conclusive) evidence for their teratogenic character exists.

The next question is, however, whether exposure conditions as they exist in the environment indeed result in increased risks of embryo-feto-toxicity. The answer is that the evidence in this area of research is increasing. It is best established for risks associated with lead and nitrates in drinking water and for people living in the neighbourhood of (hazardous) waste deposit sites. These latter show a 20% increased risk for children with chromosomal aberrations and a 33% increased risk for children with non-chromosomal congenital anomalies. The highest relative risk values have been reported in association with exposure to pesticides. Cryptorchidism is the most common congenital anomaly. Children of farmers who purchase large quantities of pesticides have a 2.3 times higher risk of cryptorchidism than the population as a whole. If the total effective xenoestrogen burden (TEXB) is taken into account, the relative risk for cryptorchidism in these children rises to 2.8 (Olea, 2006).

Of particular interest is the increase of congenital anomalies such as cryptorchidism and hypospadias, in a number of industrialised countries. Often this increase is parallel to the increase in testicular cancers and a decrease of the sperm quality in men living in the same environment. The totality of these findings might be explained by the Testicular Dysgenesis Syndrome (TDS) hypothesis, which combines genetic and environmental evidence into one mechanism in which the testosterone secretion is of key importance (Toppari, 2006).

Moreover, cryptorchidism and hypospadias are not isolated risks. They should be seen in a wider context of the social impact of decreased fertility of couples in industrial countries. To address this wider problem, a coordinated strategy between scientists and other stakeholders is imperative.

The accumulating evidence for pollutants that are teratogens should also be evaluated in the context of changing paradigmatic shifts on evaluating the mechanisms underlying congenital anomalies. They include different aspects:

- not only high, laboratory, concentrations matter. Also important are everyday concentrations of bisphenol A, phthalates, DDE and other (potential) teratogens, administered as daily doses over a period that covers both the preconceptual and the postconceptual period,

- for environmental teratogens, both situations with a threshold level and a sigmoidal dose-response curve, and situations with a linear dose-effect relationship without threshold dose, apply. The dose–effect discussion has an extra dimension, as many environmental teratogens also act as mutagens, carcinogens and as substances that impair fertility,

- for both mutagenic and epigenetic receptor binding chemicals, the dose-response might be supralinear in the very low dose range, the response being greater than expected on the basis of a linear extrapolation from a higher dose. For mutagenic substances, this might result, amongst other causes, from suboptimal induction of DNA repair. Endocrine disrupting chemicals often show non-linear dose-response relationships, and even inverted U-type dose-responses have been observed, e.g. for bisphenol A, with the response going through a maximum and then decreasing as the dose increases,

- timing of administration of teratogens is crucial. This is definitely related to fetal programming, but, moreover, there is accumulating evidence that pre-conceptual exposure of sperm and ova to teratogens also influences the state of health, both at birth and during adulthood,

- teratogens do not only result in direct effects. DES exposure is a historical example of transplacental tumourigenesis with a long latency period. This type of long latency and fetal origin of adult diseases has also been evidenced for other teratogens. These effects are rarely covered by tests for teratogenesis,

- tests for teratogenesis, even today, are typical examples of (monocausal) mechanical models of disease, whereas multicausal situations apply, particularly in environmental exposure. Although the scientific paradigm of monocausality guided medical research for many years, it is increasingly being questioned. In particular, it is difficult to apply it to environmental teratogens for which exposure to mixtures, rather than exposure to individual substances applies. Moreover, the occurrence of teratogenic outcome depends on such factors as lifestyle, social relationships, medical practice and policy. This necessitates refining our knowledge on exposure pathways in a holistic way.

The area of environmental teratogens is characterised by relatively limited scientific proof, increasing insight and understanding, uncertainties on most fundamental aspects and important public concern. This raises the question on how we should act. The answer appears to be different for the different stakeholders:

- Scientists should deal in a more professional way with risk communication. Part of this includes formulating clear points of view. These entail:

    - strong hypotheses,

    - positive lists of environmental teratogens and exposure conditions: which pesticides are the ones that influence cryptorchidism, hypospadias and congenital anomalies in general? What do we do with the finding that both chromosomal and non-chromosomal congenital anomalies are increased in populations living near waste deposit sites, when in Great Britain 20% of the population lives within 2,000 metres of a (hazardous) waste deposit site?

    - a new generation of tests and standards. Tests that allow a more holistic approach are needed. At the same time, they should be ethically sound and economically efficient. They should equally lead to a new generation of guidelines, bringing the precautionary principle on board, next to the results of the, rather unrealistic, laboratory experiments,

- General practitioners, clinicians and health advisers are an important communication forum for the population at large. Without any doubt, they should pay more attention to the existence of environmental teratogens. They should be enabled to define risks at an individual or family level and to communicate them in practical terms. They should adopt restrictive drug prescription behaviour towards pregnant women and towards couples prior to conception. Nevertheless, their task remains difficult as, for many relevant questions, we do not have clear cut answers and scientific uncertainty prevails, as regards practical suggestions. How to interpret, e.g. for a large audience, the presence of pesticides in placenta tissue?

- couples preparing for children should be aware of possible teratogenic pollutants in their lifestyle environment (medical drugs, pesticide use at home and in the garden, cosmetics). On a long term basis, it is important to be aware of the quality of the internal body environment. Some pollutants can affect the health of the children after both pre- and post-conceptional exposure. Health effects sometimes appear after a long latency period,

- NGOs are the societal counterbalance of a highly motivated industrial opposition that for ethical, perceptual or economic reasons avoids being associated with the production, promotion or sale of teratogenic products or by-products. When it comes to environmental teratogens, environmental and consumers NGOs, in particular, should continue to find examples of congenital anomalies associated with chemicals in products of daily use (e.g. packaging, disinfectants, wood-treatment substances),

- industry should continue to test, in a systematic way, new chemicals for congenital anomalies. The environmental directorate of OECD (1995, 2004) published a series of documents that provide guidance on methodological aspects, interpretation of data and strategy for testing of chemicals. The documents target a wide series of environmental and human health outcomes, including reproductive toxicology.

- policy makers are not particularly charmed by the discussion on environmental teratogens. Thus far, the general rule is to ban products that are classified as teratogenic (in a similar way as carcinogens and mutagens are "banned" for general use). Most countries also manage a congenital malformations register. EU-wide, these data are collected in EUROCAT. According to the current information on congenital anomalies causing substances and exposure conditions, this is a rather refraining attitude, that will not succeed in bringing down the teratogenicity load from, for example, pesticides or living near waste deposit sites.

An important point in this discussion is "what level of evidence is needed, to generate action at the policy level?"

Currently, different sets of criteria for answering this question exist. The criteria on association and causation developed by Sir Bradford Hill (1965) are probably the oldest ones. They relate to the strength of the association, its consistency, its specificity, its temporality, the existence of biological gradients such as a dose-effect relationship, the experimental evidence, plausibility, coherence and analogy. More recently, related sets of criteria have been developed. In its evaluation of endocrine disrupters, WHO (Damstra *et al.,* 2002) used a system in which the following evaluation criteria were handled: temporality, strength of association between exposure and health effect, biological consistency of the observations, plausibility and evidence of recovery. All these allow the determination of the overall strength of evidence, both for the hypothesis of a particular causal relation between exposure and effect, and for the mechanism explaining that relationship.

If these criteria are applied to the environmental congenital anomalies discussion, policy action is indicated for pesticides and other endocrine disrupters. In a European context, there is growing evidence for bringing endocrine disrupters under prior authorisation under REACH. This general measure should be accompanied by specific measures in the EU member countries to cope with specific problems, as revealed by their National Environmental Health Action Plans.

In the medium term, a new generation of standards, interpreting the different environmental aspects that affect congenital anomalies, needs to be developed.

It would indeed be irresponsible to delay policy actions until full scientific proof is available, for a number of reasons. The two most prominent are these:

- if we continue to build up an important burden of environmental chemicals and exposure conditions, this might affect generations to come,
- as environmentally mediated congenital anomalies are avoidable health burdens, any excess cases represent a failure of our environmental health protection system.

References

Anon (1987) Teratology Society position paper: recommendations for vitamin A use during pregnancy, *Teratology* **35**, 269–275.

Arbuckle, T.E., Sherman, G.J., Corey, P.N., Walters, D. and Lo, B. (1988) Water nitrates and CNS birth defects: a population based case–control study, *Arch. Environ. Health* **43**, 162–167.

Bellinger, D. (1994) Teratogen update: lead, *Teratology* **50**, 367–373.

Bernheim, J. (2001) The 'DES syndrome': A prototype of human teratogenesis and tumourigenesis by xenoestrogens?, in P. Nicolopoulou-Stamati, L. Hens, and C.V. Howard (eds.) *Endocrine Disrupters – Environmental health and policies*, pp. 81-118, Kluwer Academic Publishers, Dordrecht, The Netherlands.

Bishop, C.A. and Gendron, A.D. (1998) Reptiles and amphibians: shy and sensitive vertebrates of the Great Lakes basin and Saint Lawrence River, *Environ. Monit. Assess.* **53**, 225-244.

Bishop, C.A., Brooks, R.J., Carey, J.H., Ng, P., Norstrom, R.J. and Lean, D.R.S. (1991) The case for a cause-effect linkage between environmental contamination and development in eggs of the common snapping turtle (*Chelydra s. serpentina*) from Ontario, Canada. *J. Toxicol. Environ. Health* **33**, 521-547.

Blot, W.J. (1975) Growth and development following prenatal and childhood exposure to atomic radiation, *J. Radiat. Res. Suppl.* **82**, 82-88.

Botting, B. and Stone, D. (2002) Experience of dealing with the media on congenital abnormality research, *iCOT* **6**, 1-4.

Bound, J.P., Harvey, P.W., Francis, B.J., Awwad, F. and Gatrell, A.C. (1997) Involvement of deprivation and environmental lead in neural tube defects: a matched case–control study. *Arch. Dis. Child* **76**, 107-112.

Bowerman, W.W., Kubiak, T.J., Holt, J.B. Jr, Evans, D.L., Eckstein, R.G., Sindelar, C.R., Best, D.A. and Kozie, K.D. (1994) Observed abnormalities in mandibles of nestling bald eagles *Haliaeetus leucocephalis, Bull. Environ. Contam. Toxicol.* **53**, 450-457.

Brilakis, E., Fousteris, E. and Papadopulos, S.I. (2006) Congenital abnormalities in Greece - Functional evaluation of statistical data 1981-1995, in P. Nicolopoulou-Stamati, L. Hens, and C.V. Howard (eds.) *Congenital anomalies and the environment*, Springer, Dordrecht, The Netherlands.

Calabrese, E.J. and Baldwin, L.A. (2002) Defining hormesis, *Hum. Exp. Toxicol.* **21**, 91-97.

Castilla, E.E., López-Camelo, J.S., Campaña, H. and Rittler, M. (2001) Epidemiological methods to assess the correlation between industrial contaminants and rates of congenital anomalies, *Mutat. Res.* **489**, 123-145.

Cedergren, M.I., Selbing, A.J., Lofman, O. and Kallen, B.A. (2002) Chlorination byproducts and nitrate in drinking water and risk for congenital cardiac defects, *Environ. Res.* **89**, 124–130.

Clarkson, T.W. (2002) The three modern faces of mercury, *Environ. Health Perspect.* **110** (Suppl 1), 11–23.

Clementi, M., Ludwig, K. and Andrisani, A. (2006) Association of intra-uterine exposure with drugs: the thalidomide effect, in P. Nicolopoulou-Stamati, L. Hens, and C.V. Howard (eds.) *Congenital anomalies and the environment*, Springer, Dordrecht, The Netherlands.

Conover, R.M. (1984) Occurrence of supernormal clutches in Laridae, *Wilson. Bull.* **96**, 249-267.

Conover, R.M. and Hunt, G.L. (1984) Female-female pairing and sex ratios in gulls: a historical perspective, *Wilson. Bull.* **96**, 619-625.

Constable, J.D. and Hatch, M.C. (1985) Reproductive effects of herbicide exposure in Vietnam: recent studies by the Vietnamese and others, *Teratog. Carcinog. Mutagen* **5**, 231–250.

Croen, L.A., Shaw, G.M., Sanbonmatsu, L., Selvin, S. and Buffler, P.A. (1997) Maternal residential proximity to hazardous waste sites and risk of selected congenital malformations, *Epidemiology* **8**, 347-354.

Croen, L.A., Todoroff, K. and Shaw, G.M. (2001) Maternal exposure to nitrate from drinking water and diet and risk for neural tube defects, *Am. J. Epidemiol.* **153**, 325-331.

CSTEE – Scientific Committee on Toxicity, Ecotoxicity and Environment (1999) *Opinion on human and wildlife health effects of endocrine disrupting chemicals, with emphasis on wildlife and on ecotoxicity test methods*. Expressed at the 8th CSTEE plenary meeting, Brussels 4 March 1999. DCXXIV, Consumer Policy and Consumer Health Protection, European Commission.

Damstra, T., Barlow, S., Bergman, A., Kavlock, R. and Van der Kraak, G. (2002) *Global Assessment of the State-of-the-Science of Endocrine Disrupters*, WHO publication no. WHO/PCS/EDC/02.2, World Health Organization, Geneva, Switzerland.

Di Tanna, G.L., Rosano, A. and Mastroiacovo, P. (2002) Prevalence of gastroschisis at birth: retrospective study, *Br. Med. J.* **325**, 1389-1390.

Dolk, H. and Vrijheid, M. (2003) The impact of environmental pollution on congenital anomalies, *British Medical Bulletin* **68**, 25-45.

Dorsch, M.M., Scragg, R.K., McMichael, A.J., Baghurst, P.A. and Dyer, K.F. (1984) Congenital malformations and maternal drinking water supply in rural south Australia: a case–control study, *Am. J. Epidemiol.* **119**, 473-486.

Dummer, T.J., Dickinson, H.O. and Parker, L. (2003) Adverse pregnancy outcomes around incinerators and crematoriums in Cumbria, north west England, 1956–1993, *J. Epidemiol. Comm. Health* **57**, 456-461.

EEA – European Environmental Agency (2005) Personal communication.

Elliott, P., Briggs, D., Morris, S., de Hoogh, C., Hurt, C., Jensen, T.K., Maitland, I., Richardson, S., Wakefield, J. and Jarup, L. (2001) Risk of adverse birth outcome s in populations living near landfill sites, *Br. Med. J.* **829**, 363-368.

Erickson, J.D. (1980) Down syndrome, water fluoridation, and maternal age, *Teratology* **21**, 177-180.

Erickson, J.D., Mulinare, J. and McClain, P.W. (1984a) Vietnam veterans' risk for fathering babies with birth defects, *JAMA* **252**, 903-912.

Erickson, J.D., Mulinare, J. and McClain, P.W. (1984b) *Vietnam veterans' risk of fathering babies with birth defects*, Centers for Disease Control, Atlanta, USA.

EUROCAT Working Group (2002) EUROCAT Report 8: *Surveillance of Congenital Anomalies in Europe 1980–1999*, (online) www.eurocat.ulster.ac.uk/pubdata (May 10th 2005).

Ferguson, L.R. and Ford, J.H. (1997) Overlap between mutagens and teratogens, *Mutat. Res.* **396**, 1- 8.

Fox, G.A. (1992) Epidemiological and pathobiological evidence of contaminant-induced alterations in sexual development in free-living wildlife, in T. Colborn, and C. Clement, (eds.) *Chemically-Induced Alterations in Sexual and Functional Development: The Wildlife/Human Connection*. Princeton, NJ: Princeton Science Publishing Co.

Fox, G.A. (1993) What have biomarkers told us about the effects of contaminants on the health fish-eating birds in the Great Lakes? The theory and a literature review, *J. Gt. Lakes Res.* **19**, 722-736.

Fox, G.A. (2001) Wildlife as sentinels of human health effects in the Great Lakes Saint Lawrence basin, *Environ. Health Perspect.* **109**, 853-861.

Frankenberg, D., Kelnhofer, K., Bär, K. and Frankenberg-Schwager, M. (2002) Enhanced neoplastic transformation by mammography X-rays relative to 200 kVp X-rays: indication for a strong dependence on photon energy of the RBE*M* for various endpoints, *Radiat. Res.* **157**, 99–105.

Garcia, A.M. (1998) Occupational exposure to pesticides and congenital malformations: A review of mechanisms, methods, and results, *Am. J. Ind. Med.* **33**, 232–240.

Geschwind, S.A., Stolwijk, J.A., Bracken, M., Fitzgerald, E., Stark, A., Olsen, C. and Melius, J. (1992) Risk of congenital malformations associated with proximity to hazardous waste sites, *Am. J. Epidemiol.* **135**, 1197–1207.

Gilbertson, M. (1988) Epidemics in birds and mammals caused by chemicals in the Great Lakes, in M.S. Evans (ed.) *Toxic Contaminants and Ecosystem Health; A Great Lakes Focus*, New York: Wiley, 133-152.

Goldman, L.R., Paigen, B., Magnant, M.M. and Highland, J.H. (1985) Low birth weight, prematurity and birth defects in children living near the hazardous waste site, Love Canal. *Hazardous Waste Hazardous Mater* **2**, 209–223.

Grasman, K.A., Scanlon, P.F. and Fox, G.A. (1998) Reproductive and physiological effects of environmental contaminants in fish-eating birds of the Great Lakes: a review of historical trends, *Environ. Monit. Assess.* **53**,117-145.

Greenwald, P., Barlow, J.J., Nasca, P.C. and Burnett, W.S. (1971) Vaginal cancer after maternal treatment with synthetic oestrogens, *N. Engl. J. Med.* **285**, 390 – 392.

Hanselaar, A.G.J.M., Van Leusen, N.D.M., De Wilde, P.C.M. and Vooijs, G.P. (1991) Clear cell adenocarcinoma of the vagina and cervix. A report of the Central Netherlands Registry with emphasis on early detection and prognosis, *Cancer* **67**, 1971-1978.

Harada, M. (1978) Congenital Minamata disease: intrauterine methylmercury poisoning, *Teratology* **18**, 285–288.

Heinonen, O.P. (1973) Diethylstilboestrol in pregnancy. Frequency of exposure and usage patterns, *Cancer* **31**, 573-577.

Hens, L. (2000) Health and environmental impact assessment, in P. Nicolopoulou-Stamati, L. Hens, and C.V. Howard (eds.) *Health impacts of waste management policies*, Kluwer Academic Publishers, Dordrecht, The Netherlands.

Herbst, A.L., Ulfelder, H. and Poskanzer, D.C. (1971) Adenocarcinoma of the vagina. Association of maternal stilboestrol therapy with tumour. Appearance in young women, *N. Engl. J. Med.* **284**, 878-881.

Hoffman, D.J., Smith, G.J. and Rattner, B.A. (1993) Biomarkers of contaminant exposure in common terns and black-crowned night herons in the Great Lakes, *Environ. Toxicol. Chem.* **12**, 1095-1103.

Hughes, T.E. and Flood, T. (1999) *Arizona gastroschisis report, 1986-1996*, Arizona Department of Health Services. Series 1998:4, 12pp.

Infante, P.F., Wagoner, J.K. and Waxweiler, R.J. (1976) Carcinogenic, mutagenic and teratogenic risks associated with vinyl chloride, *Mutat. Res.* **41**, 131–142.

Kalter, H. (2003) Teratology in the 20th century. Environmental causes of congenital malformations in humans and how they were established, *Neurotoxicol. Teratol.* **25**, 131-282.

Kelsey, F.O. (1982) Regulatory aspects of teratology: role of the Food and Drug Administration, *Teratology* **25**, 193 – 199.

Kirstensen, P., Irgens, L.M., Andersen, A., Bye, A.S. and Sundheim, L. (1997) Birth defects among offspring of Norwegian farmers, 1967-1991, *Epidemiology* **8**, 537-544.

Koppe, J. (2006) Congenital diseases related to environmental exposure to dioxins, in P. Nicolopoulou-Stamati, L. Hens, and C.V. Howard (eds.) *Congenital anomalies and the environment*, Springer, Dordrecht, The Netherlands.

Koren, G. (1990) *Maternal-fetal toxicology. A clinician's guide*, Marcel Dekker Inc. New York.

Lagakos, S.W., Wessen, B.J. and Zelen, M. (1986) An analysis of contaminated well water and health effects in Woburn, Massachusetts, *J. Am. Stat. Assoc.* **8**, 583-596.

Lilienfeld, D.E. and Gallo, M.A. (1989) 2,4-D, 2,4,5-T, and 2,3,7,8-TCDD: an overview. *Epidemiol. Rev.* **11**, 28–58.

Little, A.B. (1976) Proceedings US District Court, District of Massachusetts, pp. 1514-1515.

Ludwig, J.P., Kurita-Matsuba, H., Auman, H.J., Ludwig, M.E., Summer, C.L., Giesy, J.P., Tillitt, D.E. and Jones, P.D. (1996) Deformities, PCBs, and TCDD-equivalents in double-crested cormorants (*Phalacrocorax auritus*) and Caspian terns (*Hydroprogne caspia*) of the upper Great Lakes 1986-1991: testing the cause-effect hypothesis, *J. Gt. Lakes Res.* **22**, 172-197.

Macdonell, J.E., Campbell, H. and Stone, D.H. (2000) Lead levels in domestic water supplies and neural tube defects in Glasgow, *Arch. Dis. Child.* **82**, 50-53.

Mantovani, A. and Maranghi, F. (2006) Endpoints for prenatal exposures in toxicological studies, in P. Nicolopoulou-Stamati, L. Hens, and C.V. Howard (eds.) *Congenital anomalies and the environment*, Springer, Dordrecht, The Netherlands.

Mastroiacovo, P., Spagnolo, A., Marni, E., Meazza, L., Bertollini, R., Segni, G. and Borgna-Pignatti, C. (1988) Birth defects in the Seveso area after TCDD contamination, *JAMA* **259**, 1668-1672.

Miller, R.W. (1985) Congenital PCB poisoning: a reevaluation, *Environ. Health Perspect.* **60**, 211-214.

Miller, R.W. and Mulvihill, J.J (1976) Small head size after atomic irradiation, *Teratology* **14**, 355-358.

Monteleone Neto, R. and Castilla, E.E. (1994) Apparently normal frequency of congenital anomalies in the highly polluted town of Cubatao, Brazil, *Am. J. Med. Genet.* **52**, 319-323.

Needleman, H.L., Pueschel, S.M. and Rothman, K.J. (1974) Fluoridation and the occurrence of Down's syndrome, *N. Engl. J. Med.* **291**, 821-823.

Nicolopoulou-Stamati, P., Hens, L. and Howard, C.V. (eds.) (2001) *Endocrine Disrupters: Environmental Health and Policies*, Kluwer Academic Publishers, Dordrecht, The Netherlands.

Nurminen, T. (2001) The epidemiologic study of birth defects and pesticides, *Epidemiology* **12**, 145-146.

OECD (1995) *Guidance document for the development of OECD guidelines for testing of chemicals (No 1)*, OECD Environment, Health and Safety Publication Series on Testing and Assessment. Environmental Directorate, OECD, Paris (France).

OECD (2004a) *Draft guidance document on the use of multimedia models for estimating overall environmental persistence and long range transport (No 45),* OECD Environment, Health and Safety Publication Series on Testing and Assessment. Environmental Directorate, OECD, Paris (France).

OECD (2004b) *Draft guidance document on reproductive toxicity testing and assessment (No 43)*, OECD Environment, Health and Safety Publication Series on Testing and Assessment. Environmental Directorate, OECD, Paris (France).

Olavi Aho, M., Koivisto, A.M., Juhani Tammela, T.L. and Auvinen, A.P. (2005) Geographical differences in the prevalence of hypospadias in Finland, *Environ. Res.* **92**, 118-123.

Olea, N. (2006) Endocrine disrupter exposure and male congenital malformations, in P. Nicolopoulou-Stamati, L. Hens, and C.V. Howard (eds.) *Congenital anomalies and the environment*, Springer, Dordrecht, The Netherlands.

Palmer, S.R., Dunstan, F.D.J., Fielder, H., Fone, D.L., Higgs, G. and Senior, M.L. (2005) Risk of congenital anomalies after the opening of landfill sites, *Environ. Health Perspect.* **113**, 1362-1365.

Paulozzi, L.J. (1999) International trends in rates of hypospadias and cryptorchidism. *Environ. Health Perspect.* **107**, 297-302

Perterka, M., Likovsky, Z. and Perterkove, R. (2006) Environmental risk and sex ratio in newborns. in P. Nicolopoulou-Stamati, L. Hens, and C.V. Howard (eds.) *Congenital anomalies and the environment*, Springer, Dordrecht, The Netherlands.

Plummer, G. (1952) Anomalies occurring in children exposed *in utero* to the atomic bomb in Hiroshima, *Pediatrics* **10**, 687-693.

Rogan, W.J. (1982) PCBs and cola-colored babies: Japan, 1968, and Taiwan, 1979, *Teratology* **26**, 259-261.

Rogan, W.J., Gladen, B.C., Hung, K.L., Koong, S.L., Shih, L.Y., Taylor, J.S., Wu, Y.C., Yang, D., Ragan, N.B. and Hsu, C.C. (1988) Congenital poisoning by polychlorinated biphenyls and the contaminants in Taiwan, *Science* **241**, 334-336.

Seller M. (2004) Genetic causes of congenital anomalies and their interaction with environmental factors, in Eurocat Special Report: *A review of environmental risk factors for congenital anomalies* (online) www.eurocat.ulster.ac.uk/pubdata (May 10th 2005).

Sharpe, R.M. (2006) Phthalate exposure during pregnancy and lower anogenital index in boys: wider implications for the general population, *Environ. Health Perspect.* **113**, A504-A505.

Sharpe, R.M. and Skakkebaek, N.E. (1993) Are oestrogens involved in falling sperm counts and disorders of the male reproductive tract? *Lancet* **341**, 1392-1395.

Silbergeld, E.K. and Mattison, D.R. (1987) Experimental and clinical studies on the reproductive toxicology of 2,3,7,8-tetrachlorodibenzo-p-dioxin, *Am. J. Ind. Med.* **11**, 131 – 144.

Silva, M.J., Barr, D.B., Reidy, J.A., Malek, N.A., Hodge, C.C., Caudill, S.P., Brock, J.W., Needham, L.L. and Calafat, A.M. (2004) Urinary levels of seven phthalate metabolites in the U.S. population from the National Health and Nutrition Examination Survey (NHANES) 1999-2000, *Environ. Health Perspect.* **112**, 331-338.

Skene, S.A., Dewhurst, I.C. and Greenberg, M. (1989) Polychlorinated dibenzo-p-dioxins and polychlorinated dibenzofurans: The risks to human health: A review, *Human Toxicology* **8**, 173-203.

Smith, O.W. and Smith, G.V. (1949) The influence of diethylstilboestrol on the progress and outcome of pregnancy as based on a comparison of treated with untreated primigravidas, *Am. J. Obstet. Gynecol.* **58**, 994-1009.

Sosniak, W.A., Kaye, W.E. and Gomez, T.M. (1994) Data linkage to explore the risk of low birthweight associated with maternal proximity to hazardous waste sites from the National Priorities List, *Arch. Environ. Health* **49**, 251-255.

Spira, A., Goujard, J., Henrion, R., Lemerle, J., Robel, P. and Tchobroutsky, C. (1983) L'administration de diethylstilboestrol (DES) pendant la grossesse, un problème de santé publique. Diethylstilboestrol (DES) during pregnancy, a public health problem, *Rev. Epidem. Santé Publ.* **31**, 249-272.

Sterling, T.D. and Arundel, A. (1986) Review of recent Vietnamese studies on the carcinogenic and teratogenic effects of phenoxy herbicide exposure, *Int. J. Health Serv.* **16**, 265-278.

Sumanovic-Glamuzina, D., Saraga-Karacic, V., Roncevic, Z., Milanov, A., Bozic, T. and Boranic, M. (2003) Incidence of major congenital malformations in a region of Bosnia and Herzegovina allegedly polluted with depleted uranium, *Croatian Med. J.* **44**, 579-584.

Swan, S.H., Main, K.M., Liu, F., Stewart, S.L., Kruse, R.L., Calafat, A.M., Mao, C.S., Redmon, J.B., Ternand, C.L., Sullivan, S., Teague, J.L. Study for Future Families Research Team (2005) Decrease in anogenital distance among male infants with prenatal phthalate exposure, *Environ. Health Perspect.* **113,** 1056-1061.

Terlemesian, K. and Stoyanov, S. (2006) Congenital diseases in Bulgaria, in P. Nicolopoulou-Stamati, L. Hens, and C.V. Howard (eds.) *Congenital anomalies and the environment*, Springer, Dordrecht, The Netherlands.

Thonneau, P.F., Huyghe, E. and Mieusset, R. (2006) Environmental impact on congenital diseases: the case of cryptorchidism, in P. Nicolopoulou-Stamati, L. Hens, and C.V. Howard (eds.) *Congenital anomalies and the environment*, Springer, Dordrecht, The Netherlands.

Toppari, J. (2006) Male dysgenesis syndrome, in P. Nicolopoulou-Stamati, L. Hens, and C.V. Howard (eds.) *Congenital anomalies and the environment*, Springer, Dordrecht, The Netherlands.

Toppari, J., Larsen, J.C., Christiansen, P., Giwercman, A., Grandjean, P., Guillette, L.J., Jégou, B., Jensen, T.K., Jouannet, P., Kleiding, N., Leffers, H., McLachlan, J.A., Meyer, O., Muller, J., Rajpert-De Meyts, E., Scheike, T, Sharpe, R., Sumpter, J. and Skakkebaek, N.E. (1996) Male reproductive health and environmental xenoestrogens, *Environ. Health Perspect.* **104**, 741-803.

Van Larebeke, N. (2000) Health effects of a household waste incinerator near Wilrijk, Belgium, in P. Nicolopoulou-Stamati, L. Hens, and C.V. Howard (eds.) *Health impacts of waste management policies*, Kluwer Academic Publishers, Dordrecht, The Netherlands.

Vinceti, M., Rovesti, S., Bergomi, M., Calzolari, E., Candela, S., Campagna, A., Milan, M. and Vivoli, G. (2001) Risk of birth defects in a population exposed to environmental lead pollution, *Sci. Total Environ.* **278**, 23-30.

Vrijheid, M., Dolk, H., Armstrong, B., Abramsky, L., Bianchi, F., Fazarinc, I., Garne, E., Ide, R., Nelen, V., Robert, E., Scott, J.E., Stone, D. and Tenconi, R. (2002) Risk of chromosomal congenital anomalies in relation to residence near hazardous waste landfill sites in Europe, *Lancet* **359**, 320-322.

Vrijheid, M., Loane, M. and Dolk, H. (2003) Chemical environmental and occupational exposures, In: EUROCAT Special Report. The environmental causes of congenital anomalies: a review of the literature, online: [www.eurocat.ulster.ac.uk/pubdata] last accessed [15/09/2005].

Waldren, C., Correll, I., Sognier, M. and Puck, T. (1986) Measurement of low levels of x-ray mutagenesis in relation to human disease, *PNAS* **83**, 4839-4843.

Wilson, J.G. (1979) The evolution of teratological testing, *Teratology* **20**, 205-211.

Winder, C. (1993) Lead, reproduction and development, *Neurotoxicology* **14**, 303-318.

Wren, C.D. (1991) Cause-effect linkages between chemicals and populations of mink (Mustela vison) and otter (Lutra Canadensis) in the Great Lakes basin. *J. Toxicol. Environ. Health* **33**, 549-585.

LIST OF ABBREVIATIONS

17βHSD/17KSR	17β-hydroxysteroid dehydrogenase/17-ketosteroid reductase
3βHSD	3β-hydroxysteroid dehydrogenase
ADHD	Attention Deficit Hyperactivity Disorders
AGD	Anogenital Distance
AhR	Aryl Hydrocarbon Receptor
AIDS	Acquired Immune Deficiency Syndrome
AMH	Anti-Müllerian Hormone
AR	Androgen Receptor
AREHNA	Awareness Raising about Environment and Health of Non expert Advisors
BBB	Blood Brain Barrier
BBP	Benzyl Butyl phthalate
BINOCAR	British Isles Network of Congenital Anomaly Registers
CA	Congenital Anomaly
CARIS	Congenital Anomaly Register and Information System for Wales
CCA	Clear Cell Carcinoma
CEFIC	The European Chemical Industry Council
CI	Confidence Interval
CIS	Carcinoma in Situ
CNS	Central Nervous System
CSTEE	Commission's Scientific Committee on Toxicity, Ecotoxicity and the Environment
DBP	DiButyl Phthalate
DDD	Bis(4-chlorophenyl)-1,1-dichloroethane
DDE	1,1-Dichloro-2,2-bis(p-chlorophenyl) ethylene
DDT	Dichloro-Diphenyl-trichloroethane
DEHP	Diethylhexyl phthalate
DES	Diethylstilboestrol
DHT	Dihydrotestosterone
DNA	Deoxyribonucleic acid
ECLAMC	Estudio Latinoamericano de Colaborativo Malformaciones Congénitas
ED	Endocrine Disrupters

EDC	Endocrine Disrupting Chemical
EEA	European Environmental Agency
EEG/g	Estradiol Equivalents per gram of lipid
EF	Environmental Factor
ELISA	Enzyme-Linked Immunosorbent Assay
ENL	Erythema Nodosum Leprosum
ENTIS	European Network of Teratogen Information Services
ER	Estrogen Receptor
ESF	European Science Foundation
EU	European Union
EUROCAT	European Surveillance of Congenital Anomalies
FDA	Food and Drug Administration
FSH	Follicle Stimulating Hormone
GC/MS	Gas shromatography/ Mass Spectrometry
GD	Gestation Days
GDP	Gross Domestic Product
GNRH	Gonadotropin-Releasing Hormone
GVHD	Graft Versus Host Disease
HCB	Hexachlorobenzene
HCG	Human Chorionic Gonadotropin
HE	Heptachlor Epoxide
HIV	Human Immudodeficiency Virus
ICBDSR	International Clearinghouse for Birth Defects Surveillance and Research
ICD	International Classification of Diseases
ICM	Integrated Crop Management
IgG	Immunoglobulin G
IL-1	Interleukin-1
ILO	International Labour Organisation
IMR	Infant Mortality Rate
IPM	Integrated Pest Management
IUGR	Intrauterine Growth Retardation
LH	Luteinizing Hormone
Linurone	(3-(3,4-dichlorophenyl)-1-methoxy-1-methylurea
LPS	Lipopolysacharide
MC	Methoxychlor
MCF-7	Human Breast Cancer Cells from the Michigan Cancer Foundation
MDS	Male Dysgenesis Syndrome
MEHP	Monoethylhexyl phthalate
MTHFR	Methylene tetrahydrofolate reductase
NCAS	National Congenital Anomaly System

NEHAP	National Environmental Health Policy Plans
NGO	Non Governmental Organisation
NOEL	No-Observed-Effect-Level
NorCAS	Northern Congenital Abnormality Survey
NSSG	National Statistical Service of Greece
NTD	Neural Tube Defect
OCs	Organochlorine chemical
OECD	Organisation for Economic Cooperation and Development
ONS	Office for National Statistics (formerly OPCS)
OPCS	Office for Population, Census and Surveys
OR	Odds Ratio
OTIS	Organization of Teratogen Information Services
PCB	Polychlorinated Biphenyl
PCDD	Polychlorinated-dibenzo-p-dioxin
PDP	Plant Defense Products
PHA	Polycyclic Aromatic Hydrocarbon
RIVM	National Institute of Public Health and Environmental Protection
RR	Relative Risk
SCF	Scientific Committee for Food
SIDS	Sudden Infant Death Syndrome
SORCA	Sofia Registry of Congenital Anomalies
SPE	Study center for Perinatal Epidemiology
SR-B1	Scavenger Receptor B1
StAR	Steroidogenic Acute Regulatory Protein
STEPS	System for Thalidomide Education and Prescribing Safety
T4	Thyroxine
TCDD	Tetrachloro-dibenzo-p-dioxin
TDS	Testicular Dygenesis Syndrome
TEXB	Total Effective Xenoestrogen Burden
TG	Test Guideline
TIS	Teratogen Information Services
TNF	Tumor Necrosis Factor
TSD	Tay Sachs Disease
TSH	Thyroid Stimulating Hormone
UK	United Kingdom
UN	United Nations
UNEP	United Nations Environment Programme
US	United States
UV	Ultra-violet
VSD	Ventricular Septal Defect

VTG	Vitellogenin
WHO	World Health Organization

LIST OF UNITS

Prefixes to Units

da	deca	(10^1)	d	deci	(10^{-1})
h	hecto	(10^2)	c	centi	(10^{-2})
k	kilo	(10^3)	m	milli	(10^{-3})
M	Mega	(10^6)	μ	micro	(10^{-6})
G	Giga	(10^9)	n	nano	(10^{-9})
T	Tera	(10^{12})	p	pico	(10^{-12})
P	Peta	(10^{15})	f	femto	(10^{-15})

Units

°C	degree Celcius or centigrade	m	metre
		Nm^3	Normalised cubic metre
d	day	pa	per annum
Drachme	Former Greek currency unit	pH	acidity
		ppb	parts per billion
Euro	European currency unit	ppm	parts per million
g	gram	s	second
h	hour	t	ton
kg_{bw}	kilogram body weight	te	ton emission gas
kgpa	kilogram per annum	tpa	ton per annum
l	litre	US$	US Dolar

INDEX

B

C

D

F

G

H

I

O

P

T

U

V

W

X

Y

Z

Environmental Science and Technology Library

1. A. Caetano, M.N. De Pinho, E. Drioli and H. Muntau (eds.), *Membrane Technology: Applications to Industrial Wastewater Treatment.* 1995 ISBN 0-7923-3209-1
2. Z. Zlatev: *Computer Treatment of Large Air Pollution Models.* 1995 ISBN 0-7923-3328-4
3. J. Lemons and D.A. Brown (eds.): *Sustainable Development: Science, Ethics, and Public Policy.* 1995 ISBN 0-7923-3500-7
4. A.V. Gheorghe and M. Nicolet-Monnier: *Integrated Regional Risk Assessment.*
   Volume I: Continuous and Non-Point Source Emissions: Air, Water, Soil. 1995 ISBN 0-7923-3717-4
   Volume II: Consequence Assessment of Accidental Releases. 1995 ISBN 0-7923-3718-2
   Set: ISBN 0-7923-3719-0
5. L. Westra and J. Lemons (eds.): *Perspectives on Ecological Integrity.* 1995 ISBN 0-7923-3734-4
6. J. Sathaye and S. Meyers: *Greenhouse Gas Mitigation Assessment: A Guidebook.* 1995 ISBN 0-7923-3781-6
7. R. Benioff, S. Guill and J. Lee (eds.): *Vulnerability and Adaptation Assessments.* An International Handbook. 1996 ISBN 0-7923-4140-6
8. J.B. Smith, S. Huq, S. Lenhart, L.J. Mata, I. Nemošová and S. Toure (eds.): *Vulnerability and Adaptation to Climate Change.* Interim Results from the U.S. Country Studies Program. 1996 ISBN 0-7923-4141-4
9. B.V. Braatz, B.P. Jallow, S. Molnár, D. Murdiyarso, M. Perdomo and J.F. Fitzgerald (eds.): *Greenhouse Gas Emission Inventories.* Interim Results from the U.S. Country Studies Program. 1996 ISBN 0-7923-4142-2
10. M. Palo and G. Mery (eds.): *Sustainable Forestry Challenges for Developing Countries.* 1996 ISBN 0-7923-3738-7
11. S. Guerzoni and R. Chester (eds.): *The Impact of Desert Dust Across the Mediterranean.* 1996 ISBN 0-7923-4294-1
12. J.J.C. Picot and D.D. Kristmanson: *Forestry Pesticide Aerial Spraying.* Spray Droplet Generation, Dispersion, and Deposition. 1997 ISBN 0-7923-4371-9
13. J. Lemons, L. Westra and R. Goodland (eds.): *Ecological Sustainability and Integrity.* Concepts and Approaches. 1998 ISBN 0-7923-4909-1
14. V. Kleinschmidt and D. Wagner (eds.): *Strategic Environmental Assessment in Europe.* 4th European Workshop on Environmental Impact Assessment. 1998 ISBN 0-7923-5256-4
15. A. Bejan, P. Vadász and D.G. Kröger (eds.): *Energy and the Environment.* 1999 ISBN 0-7923-5596-2
16. P. Nicolopoulou-Stamati, L. Hens and C.V. Howard (eds.): *Health Impacts of Waste Management Policies.* 2000 ISBN 0-7923-6362-0
17. U. Albarella (ed.): *Environmental Archaeology: Meaning and Purpose.* 2001 ISBN 0-7923-6763-4
18. P. Nicolopoulou-Stamati, L. Hens and C.V. Howard (eds.): *Endocrine Disrupters.* Environmental Health and Policies. 2001 ISBN 0-7923-7056-2
19. S.D. Schery: *Understanding Radioactive Aerosols and Their Measurement.* 2001 Hb ISBN 0-7923-7068-6; Pb ISBN 0-7923-7176-3
20. P. Nicolopoulou-Stamati, L. Hens, C.V. Howard and N. Van Larebeke (eds.): *Cancer as an Environmental Disease.* 2004 ISBN Hb 1-4020-2019-8; Pb 1-4020-2020-1

Environmental Science and Technology Library

21. P. Nicolopoulou-Stamati, L. Hens and C.V. Howard (eds.): *Environmental Health Impacts of Transport and Mobility.* 2005 ISBN 1-4020-4304-X
22. P. Nicolopoulou-Stamati, L. Hens and C.V. Howard (eds.): *Reproductive Health and the Environment.* 2007 ISBN 1-4020-4828-9
23. P. Nicolopoulou-Stamati, L. Hens and C.V. Howard (eds.): *Congenital Diseases and the Environment.* 2007 ISBN 978-1-4020-4830-2